AF389446

Feature Papers of Forecasting 2021

Feature Papers of Forecasting 2021

Editor

Sonia Leva

MDPI • Basel • Beijing • Wuhan • Barcelona • Belgrade • Manchester • Tokyo • Cluj • Tianjin

Editor
Sonia Leva
Department of Energy,
Politecnico Di Milano
Italy

Editorial Office
MDPI
St. Alban-Anlage 66
4052 Basel, Switzerland

This is a reprint of articles from the Special Issue published online in the open access journal *Forecasting* (ISSN 2571-9394) (available at: https://www.mdpi.com/journal/forecasting/special_issues/FP_2021).

For citation purposes, cite each article independently as indicated on the article page online and as indicated below:

LastName, A.A.; LastName, B.B.; LastName, C.C. Article Title. *Journal Name* **Year**, *Volume Number*, Page Range.

ISBN 978-3-0365-5571-3 (Hbk)
ISBN 978-3-0365-5572-0 (PDF)

Contents

About the Editor

Sonia Leva

Sonia Leva is Full Professor in "Elettrotecnica" (Electrical Engineering-Circuit Theory) in Politecnico di Milano (Italy). She received a Ph.D. degree in 2001 in Electrical Engineering from the Faculty of Engineering, Politecnico di Milano, Italy. She was Research Associate and after Associate Professor of Electrical Engineering at Politecnico di Milano. Since 2016, Sonia Leva is a Full Professor in "Elettrotecnica" (Electrical Engineering-Circuit Theory), starting her professor activity on January 06, 2016.

She has been an IEEE member since 2000 and a senior member since 2013. She served as a Chairperson of sessions in an international conference organized by Institute of Electrical and Electronic Engineers. She is the author of about 300 papers mainly published on international and national journal or conference proceedings. She served as Editor-in-Chief for *Forecasting* from 2019.

Sonia Leva is founder and coordinator of SolartechLab and MultyGood Microgrid Lab ad Department of Energy, Politecnico di Milano.

Editorial

Editorial for Special Issue: "Feature Papers of Forecasting 2021"

Sonia Leva

Department of Energy, Politecnico di Milano, 20156 Milano, Italy; sonia.leva@polimi.it

Citation: Leva, S. Editorial for Special Issue: "Feature Papers of Forecasting 2021". *Forecasting* **2022**, *4*, 335–337. https://doi.org/10.3390/forecast4010018

Received: 24 January 2022
Accepted: 4 February 2022
Published: 3 March 2022

Publisher's Note: MDPI stays neutral with regard to jurisdictional claims in published maps and institutional affiliations.

The human capability to react or adapt to upcoming changes strongly relies on the ability to forecast them. Forecasting and its applications are increasingly important because they allow to improve decision-making processes by providing useful insights about the future. Scientific research is giving unprecedent attention to forecasting methods and applications, with a continuously growing number of articles about novel forecast approaches being published.

In this Special Issue, as well as in the one published in 2020 [1], high-quality papers in *Forecasting* spread into topics such as power and energy forecasting, forecasting in economics and management, forecasting in computer science, weather and forecasting and environmental forecasting have been selected and published. In particular, in this Special Issue, the most recent and high-quality research about forecasting is collected. Eleven papers are selected to represent a wide range of research fields where forecasting applications are playing a crucial role.

Nikolaidis et al. [2] propose a dynamical forecaster capable of estimating the required spinning reserves on the basis of a real-time load forecast. A neural network is trained via non-linear regression to accurately predict the load ahead starting from eight predictors, divided into constant and variable inputs by exploiting a model predictive control. The results provided demonstrate that the adoption of the proposed dynamical forecaster allows for significant improvements in terms of decreasing operating reserve requirements: Based on real-time updates, the load forecasting can achieve lower costs while the system security is preserved.

Ramos et al. [3] present a methodology designed for office buildings and aimed at improving the accuracy in electricity consumption forecasting on a 5-min time interval, providing proper support to decisions related to energy management towards higher efficiency. The prediction, based on data measured by different devices including presence, temperature, consumption and humidity, is carried out by means of two different forecasting algorithms, namely, Artificial Neural Network (ANN) and K-Nearest Neighbor (KNN) algorithms. The present research demonstrated that in order to achieve the maximum forecast accuracy in different periods of the day, hence in different contexts regarding consumption patterns, different forecasting algorithms must be used.

Chaiton et al. [4] present the outcomes of simulations forecasting the impact of five possible Tobacco Endgame policies on smoking prevalence and on tax revenues in Ontario by 2035. The Ontario SimSmoke simulation is exploited for modeling the expected effect of the first four strategies, namely: plain packaging, free cessation services, decreasing the number of tobacco outlets and increasing tobacco taxes. On the other hand, different models are involved in the evaluation of the impact of increasing the minimum required age to legally purchase tobacco to 21 years. Simulations predict that an increase in tobacco taxes will determine the greatest decrease in smoking prevalence, and that reducing smoking prevalence to "less than 5 by 35" by combining non-tax interventions and excise tax increase will result in a minimal impact on tax revenues.

Petropoulos et al. [5] focus on univariate time series forecasting and provide an overview of five different approaches allowing an improvement in the performances achievable with standard extrapolation methods. In further detail, the Theta method (manipulation of local curvatures), Multiple Temporal Aggregation (MTA), bootstrapping,

Forecasting with Sub-seasonal Series (FOSS) and forecasting with multiple starting points are discussed and compared in terms of how information is extracted from data, the computational cost and the performance. Moreover, the concept of the "wisdom of the data" is presented, explaining how a proper data manipulation can translate into improved forecast accuracy by combining forecasts carried out from different perspectives on the same data.

Watson et al. [6] investigate how the quality of weather data derived from thunderstorm simulations influences the outcome of power outage models. A comparative analysis is conducted using two different Numerical Weather Prediction (NWP) systems with various levels of data assimilation, determining how outage models trained on these different sets of weather data differ in terms of performance. It is demonstrated that erroneous estimations in weather simulations propagate into the outage models in specific and quantifiable ways, suggesting how improved weather representations can possibly improve the quality of the power outage insights obtained.

Nespoli et al. [7] propose a preliminary forecast procedure with the objective to predict a family of batteries which is suitable, from both a technical and a financial point of view, for coupling with a certain PV plant configuration. The procedure is applied to hypothetical plants aimed at fulfilling the energy requirements of a commercial and an industrial loads. The amount of energy produced by the PV system is estimated on the basis of a performance analysis carried out on real plants with similar characteristics, while the battery operations are determined by two distinct control logics regulating charge and discharge, respectively. Finally, an unsupervised clustering based on k-means algorithm applied to all possible PV+BESS (Battery Energy Storage System) configurations allowed the researchers to identify the family of feasible solutions which, as expected, was characterized by a low payback time and a low number of residual cycles.

Boudhaouia et al. [8] describe a novel web-oriented data analysis platform capable of forecasting water consumption in real-time by exploiting Machine Learning techniques. The prediction is carried out with no prior and contextual information, relying only on past water consumption data recorded by smart meters as unevenly spaced time series with high-resolution and based on two different algorithms, namely, a Long Short-Term Memory (LSTM) and a Back-Propagation Neural Network (BPNN). The two models are tested on forecasting the water consumption in a private building: By evaluating their performance, it is observed that LSTM outperforms BPNN, providing more accurate predictions. According to the authors, the developed model can even be generalized to different types of consumption, such as electricity and gas.

Bas et al. [9] introduce a novel time series forecasting approach based on the Holt method modified by using time-varying smoothing parameters instead of fixed ones. Holt's smoothing parameters are obtained for each observation exploiting first-order autoregressive models whose parameters, in turn, are assessed through a Harmony Search Algorithm (HSA). The proposed method is tested on Istanbul Stock Exchange datasets covering the years between 2000 and 2017: The forecasts are obtained with a subsampling bootstrap approach, and different test lengths are considered during this analysis.

Wu et al. [10] deal with the topic of forecasting volatility from econometric datasets, a crucial task in finance. First, they assess the robustness of state-of-art Normalizing and Variance-Stabilizing (NoVaS) methods for long-term time-aggregated predictions, addressing the lack of experimental results in current NoVaS-related studies. Then, they develop a novel model-free method that, after an extensive analysis, demonstrated improved and more stable performance with respect to state-of-art NoVaS and standard GARCH-type methods in both the short and long term, regardless of whether simulation or real-world data are used.

Ali et al. [11] propose a novel approach aimed at predicting ocean currents by means of deep learning. In detail, a LSTM model is applied to the prediction of the three-dimensional tensors representing water column velocity. The proposed method is tested on estimating the Loop Current (LC) measured in the Gulf of Mexico between 2009 and 2011 at multiple spatial and temporal scales, where an RMSE (Root Mean Square Error) lower than 0.05 cm/s and a correlation coefficient of 0.6 were presented. Moreover, the model presented a useful forecast period, hence the time interval after which the forecast significantly diverges from the observed motion field, larger than 4 days.

Vega et al. [12] face the challenge of forecasting the number of new COVID-19 infections in the short and medium term by proposing the SIMLR model, incorporating Machine Learning (ML) into the epidemiological SIR model. By combining these two components, it is substantially possible to reduce the amount of data required by Machine Learning in order to produce accurate predictions and to estimate the time-varying parameters of a SIR model to produce forecasts with an advance of one to four weeks. The proposed SIMLR model is applied to study cases from Canada and the United States, demonstrating state-of-the-art forecasting performance with the additional advantage of providing probabilistic and interpretable outcomes. The authors expect this approach to be involved not only in COVID-19 modeling and for other infectious diseases as well.

Funding: This research received no external funding.

Institutional Review Board Statement: Not applicable.

Informed Consent Statement: Not applicable.

Data Availability Statement: Not applicable.

Conflicts of Interest: The authors declare no conflict of interest.

References

1. Leva, S. Editorial for Special Issue: "Feature Papers of Forecasting". *Forecasting* **2021**, *3*, 135–137. [CrossRef]
2. Nikolaidis, P.; Partaourides, H. A Model Predictive Control for the Dynamical Forecast of Operating Reserves in Frequency Regulation Services. *Forecasting* **2021**, *3*, 228–241. [CrossRef]
3. Ramos, D.; Khorram, M.; Faria, P.; Vale, Z. Load Forecasting in an Office Building with Different Data Structure and Learning Parameters. *Forecasting* **2021**, *3*, 242–254. [CrossRef]
4. Chaiton, M.; Dubray, J.; Guindon, G.E.; Schwartz, R. Tobacco Endgame Simulation Modelling: Assessing the Impact of Policy Changes on Smoking Prevalence in 2035. *Forecasting* **2021**, *3*, 267–275. [CrossRef]
5. Petropoulos, F.; Spiliotis, E. The Wisdom of the Data: Getting the Most Out of Univariate Time Series Forecasting. *Forecasting* **2021**, *3*, 478–497. [CrossRef]
6. Watson, P.L.; Koukoula, M.; Anagnostou, E. Influence of the Characteristics of Weather Information in a Thunderstorm-Related Power Outage Prediction System. *Forecasting* **2021**, *3*, 541–560. [CrossRef]
7. Nespoli, A.; Matteri, A.; Pretto, S.; De Ciechi, L.; Ogliari, E. Battery Sizing for Different Loads and RES Production Scenarios through Unsupervised Clustering Methods. *Forecasting* **2021**, *3*, 663–681. [CrossRef]
8. Boudhaouia, A.; Wira, P. A Real-Time Data Analysis Platform for Short-Term Water Consumption Forecasting with Machine Learning. *Forecasting* **2021**, *3*, 682–694. [CrossRef]
9. Bas, E.; Egrioglu, E.; Yolcu, U. Bootstrapped Holt Method with Autoregressive Coefficients Based on Harmony Search Algorithm. *Forecasting* **2021**, *3*, 839–849. [CrossRef]
10. Wu, K.; Karmakar, S. Model-Free Time-Aggregated Predictions for Econometric Datasets. *Forecasting* **2021**, *3*, 920–933. [CrossRef]
11. Ali, A.M.; Zhuang, H.; Vanzwieten, J.; Ibrahim, A.K.; Chérubin, L. A Deep Learning Model for Forecasting Velocity Structures of the Loop Current System in the Gulf of Mexico. *Forecasting* **2021**, *3*, 934–953.
12. Vega, R.; Flores, L. SIMLR: Machine Learning inside the SIR Model for COVID-19. *Forecasting* **2022**, *4*, 72–94. [CrossRef]

Article

SIMLR: Machine Learning inside the SIR Model for COVID-19 Forecasting

Roberto Vega [1,2,*], Leonardo Flores [3] and Russell Greiner [1,2]

[1] Department of Computing Science, University of Alberta, Edmonton, AB T6G 2R3, Canada; rgreiner@ualberta.ca

[2] Alberta Machine Intelligence Institute, Edmonton, AB T5J 3B1, Canada

[3] Independent Researcher, San Luis Potosi 78170, Mexico; leonardo.flores.q@gmail.com

* Correspondence: rvega@ualberta.ca

Abstract: Accurate forecasts of the number of newly infected people during an epidemic are critical for making effective timely decisions. This paper addresses this challenge using the SIMLR model, which incorporates machine learning (ML) into the epidemiological SIR model. For each region, SIMLR tracks the changes in the policies implemented at the government level, which it uses to estimate the time-varying parameters of an SIR model for forecasting the number of new infections one to four weeks in advance. It also forecasts the probability of changes in those government policies at each of these future times, which is essential for the longer-range forecasts. We applied SIMLR to data from in Canada and the United States, and show that its mean average percentage error is as good as state-of-the-art forecasting models, with the added advantage of being an interpretable model. We expect that this approach will be useful not only for forecasting COVID-19 infections, but also in predicting the evolution of other infectious diseases.

Keywords: COVID-19; probabilistic graphical models; interpretable machine learning

Citation: Vega, R.; Flores, L.; Greiner, R. SIMLR: Machine Learning inside the SIR Model for COVID-19 Forecasting. *Forecasting* **2022**, *4*, 72–94. https://doi.org/10.3390/forecast4010005

Academic Editor: Sonia Leva

Received: 9 December 2021
Accepted: 10 January 2022
Published: 13 January 2022

Publisher's Note: MDPI stays neutral with regard to jurisdictional claims in published maps and institutional affiliations.

1. Introduction

Since its identification in December 2019, COVID-19 has posed critical challenges for the public health and economies of essentially every country in the world [1–3]. Government officials have taken a wide range of measures in an effort to contain this pandemic, including closing schools and workplaces, setting restrictions on air travel, and establishing stay at home requirements [4]. Accurately forecasting the number of new infected people in the short and medium term is critical for the timely decisions about policies and for the proper allocation of medical resources [5,6].

There are three basic approaches for predicting the dynamics of an epidemic: compartmental models, statistical methods, and ML-based methods [5,7]. Compartmental models subdivide a population into mutually exclusive categories, with a set of dynamical equations that explain the transitions among categories [8]. The Susceptible-Infected-Removed (SIR) model [9] is a common choice for the modelling of infectious diseases. Statistical methods extract general statistics from the data to fit mathematical models that explain the evolution of the epidemic [6]. Finally, ML-based methods use machine learning algorithms to analyze historical data and find patterns that lead to accurate predictions of the number of new infected people [7,10].

Arguably, when any approach is used to make high-stake decisions, it is important that it be not just accurate, but also interpretable: It should give the decision-maker enough information to justify the recommendation [11]. Here, we propose SIMLR, which is an interpretable probabilistic graphical model (PGM) that combines compartmental models and ML-based methods. As its name suggests, it incorporates machine learning (ML) within an SIR model. This combines the strength of curve fitting models that allow accurate predictions in the short-term, involving many features, with mechanistic models that allow to extend the range to predictions in the medium and long terms [12].

SIMLR uses a mixture of experts approach [13], where the contribution of each expert to the final forecast depends on the changes in the government policies implemented at various earlier time points. When there is no recent change in policies (two to four weeks before the week to be predicted), SIMLR relies on an SIR model with time-varying parameters that are fitted using machine learning methods. When a change in policy occurs, SIMLR instead relies on a simpler model that predicts that the new number of infections will remain constant. Note that forecasting the number of new infections one and two weeks in advance (ΔI_1 and ΔI_2) is relatively easy as SIMLR knows, at the time of the prediction, whether the policy has changed recently. However, for three- or four-week forecasts (ΔI_3 and ΔI_4), our model needs to estimate the likelihood of a future change of policy. SIMLR incorporates prior domain knowledge to estimate such policy-change probabilities.

The use of such prior models—here epidemiological models—is particularly important when the available data is scarce [14]. At the same time, machine learning models need to acknowledge that the reported data on COVID-19 is imperfect [15,16]. The use of probabilistic graphical models allows SIMLR to account for this uncertainty on the data. At the same time, the probability tables associated with this graphical model can be manually modified to adapt SIMLR to the specific characteristics of a region.

This work makes three important contributions. (1) It empirically shows that an SIR model with time-varying parameters can describe the complex dynamics of COVID-19. (2) It describes an interpretable model that predicts the new number of infections one to four weeks in advance, achieving state-of-the-art results, in terms of mean absolute percentage error (MAPE), on data from Canada and the United States. (3) It presents a machine learning model that incorporates the uncertainty of the input data and can be tailored to the specific situations of a particular region.

The rest of Section 1 describes the related work and the basics of the SIR compartmental model. Section 2 then describes in detail our proposed SIMLR approach. Section 3 shows the results of the predicting the number of new infections in the United States and provinces of Canada. Finally, Section 4 presents our final remarks.

1.1. Basic SIR Model

The Susceptible-Infected-Removed (SIR) compartmental model [9] is a mathematical model of infectious disease dynamics that divide the population into three disjoint groups [8]. Susceptible (S) refers to the set of people who have never been infected but can acquire the disease. Infected (I) refers to the set of people who have and can transmit the infection. Removed (R) refers to the people who have either recovered or died from the infection and cannot transmit the disease anymore. This model is defined by the differential equations:

$$\frac{dS}{dt} = -\frac{\beta S(t)I(t)}{N}, \quad \frac{dI}{dt} = \frac{\beta S(t)I(t)}{N} - \gamma I(t), \quad \frac{dR}{dt} = \gamma I(t) \tag{1}$$

SIR assumes an homogeneous and constant population, and it is fully defined by the parameters β (transmission rate) and γ (recovery rate). The intuition behind this model is that every infected patient gets in contact with β people. Since only the susceptible people can become infected, the chance of interacting with a susceptible person is simply the proportion of susceptible people in the entire population, $N = S + I + R$. Likewise, at every time point, γ proportion of the infected people is removed from the system. Figure 1a depicts the general behaviour of an SIR model.

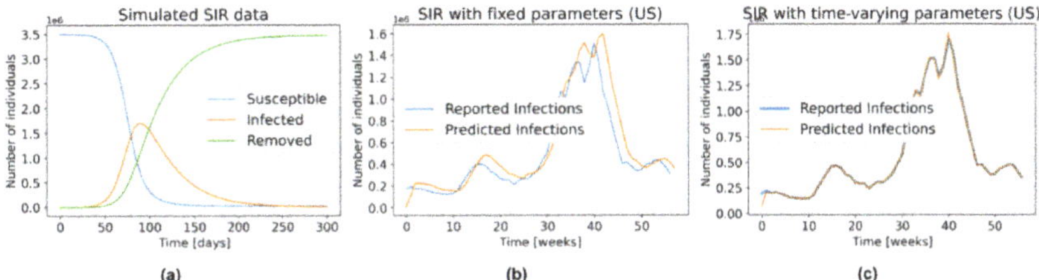

Figure 1. (**a**) General behaviour of the SIR model. (**b**) The number of infections predicted by the SIR model with fixed parameters, fitted to the US data for 1 week prediction. (**c**) Similar to (**b**), but with time-varying parameters.

1.2. Related Work

The main idea behind combining compartmental models with machine learning is to replace the fixed parameters of the former with time-varying parameters that can be learned from data [6,17–19]. However, most of the approaches focus on finding the parameters that can explain the past data, and not on predicting the number of newly infected people. Although those approaches are useful for obtaining insight into the dynamics of the disease, it does not mean that those parameters will accurately predict the behaviour in the future.

Particularly relevant to our approach is the work by Arik et al. [5], who used latent variables and autoencoders to model extra compartments in an extended Susceptible-Exposed Infected Removed (SEIR) model. Those additional compartments bring further insight into how the disease impacts the population [20,21]; however, our experiments suggest that they are not needed for an accurate prediction of the number of new infections. One limitation of their model is a decrease in performance when the trend in the number of new infections changes. We hypothesize that those changes in trend are related to the government policies that are in place at a specific point in time. SIMLR is able to capture those changes by tracking the policies implemented at the government level.

A different line of work replaces epidemiological models with machine learning methods to directly predict the number of new infections [22–25]. Importantly, Yeung et al. [26] added non-pharmaceutical interventions (policies) as features in their models; however, their approach is limited to make predictions up to two weeks in advance, since information about the policies that will be implemented in the future is not available at inference time. Our SIMLR approach differs by being interpretable and also by forecasting policy changes, which allows it to extend the horizon of the ΔI predictions.

There are many models that attempt to predict the evolution of the COVID-19 epidemic. The Center for Disease Control and Prevention (CDC) in the United States allows different research teams across the globe to submit their forecasts of the number of cases and deaths 1 to 8 weeks in advance [27]. More than 100 teams have submitted at least one prediction to this competition. We compare SIMLR with all of the models that made predictions 1 to 4 weeks in advance in the same time span as our study.

2. Materials and Methods

We view SIMLR as a probabilistic graphical model that uses a mixture of experts approach to forecast the number of new COVID-19 infections, 1 to 4 weeks in advance. Figure 2 shows the intuition behind SIMLR. Changes in the government policies are likely to modify the trend of the number of new infections. We assume that stronger policies are likely to decrease the number of new infections, while the opposite effect is likely to occur when relaxing the policies. These changes are reflected as a change in the parameters of the SIR model. Using those parameters, we can then predict the number of new infections, then use that to compute the likelihood of observing other new policy changes in the short term.

While Figure 2 is an schematic diagram used for pedagogical purposes; Figure 3 depicts the formal probabilistic graphical model, as a plate model, that we use to estimate the parameters of the SIR model, the number of new infections, and the likelihood of observing changes in policies 1 to 4 weeks in advance. The blue nodes are estimated at every time point, while the values of the green nodes are either known as part of the historical data, or inferred in a previous time point. The random variables are assumed to have the following distributions:

$$
\begin{aligned}
CT_{t+1} \mid & \quad \{CP_{t-\tau}\}_{\tau \in \{1,2,3\}} & \sim & \quad Cat_{K \in \{-1,0,1\}}(\theta_{CT}) \\
\beta_{t+1} \mid & \{\beta_{t-\tau}\}_{\tau \in \{0,1,2\}}, CT_{t+1} & \sim & \quad \mathcal{N}(\mu_\beta, \Sigma_\beta) \\
\gamma_{t+1} \mid & \{\gamma_{t-\tau}\}_{\tau \in \{0,1,2\}}, CT_{t+1} & \sim & \quad \mathcal{N}(\mu_\gamma, \Sigma_\gamma) \\
SIR_{t+1} \mid & \quad \beta_{t+1}, \gamma_{t+1} & \sim & \quad \mathcal{N}(\mu_{SIR}, \Sigma_{SIR}) \\
U_t \mid & \quad \{SIR_{t-\tau}\}_{\tau \in \{0,1,2\}} & \sim & \quad Cat_{K \in \{-1,0,1\}}(\theta_U) \\
O_t \mid & \quad W_t & \sim & \quad Cat_{K \in \{0,1\}}(\theta_O) \\
CP_{t+1} \mid & \quad O_t, U_t & \sim & \quad Cat_{K \in \{-1,0,1\}}(\theta_{CP})
\end{aligned}
\tag{2}
$$

where t indexes the current week, $SIR_t = [S_t, I_t, R_t]$, $\mu_{SIR} \in \mathbb{R}^3$ is given below by Equation (3), $\mu_\beta = (\alpha_{0,CT_{t+1}}) + (\alpha_{1,CT_{t+1}})\beta_{t-1} + (\alpha_{2,CT_{t+1}})\beta_{t-2} + (\alpha_{3,CT_{t+1}})\beta_{t-3}$ and $\mu_\gamma = (\omega_{0,CT_{t+1}}) + (\omega_{1,CT_{t+1}})\gamma_{t-1} + (\omega_{2,CT_{t+1}})\gamma_{t-2} + (\omega_{3,CT_{t+1}})\gamma_{t-3}$ are linear combinations of the three previous values of β and γ, (respectively). The coefficients of those linear combinations depend on the value of the random variable CT_{t+1}. We did not specify a distribution for the node New_infections$_{t+1}$ because its value is deterministically computed as $S_t - S_{t+1}$.

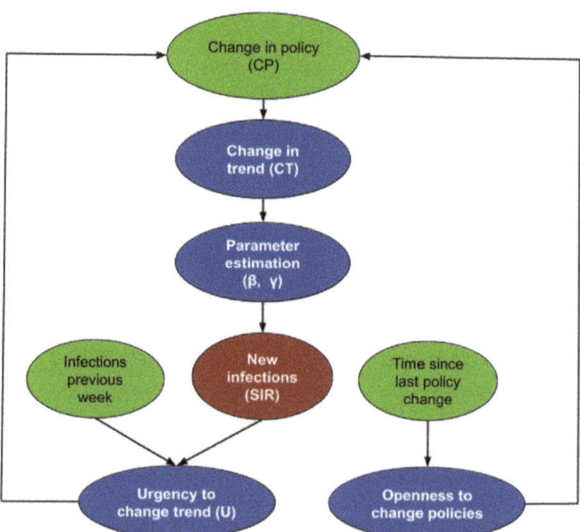

Figure 2. Intuition behind SIMLR. The policies currently in place determine the value of the parameters needed to infer the next values, using an SIR model. Those predictions are then used to estimate how the policies might change in the future.

Informally, the assignment $CT_t = -1$ means that we expect a change in trend from an increasing number of infections to a decreasing one. The opposite happens when $CT_t = 1$, while $CT_t = 0$ means that we expect the population to follow the current trend (either increasing or decreasing). We assume these changes in trend depend on changes in the government policies 2 to 4 weeks prior to the week of our forecast—e.g., we use $\{CT_{t-3}, CT_{t-2}, CT_{t-1}\}$ when predicting the number of new infections at $t+1$, ΔI_{t+1}, and we need $\{CT_t, CT_{t+1}, CT_{t+2}\}$ when predicting ΔI_{t+4}. Note that, at time t, we will not know CT_{t+1} nor CT_{t+2}. We chose this interval based on the assumption that the incubation period of the virus is 2 weeks.

The status of CT_{t+1} defines the coefficients that relate β_{t+1} and γ_{t+1} with their three previous values $\beta_t, \beta_{t-1}, \beta_{t-2}$ and $\gamma_t, \gamma_{t-1}, \gamma_{t-2}$, respectively. Since β_{t+1} and γ_{t+1} fully parameterize the SIR model in Equation (1), we can estimate the new number of infected people, ΔI_{t+1}, from these parameters (as well as the SIR values at time t).

The random variables $U_t \in \{-1, 0, 1\}$ and $O_t \in \{0, 1\}$ are auxiliary variables designed to predict the probability of observing a change in policy at time $t + 1$. Intuitively, U_t represents the "urgency" of modifying a policy. As the number of cases per 100K inhabitants and the rate of change between the number of cases in two consecutive time points increases, the urgency to set stricter government policies increases. As the number (and rate of change) of cases decreases, the urgency to relax the policies increases. Finally, O_t models the "willingness" to execute a change in government policies. As the number of time points without a change increases, so does this "willingness".

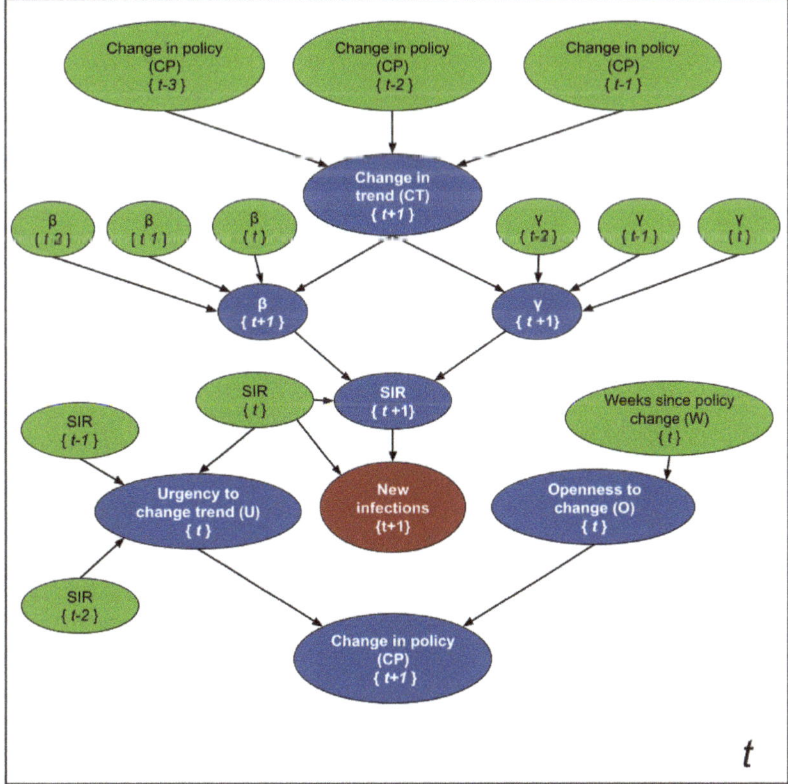

Figure 3. Modeling SIMLR as a PGM for forecasting new cases of COVID-19. The blue nodes are estimated at each time point, while the green ones are either based on past information, or where estimated in a previous iteration.

2.1. SIR with Time-Varying Parameters

We can approximate an SIR model by transforming the differential Equation (1) into the equations of differences:

$$
\begin{aligned}
S_t &= -\beta \frac{S_{t-1} I_{t-1}}{N} + S_{t-1} \\
I_t &= \beta \frac{S_{t-1} I_{t-1}}{N} - \gamma I_{t-1} + I_{t-1} \\
R_t &= \gamma I_{t-1} + R_{t-1}
\end{aligned}
\tag{3}
$$

where S_t, I_t, R_t are the number of individuals in the groups Susceptible, Infected and Removed, respectively, at time t. Similarly $S_{t-1}, I_{t-1}, R_{t-1}$ represent the number individuals in each group at time $t-1$. β is the transmission rate, and γ is the recovery rate.

While the SIR model is non-linear with respect to the states (S, I, R), it is linear with respect to the parameters β and γ. Therefore, under the assumption of constant and known population size (i.e., $N = S_t + I_t + R_t$) we can re-write the set of Equation (3) as:

$$
\begin{aligned}
\begin{bmatrix} S_t \\ I_t \end{bmatrix} &= \begin{bmatrix} -\frac{S_{t-1} I_{t-1}}{N} & 0 \\ \frac{S_{t-1} I_{t-1}}{N} & -I_{t-1} \end{bmatrix} \begin{bmatrix} \beta \\ \gamma \end{bmatrix} + \begin{bmatrix} S_{t-1} \\ I_{t-1} \end{bmatrix} \\
R_t &= N - S_t - I_t
\end{aligned}
\tag{4}
$$

Given a sequence of states $x_1, \ldots, x_n$, where $x_t = [S_t \ I_t]^T$, it is possible to estimate the optimal parameters of the SIR model as:

$$
(\beta^*, \ \gamma^*) = \underset{\beta, \gamma}{\arg\min} \sum_{i=1}^{n} ||x_i - \hat{x}_i||^2 + \lambda_1 (\beta - \beta_0)^2 + \lambda_2 (\gamma - \gamma_0)^2
\tag{5}
$$

where $\hat{x}_i$ is computed using Equation (4), and λ_1 and λ_2 are optional regularization parameters that allow the incorporation of the priors β_0 and γ_0. For the case of Gaussian priors—i.e., $\beta \sim \mathcal{N}(\beta_0, \sigma_\beta^2)$ and $\gamma \sim \mathcal{N}(\gamma_0, \sigma_\gamma^2)$—we use $\lambda_1 = \frac{1}{2\sigma_\beta^2}$ and $\lambda_2 = \frac{1}{2\sigma_\gamma^2}$ [28]. Intuitively, Equation (5) computes the transmission rate (β^*) and the recovery rate (γ^*) that best explain the number of new infections, deaths, and recovered people in a fixed time frame. If we know a standard recovery rate and transmission rate a priori (β_0, γ_0), it is possible to incorporate them into the Equation (5) as regularization parameters. The weights λ_1 and λ_2 control how much to weight those prior parameters. Small weights means we basically use the parameters learned by the data, and large weights mean more emphasis on the prior information.

In the traditional SIR model, we set $\lambda_1 = \lambda_2 = 0$ and fit a single β and γ to the entire time series. However, as shown in Figure 1a, an SIR model with fixed parameters is unable to accurately model several waves of infections. As illustration, Figure 1b shows the predictions produced by fitting an SIR with fixed parameters (Equation (5)) to the US data from 29 March 2020 to 3 May 2021, and then using those parameters to make predictions one week in advance, over this same interval. That is, using this learned (β, γ), and the number of people in the S, I, and R compartments on 28 March 2020, we predicted the number of observed cases during the week of 29 March 2020 to 4 April 2020. We repeated the same procedure for the entire time series. Note that even though the parameters β and γ were found using the entire time series – i.e., using information that was not available at the time of prediction—the resulting model still does a poor job fitting the reported data.

Figure 1c, on the other hand, was created by allowing β and γ to change every week. Here, we first found the parameters that fit the data from 29 March 2020 to 4 April 2020—call them β_1 and γ_1—then used those parameters along with the SIR state on 28 March 2020 to predict the number of new infections one week ahead—i.e., the sampled week of 29 March 2020 to 4 April 2020. By repeating this procedure during the entire time series we obtained an almost perfect fit to the data. Of course, these are also not "legal" predictions since they too use information that is not available at prediction time—i.e., they used the number of reported infections during this first week to find the parameters, which were then used to estimate the number of cases over this time. However, this "cheating" example shows that an SIR model, with the optimal time-varying parameters, can model the complex dynamics of COVID-19. Recall from Figure 1b that this is not the case in the SIR model with fixed parameters, which cannot even properly fit the training data.

2.2. Estimating β_{t+1} and γ_{t+1}

Naturally, the challenge is "legally" computing the appropriate values of β_{t+1} and γ_{t+1}, for each week, using only the data that is known at time t. Figure 3 shows that computing β_{t+1} and γ_{t+1} depends on the status of the random variable CT_{t+1}. When $CT_{t+1} = 0$—i.e., there is no change in the current trend—we assume that:

$$
\begin{aligned}
\beta_{t+1} &\sim \mathcal{N}(\alpha_0 + \alpha_1\beta_t + \alpha_2\beta_{t-1} + \alpha_3\beta_{t-2},\ \sigma_\beta^2) \\
\gamma_{t+1} &\sim \mathcal{N}(\omega_0 + \omega_1\gamma_t + \omega_2\gamma_{t-1} + \omega_3\gamma_{t-2},\ \sigma_\gamma^2)
\end{aligned}
\tag{6}
$$

At time t, we can use the historical daily data $x_1, x_2, \ldots, x_t$ to find the weekly parameters $\beta_1, \beta_2, \ldots, \beta_{t/7}$ and $\gamma_1, \gamma_2, \ldots, \gamma_{t/7}$. Note that the is just one value for each week, so is there are 140 days, there are $140/7 = 20$ weeks. The first weekly pair (β_1, γ_1) is found by fitting Equation (5) to $x_1, \ldots, x_7$; (β_2, γ_2) to $x_8, \ldots, x_{14}$; and so on. Finally, we find the parameters α and ω in Equation (6) by maximizing the likelihood of the computed pairs. After finding those parameters, it is straightforward to infer $(\beta_{t+1}, \gamma_{t+1})$. Note that this approach is the probabilistic version of linear regression. To estimate the parameters σ_β^2 and σ_γ^2 we can simply estimate the variance of the residuals. An advantage of also computing these variances is that it is possible to obtain confidence intervals by sampling from the distribution in Equation (6) and then using those samples along with Equation (3) to estimate the distribution of the new infected people.

We estimated β_{t+1} and γ_{t+1} as a function of the 3 previous values of those parameters since this allows them to incorporate the velocity and acceleration at which the parameters change. We computed the velocity of β as $v_{\beta,t} = \beta_t - \beta_{t-1}$ and its acceleration as $a_{\beta,t} = v_{\beta,t} - v_{\beta,t-1}$. Then, estimating $\beta_t = \theta_0 + \theta_1\beta_{t-1} + \theta_2 v_{\beta,t-1} + \theta_3 a_{\beta,t-1}$ is equivalent to the model in Equation (6). The same reasoning applies to the computation of γ_t. We call this approach the "trend-following varying-time parameters SIR", tf-v-SIR.

For the case of $CT_t = -1$ and $CT_t = 1$ (which represents a change in trend from increasing number of infections to decreasing number of infections or vice-versa), we set β_{t+1} and γ_{t+1} to values such that the predicted number of new cases at week $t + 1$ is identical to the one at week t. We call this the "Same as the Last Observed Week" (SLOW) model. As shown in Section 3, SLOW is a baseline with very good performance despite its simplicity. Given that the pandemic is a physical phenomenon that changes relatively slowly from one week to the next, making a prediction that assumes that the new number of cases will remain constant is not a bad prediction.

2.3. Estimating CT_{t+1}, CP_{t+1}, O_t

The random variables CT_{t+1}, CP_{t+1} and O_t in Figure 3 are all discrete nodes with discrete parents, meaning their probability mass functions are fully defined by conditional probability tables (CPTs). Learning the parameters of such CPTs from data is challenging due to the scarcity of historical information. The random variable CT_{t+1} depends on the random variable changes in policy (CP) at times $t - 1, t - 2, t - 3$; however, there are very few changes in policy in a given region, meaning it is difficult to accurately estimate those probabilities from data. For the random variable O, which represents the "willingness" of the government to implement a change in policy, there is no observable data at all. We therefore relied on prior expert knowledge to set the parameters of the conditional probability tables for these random variables. Figure 4 shows the conditional probability tables (CPT) for the random variables CT_{t+1}, CP_{t+1}, O_t. The intuition used to generate the CPT's is as follows:

We considered that a change in trend in the current week depends on changes in policies during the previous three weeks. We chose 3 weeks using the hypothesis that the incubation period for the virus is 2 weeks. Then the effects of a policy will be reflected approximately 2 weeks after a change. We decided to analyze also one week after, and one week before this period, giving as a result the tracking of CP_{t-3} to CP_{t-1}. Secondly, we also assume that whenever we observe a change of policy that will move the trend from going up to going down, then that event will most likely happen. This is why most of the probability mass is located in a single column. For example, if we observe that the policies are relaxed at any point during the weeks $t-3$, $t-2$, or $t-1$, then we assume that we will observe a change in trend with 99.9% probability.

The rationale for the CPT $P(O_t \mid W_t)$ is that the government becomes more open to implement changes after long periods of 'inactivity'. For example, if they implement a change in policy this week ($W_t = 0$), then the probability of considering a second change of policy during the same week is very small (0.01%). We are assuming that, after a change in policy, the government will wait to see the effect of that change before taking further action. If 4 weeks have passed since the last change in policy, we estimated the probability of considering a change in the policy as 50%, while if more than 7 weeks have passed, then they are fully open to the possibility of implementing a new change.

$P(O_t \mid W_t)$ estimates the probability of considering a change in the policy. The probability of actually implementing a change, $P(CP_{t+1} \mid O_t, U_t)$ depends not only on how willing the government is, but also on how urgent it is to make a change. In general, if the government is open to implement a change, and the urgency is "high", then the probability of changing a policy is high. We also considered that the government "prefers" to either not make changes in policy or relax the policies, rather than to implement more strict policies.

CP $_{t-3}$	CP $_{t-2}$	CP $_{t-1}$	$P(CT_{t+1} = x \mid CP_{t-3}, CP_{t-2}, CP_{t-1})$		
			x = -1	x = 0	x = 1
-1	-1	-1	0.999	0.0005	0.0005
-1	-1	0	0.999	0.0005	0.0005
-1	-1	1	0.999	0.0005	0.0005
-1	0	-1	0.999	0.0005	0.0005
-1	0	0	0.999	0.0005	0.0005
-1	0	1	0.0005	0.0005	0.999
-1	1	-1	0.999	0.0005	0.0005
-1	1	0	0.0005	0.0005	0.999
-1	1	1	0.0005	0.0005	0.999
0	-1	-1	0.999	0.0005	0.0005
0	-1	0	0.999	0.0005	0.0005
0	-1	1	0.0005	0.0005	0.999
0	0	-1	0.999	0.0005	0.0005
0	0	0	0.0005	0.999	0.0005
0	0	1	0.0005	0.0005	0.999
0	1	-1	0.999	0.0005	0.0005
0	1	0	0.0005	0.0005	0.999
0	1	1	0.0005	0.0005	0.999
1	-1	-1	0.999	0.0005	0.0005
1	-1	0	0.999	0.0005	0.0005
1	-1	1	0.0005	0.0005	0.999
1	0	-1	0.999	0.0005	0.0005
1	0	0	0.0005	0.0005	0.999
1	0	1	0.0005	0.0005	0.999
1	1	-1	0.0005	0.0005	0.999
1	1	0	0.0005	0.0005	0.999
1	1	1	0.0005	0.0005	0.999

W_t	$P(O_t = x \mid W_t)$	
	x = 0	x = 1
0	0.9999	0.0001
1	0.9	0.1
2	0.85	0.15
3	0.75	0.25
4	0.5	0.5
5	0.25	0.75
6	0.0001	0.9999
7+	0.0001	0.9999

O_t	U_t	$P(CP_{t+1} = x \mid O_t, U_t)$		
		x = -1	x = 0	x = 1
0	-1	0.02	0.97	0.01
0	0	0.005	0.99	0.005
0	1	0.01	0.97	0.02
1	-1	0.8	0.19	0.01
1	0	0.09	0.9	0.01
1	1	0.01	0.24	0.75

Figure 4. Conditional probability tables used by SIMLR. The names of the variables refer to the nodes that appear on Figure 2 on the main text.

2.4. Estimating U_t

For modelling the random variable U_t, which represents the "Urgency to change the trend", we use an NN-CPD (neural-network conditional probability distribution), which is a modified version of the multinomial logistic conditional probability distribution [29].

Definition 1 (NN-CPD). *Let $Y \in \{1, \dots, m\}$ be an m-valued random variable with k parents $X_1, \dots, X_k$ that each take on numerical values. The conditional probability distribution $P(Y \mid X_1, \dots, X_k)$ is an NN-CPD if there is an function $z = f_\theta(X_1, \dots, X_k) \in \mathbb{R}^m$, represented as a neural network with parameters θ, such that $p(Y = i \mid x_1, \dots, x_k) = \exp(z_i) / \sum_j \exp(z_j)$, where z_i represents the i-th entry of z.*

Note U_t is a latent variable, so there is no observable data at all. We again rely on domain knowledge to estimate its probabilities. To compute $P(U_t \mid \text{SIR}_{t-2}, \text{SIR}_{t-1}, \text{SIR}_t)$, we extract two features: $c_t = 10 \times 10^5 (S_{t-1} - S_t)/N$, which represents the number of new reported infections per 100K inhabitants; and $v_t = c_t - c_{t-1}$, which estimates the rate of change of c_t. Then define $P(U_t \mid \text{SIR}_{t-2}, \text{SIR}_{t-1}, \text{SIR}_t) = P(U_t \mid c_t, v_t)$.

To learn the parameters θ we created the dataset shown in Figure 5. Note that the targets in such dataset are probabilities. We relied on the probabilistic labels approach proposed by Vega et al. [30] to use a dataset with few training instances along with their probabilities to learn the parameters of a neural network more efficiently. We trained and a simple neural network with a single hidden layers with 64 units, and 3 output units with softmax activation.

The random variables $U_t \in \{-1, 0, 1\}$ and $O_t \in \{0, 1\}$ are auxiliary variables designed to predict the probability of observing a change in policy at time $t + 1$. Intuitively, U_t represents the "urgency" of modifying a policy. As the number of cases per 100 K inhabitants and the rate of change between the number of cases in two consecutive time points increases, the urgency to set stricter government policies increases. As the number (and rate of change) of cases decreases, the urgency to relax the policies increases. Most of the parameters in both NN-CPD tables are similar for the US and Canada, the difference arises from a perceived preference for not setting very strict policies in the US during the first year of the pandemic.

NN- CPD for US

C_t	V_t	$P(U_t = x \mid C_t, V_t)$		
		x = -1	x = 0	x = 1
0	-40	1	0	0
50	-20	0.9	0.1	0
50	0	0.05	0.95	0
50	20	0	1	0
125	-20	0.05	0.95	0
125	0	0.01	0.98	0.01
125	20	0	0.95	0.05
200	-20	0	1	0
200	0	0	0.95	0.05
200	20	0	0.1	0.9
200	40	0	0	1
250	40	0	0	1

NN- CPD for Canada (All provinces)

C_t	V_t	$P(U_t = x \mid C_t, V_t)$		
		x = -1	x = 0	x = 1
0	-40	1	0	0
50	-20	0.9	0.1	0
50	0	0.5	0.5	0
50	20	0	1	0
125	-20	0.5	0.5	0
125	0	0.01	0.98	0.01
125	20	0	0.5	0.5
200	-20	0	1	0
200	0	0	0.5	0.5
200	20	0	0.1	0.9
200	40	0	0	1
250	40	0	0	1

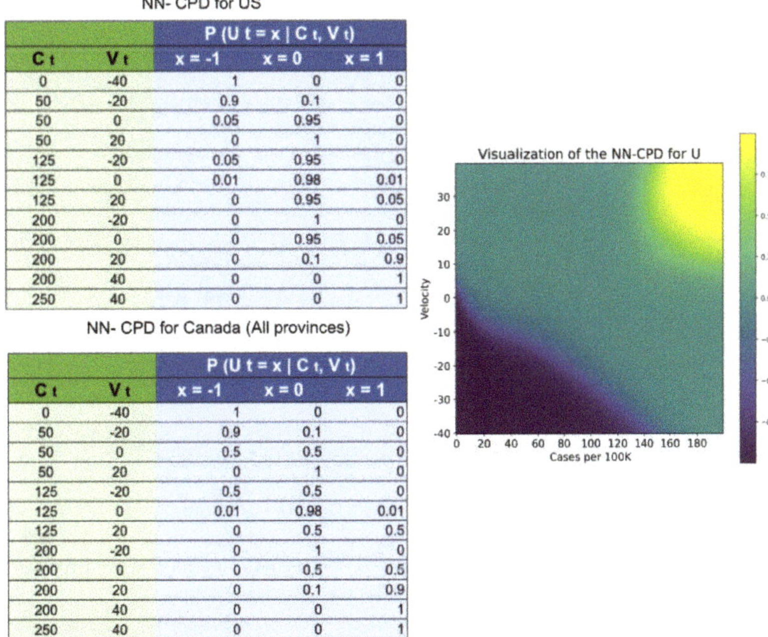

Figure 5. Dataset used to create the NN-CPD for the variable U_t and its visualization. Values closer to 1 (yellow) increase $p(U_t = 1 \mid C_t, V_t)$. Values closer to 0 (green) increase $p(U_t = 0 \mid C_t, V_t)$. Values closer to -1 (blue) increase $p(U_t = -1 \mid C_t, V_t)$.

2.5. Evaluation

We evaluated the performance of SIMLR, in terms of the mean absolute percentage error (MAPE) and mean absolute error (MAE), for forecasting the number of new infections one to four weeks in advance, in data from United States (as a country and individually for every state) and the six biggest provinces of Canada: Alberta (AB), British Columbia (BC), Manitoba (MN), Ontario (ON), Quebec (QB), and Saskatchewan (SK). For each of the regions, the predictions are done on a weekly basis, over the 39 weeks from 26 July 2020 to 1 May 2021. This time span captures different waves of infections. Equation (7) show the computation of the metrics used for evaluating our approach.

$$
\begin{aligned}
\text{MAPE} &= \frac{1}{n} \sum_{t=1}^{n} \left| \frac{y_t - \hat{y}_t}{y_t} \right| \\
\text{MAE} &= \frac{1}{n} \sum_{t=1}^{n} |y_t - \hat{y}_t|
\end{aligned}
\tag{7}
$$

At the end of every week, we fitted the SIMLR parameters using the data that was available until that week. For example, on 25 July 2020, we used all the data available from 1 January 2020 to 25 July 2020 to fit the parameters of SIMLR. Then, we made the predictions for the number of new infections during the weeks: 26 July 2020–1 August 2020 (one week in advance), 2 August 2020–8 August 2020 (two weeks in advance), 9 August 2020–15 August 2020 (three weeks in advance), and 16 August 2020–22 August 2020 (four weeks in advance). After this, we then fitted the parameters with data up to 1 August 2020 and repeated the same process, for 38 more iterations, until we covered the entire range of predictions.

We compared the performance of SIMLR with the SIR compartmental model with time-varying parameters learned using Equation (6) but no other random variable (tf-v-SIR), and with the simple model that forecasts that the number of cases one to four weeks in advance is the "Same as the Last Observed Week" (SLOW). For the United States data, we also compared the performance of SIMLR against the publicly available predictions at the COVID-19 Forecast Hub, which are the predictions submitted to the Center for Disease Control and Prevention (CDC) [31].

For training, we used the publicly available dataset OxCGRT [4], which contains the policies implemented by different regions, as well as the time period over which they were implemented. We limited our analysis to three policy decisions: `Workplace closing`, `Stay at home requirements`, and `Cancellation of public events` in the case of Canada. For the case of the United States we used `Restrictions on gatherings`, `Vaccination policy`, and `Cancellation of public events`. For information about the new number of reported cases and deaths, we used the publicly available COVID-19 Data Repository by the Center for Systems Science and Engineering at Johns Hopkins University [1]. The code for reproducing the results presented here are discussed in Appendix A.

3. Results

3.1. Data Preprocessing

Before inputting the time-series data to SIMLR, we performed some basic preprocessing during the training phase, and exclusively on the training data. We evaluated of our models by comparing its predictions with the results reported by the different health agencies –i.e., we did not fill in the data on the test sets:

1. The original data contains the cumulative number of reported infections/deaths on a daily basis. We trivially transformed this time-series into the number of new daily infections/deaths.

2. We considered negative values from the new daily infections/deaths time-series as missing, assuming these negative values arose due to inconsistencies during the data reporting procedure.
3. We "filled-in" the missing values. When the number of new infections/deaths was missing at day d, we assumed that the entry at $d + 1$ contained the cases for both d and $d + 1$, and divided the number of new infections/deaths evenly between both days.
4. We eliminated outliers. For each day d, with number of reported new infections, ΔI_d, we computed the mean (μ_d) and standard deviation (σ_d of the set $\Delta I_{d-10}, \ldots, \Delta I_{d-1}$; we then set $\Delta I_d := \min\{\Delta I_d, \mu_d + 4\sigma_d\}$.
5. We used the number of new infections and new deaths to produce the SIR vector $SIR_t = [S_t, I_t, R_t]$.

In step 5, we assumed that everyone in a given region was susceptible at the start time—i.e., $S_0 = N$. At each new time point, we transfer the number of new infections from S to I, and the number of new deaths and recovered from I to R. If the number of new recovered people is not reported, we used the surveillance definition of recovered used by Canadian health agencies. This definition is based on the assumption that a recovered person is one who is not hospitalized and is 14 days past the day when they tested positive [32,33]:

> "Active and recovered status is a surveillance definition to try to understand the number of active cases in the population. It is not related to clinical management of cases. It is based on the assumption that a case is recovered 14 days after a particular date..."

3.2. MAPE and MAE

Figure 6 shows the MAPE of the one- to four-week forecasts for the United States as a country and the six biggest provinces of Canada. Note that SIMLR has a consistently lower MAPE than tf-v-SIR and SLOW. Figure 7 shows a similar result in terms of MAE. Tables 1 and 2 show the mean and standard deviations of the metrics corresponding to the Figures 6 and 7. In addition Table 3 show the correlation coefficient between the time series of the reported new infections every week and the predictions made by the different models.

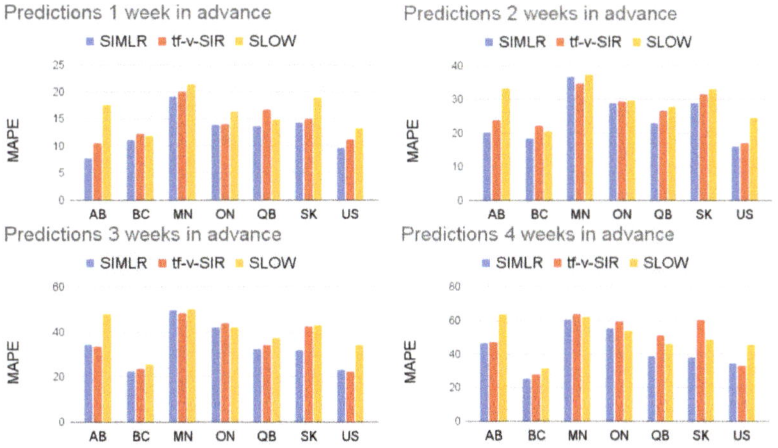

Figure 6. Comparison of SIMLR, SIR model with time-varying parameters, and SLOW. Table 1 contains the numerical information.

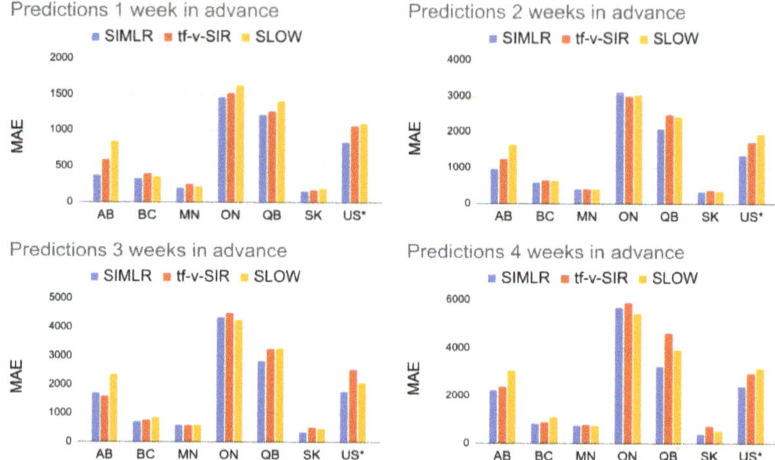

Figure 7. Comparison of SIMLR, SIR model with time-varying parameters, and SLOW in terms of MAE. To make the numbers comparable, the figures each show the US MAE values divided by 100.

Table 1. MAPE of the six biggest provinces in Canada and United States as a country, one- to four-weeks in advance. The number in parenthesis is the standard deviation.

	Week 1			Week 2		
	SIMLR	**tf-v-SIR**	**SLOW**	**SIMLR**	**tf-v-SIR**	**SLOW**
AB	7 (8)	10 (10)	17 (9)	20 (14)	23 (16)	33 (17)
BC	11 (8)	12 (10)	11 (8)	18 (10)	22 (15)	20 (13)
MN	19 (14)	20 (13)	21 (15)	36 (24)	34 (22)	37 (24)
ON	14 (9)	14 (10)	16 (10)	28 (21)	29 (24)	29 (19)
QB	13 (11)	14 (11)	16 (11)	23 (20)	26 (30)	27 (19)
SK	14 (9)	15 (12)	18 (13)	28 (17)	31 (18)	33 (18)
US	9 (6)	11 (8)	13 (9)	16 (13)	19 (16)	24 (17)
	Week 3			Week 4		
	SIMLR	**tf-v-SIR**	**SLOW**	**SIMLR**	**tf-v-SIR**	**SLOW**
AB	34 (21)	33 (22)	48 (26)	46 (35)	47 (33)	63 (35)
BC	22 (14)	23 (16)	25 (18)	25 (20)	27 (21)	31 (20)
MN	49 (31)	48 (34)	50 (27)	60 (38)	63 (42)	62 (33)
ON	42 (37)	44 (40)	42 (30)	55 (51)	59 (58)	53 (40)
QB	32 (28)	34 (36)	37 (27)	38 (41)	51 (64)	45 (35)
SK	32 (23)	42 (32)	43 (22)	38 (24)	60 (50)	49 (26)
US	23 (23)	25 (26)	34 (28)	36 (38)	38 (41)	45 (40)

Table 2. MAE of the six biggest provinces in Canada and United States as a country, one- to four-weeks in advance. The number in parenthesis is the standard deviation. For the case of the US the number of cases was divided by 100.

	Week 1			Week 2		
	SIMLR	**tf-v-SIR**	**SLOW**	**SIMLR**	**tf-v-SIR**	**SLOW**
AB	385 (559)	598 (905)	850 (724)	966 (971)	1245 (1430)	1651 (1258)
BC	339 (304)	397 (426)	361 (294)	594 (443)	661 (480)	648 (485)
MN	204 (227)	252 (271)	221 (224)	422 (371)	418 (379)	413 (346)
ON	1471 (1343)	1520 (1662)	1635 (1388)	3124 (2632)	3001 (2847)	3044 (2351)
QB	1229 (1443)	1265 (1354)	1410 (975)	2098 (2264)	2496 (3270)	2446 (1743)
SK	161 (161)	171 (203)	194 (174)	339 (294)	382 (324)	355 (264)
US*	841 (796)	1061 (1149)	1103 (913)	1361 (1398)	1729 (1979)	1933 (1580)
	Week 3			Week 4		
	SIMLR	**tf-v-SIR**	**SLOW**	**SIMLR**	**tf-v-SIR**	**SLOW**
AB	1719 (1381)	1601 (1558)	2378 (1649)	2261 (1863)	2385 (2087)	3074 (1858)
BC	731 (566)	777 (672)	853 (716)	835 (709)	883 (703)	1127 (892)
MN	609 (504)	591 (501)	602 (467)	749 (612)	775 (630)	753 (571)
ON	4357 (3672)	4511 (3983)	4266 (3053)	5702 (4427)	5910 (4988)	5447 (3417)
QB	2854 (2527)	3261 (4096)	3288 (2389)	3244 (3131)	4636 (6115)	3947 (2788)
SK	351 (320)	522 (500)	472 (306)	410 (287)	736 (733)	541 (348)
US*	1793 (2012)	2089 (2768)	2538 (2151)	2414 (2755)	2933 (4027)	3157 (2679)

Table 3. Pearson correlation coefficient between the ground truth and the predictions of the six biggest provinces in Canada and United States as a country one- to four-weeks in advance.

	Week 1			Week 2		
	SIMLR	**tf-v-SIR**	**SLOW**	**SIMLR**	**tf-v-SIR**	**SLOW**
AB	0.99	0.98	0.96	0.94	0.95	0.84
BC	0.97	0.97	0.97	0.90	0.89	0.90
MN	0.96	0.96	0.95	0.85	0.87	0.86
ON	0.96	0.97	0.96	0.83	0.85	0.85
QB	0.97	0.97	0.96	0.93	0.89	0.86
SK	0.97	0.96	0.95	0.89	0.89	0.86
US	0.97	0.97	0.96	0.93	0.93	0.87
	Week 3			Week 4		
	SIMLR	**tf-v-SIR**	**SLOW**	**SIMLR**	**tf-v-SIR**	**SLOW**
AB	0.90	0.90	0.68	0.84	0.85	0.50
BC	0.84	0.83	0.83	0.80	0.81	0.75
MN	0.69	0.75	0.73	0.51	0.56	0.57
ON	0.67	0.68	0.71	0.51	0.53	0.58
QB	0.83	0.84	0.74	0.71	0.71	0.61
SK	0.82	0.81	0.78	0.73	0.69	0.71
US	0.88	0.90	0.77	0.80	0.84	0.65

Figure 8c shows how our proposed SIMLR approach compares with the 18 models that submitted predictions at the country level to the CDC during the same span of time (results at the state level are included in the Appendix B). Note that SIMLR and the model *LNQ-ens1* are the best performing models, with no statistically significant difference ($p > 0.05$ on a paired t-test) with respect to MAPE.

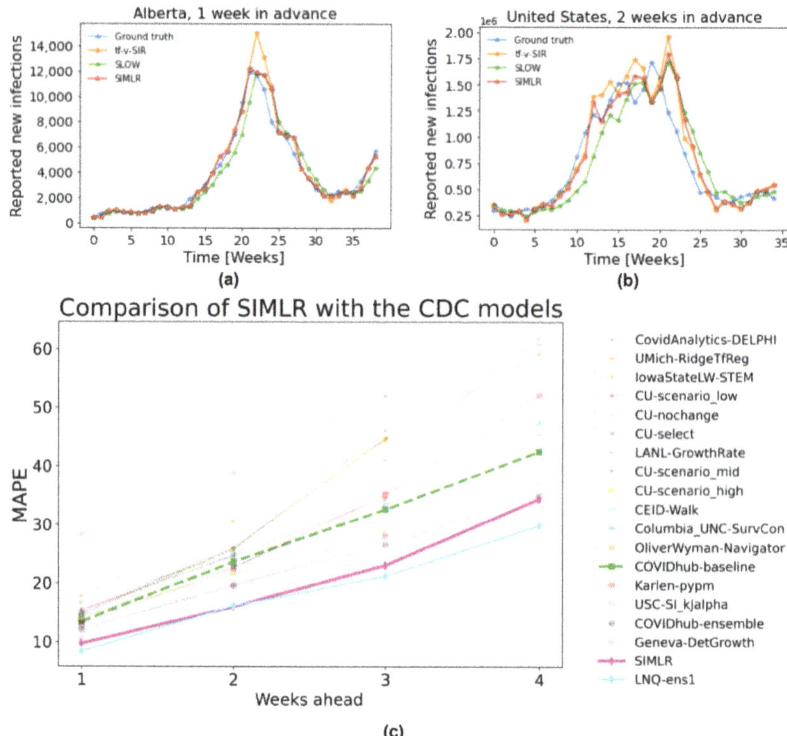

Figure 8. (**a**) 1-week forecasts SIMLR, tf-v-SIR, and SLOW, for Alberta, Canada. (**b**) 2-week forecasts, of the same models, for US data. (**c**) Comparison of SIMLR versus models submitted to the CDC (on US data).

4. Discussion

Figure 8 illustrates the actual predictions of SIMLR one week in advance for the province of Alberta, Canada; and two weeks in advance for the US as a country. These two cases exemplify the behaviour of SIMLR. As noted above, there is a 2- to 4-week lag after a policy changes, before we see the effects. This means the task of making 1-week forecasts is relatively simple, as the relevant policy (at times $t - 3$ to $t - 1$) is fully observable. This allows SIMLR to directly compute CT_{t+1}, which can then choose whether to continue using the SIR with time-varying parameters if no policy changed at time $t - 1$, $t - 2$, or $t - 3$, or using the SLOW predictor if the policy changed.

Figure 8a shows a change in the trend of reported new cases at week 22. However, just by looking at the evolution of number of new infections before week 22, there is no way to predict this change, which is why tf-v-SIR predicts that the number of new infections will continue growing. However, since SIMLR observed a change in the government policies at week 20, it realized it could no longer rely on its estimation of parameters and so switched to the SLOW model, which is why it was more accurate here. A similar behaviour occurs in week 34, when the third wave of cases in Alberta started. Due to a relaxation in the policies on week 31, SIMLR (at week 31) correctly predicted a change of trend around weeks 33–35.

This behavior is not exclusive for the data of Alberta and it explains why the performance of SIMLR is consistently higher than the baselines used for comparison in Figure 6 and Figure 8c. A striking result is how hard it is to beat the simple SLOW model (COVIDhub-baseline). Out of the 19 models considered here, only five (including SIMLR) do better than this simple baseline when predicting three to four weeks ahead. This brings some insight into the challenge of making accurate prediction in the medium term—probably due to the need to predict, then use, policy change information. Tables A1–A4 in the Appendix B show a comparison between our proposed SIMLR and tf-v-SIR against the models submitted to the CDC for all the states in the US. SIMLR consistently ranks among the best performers, with the advantage of being an interpretable model.

A deeper analysis of Tables A1–A4 shows that, in some states, the performance of SIMLR degrades for longer range predictions. This occurs because we are monitoring only the same three policies for all the states; however, different states might have implemented different policies and reacted differently to them. For example, closing schools might be a relevant policy in a state where there is an outbreak that involves children, but not as relevant if most of the cases are in older people.

Tracking irrelevant policies might degrade the performance of SIMLR. If the status of an irrelevant policy changes, then the dynamics of the disease will not be affected. The model however, will assume that the change in the policy will cause a change of trend and it will rely on the SLOW model, instead of the more accurate tf-v-SIR. Although SIMLR can be adapted to track different policies, the policies that are relevant for a given state must be given as an input. So while we think our overall approach applies in general, our specific model (tracking these specific policies, etc.) might not perform accurate predictions across all the regions. This is also a strength, in that it is trivial to adapt our specific model to track the policies of interest within a given region.

Predictions at the country level are more complicated, since most of the time policies are implemented at the state (or province) level instead of nationally. For making predictions for an entire country, as well as predictions three or four weeks in advance, SIMLR first predicts, then uses, the likelihood of observing a change in trend, at every week. In these cases, the random variable CT_{t+1} no longer acts like a "switch", but instead it mixes the predictions of the tf-v-SIR and SLOW models, according to the probability of observing a change in the trend.

Figure 8b shows that whenever there is a stable trend in the number of new reported infections—which suggests there have been no recent policy changes—SIMLR relies on the predictions of the tf-v-SIR model; however, as the number (and rate of change) of new infections increases, so does the probability of observing a change in the policy. Therefore, SIMLR starts giving more weight to the predictions of the SLOW model. Note this behavior in the same figure during weeks 13–20.

One limitation of SIMLR is that it relies on conditional probabilities that are hard to learn due to lack of data, which forced us to build them based on domain knowledge. If this prior knowledge is inaccurate, then the predictions might be also misleading. Also, different regions might have different "thresholds" for taking action. Despite this limitation, SIMLR produced state-of-the-art results in both forecasting in the US as a country and at the provincial level in Canada, as well as very competitive results in predictions at the state level in the US.

Note that modelling SIMLR as a PGM does not imply causality. Although changes in the observed policy influence changes in the trend of new reported cases, the opposite is also true in reality. However, using probabilistic graphical models does makes it interpretable. It also allows us to incorporate domain knowledge that compensates for the relatively scarce data. SIMLR's excellent performance—comparable to state-of-the-art systems in this competitive task—show that it is possible to design interpretable machine learning models without sacrificing performance.

5. Conclusions

Forecasting the number of new COVID-19 infections is a very challenging task. Many factors play a role on how the disease spreads, including the government policies and the adherence of citizens to such policies. These elements are difficult to model mathematically; however, the collected data (number of new infections and deaths, for example) are a reflection of all those complex interactions.

Machine learning, on the other side, excels at learning patterns directly from the data. Unfortunately, training many models from scratch can require a great deal of data, especially to learn complex patterns, such as the evolution of a pandemic.

We proposed SIMLR, a methodology that uses machine learning (ML) techniques to learn a model that can set, and adjust, the parameters of mathematical model for epidemiology (SIR). SIMLR augments that SIR model by incorporating expert knowledge in the form of a probabilistic graphical model. In this way, human experts can incorporate their believes in the likelihood that a policy will change, and when. By combining both components we substantially reduce the data that machine learning usually requires to produce models that can make accurate predictions.

Importantly, besides providing state-of-the-art predictions in terms of MAPE in the short and medium term, the resulting SIMLR model is interpretable and probabilistic. The first means that we can justify the predictions given by the algorithm—e.g., "SIMLR predicts 1000 cases for the next week due to a change in the government policies that will decrease the transmission rate". The second means we can produce probabilistic values—so instead of predicting a single value, it can predict the entire probability distribution—e.g., the probability of 100 cases next week, or of 200 cases or of 1000, etc.

This paper demonstrated that a model that explicitly models and incorporates government policy decisions can accurately produce one- to four-week forecasts of the number of COVID-19 infections. This involved showing that an SIR model with time-varying parameters is enough to describe the complex dynamics of this pandemic, including the different waves of infections. We expect that this approach will be useful not only for modelling COVID-19, but other infectious diseases as well. We also hope that its interpretability will leads to its adoption by researchers, and users, in epidemiology and other non-ML fields.

Author Contributions: Conceptualization, R.V., L.F. and R.G.; methodology, R.V., L.F. and R.G.; software, R.V.; validation, R.V.; formal analysis, R.V., L.F. and R.G.; investigation, R.V., L.F. and R.G.; resources, R.G.; data curation, R.V.; writing—original draft preparation, R.V.; writing—review and editing, R.V., L.F. and R.G.; visualization, R.V.; supervision, R.G.; project administration, R.G.; funding acquisition, R.G. All authors have read and agreed to the published version of the manuscript.

Funding: This research was funded by Alberta Machine Intelligence Institute.

Institutional Review Board Statement: Not applicable.

Informed Consent Statement: Not applicable.

Data Availability Statement: All the datasets used for this manuscript are publicly available. For information about the new number of reported cases and deaths, we used the publicly available COVID-19 Data Repository by the Center for Systems Science and Engineering at Johns Hopkins University [1] https://github.com/CSSEGISandData/COVID-19, accessed on 1 September 2020. For policy tracking we used the OxCGRT [4] https://github.com/OxCGRT/covid-policy-tracker, accessed on 1 September 2020. For comparing our approach with other models we used the publicly available predictions at the COVID-19 Forecast Hub [31] https://github.com/reichlab/covid19-forecast-hub, accessed on 1 September 2020.

Acknowledgments: We thank the Google Cloud Research Credits program and Compute Canada for providing computational support. We also benefited from our many meetings with our colleagues of the greater University of Alberta Covid-Team.

Conflicts of Interest: The authors declare no conflict of interest.

Appendix A. Code Availability

The code for reproducing the main results of this manuscript are publicly available at: https://github.com/rvegaml/SIMLR, accessed on 7 December 2021.

There are six jupyter notebooks on that repository. All the experiments were run using an e2-standard-4 (4 vCPUs, 16 GB memory) computer in the Google Cloud Platform.

- CDC_models.ipynb: It contains the code used to compile the predictions of the models submitted to the CDC. The dataset required to run this script was not included due to the size, but it is publicly available.
- Comparison_CDC.ipynb: It contains the code to create the graphs that compare SIMLR with the models submitted to the CDC. It uses the files created by the previous notebook.
- Model_Canada_Provinces.ipynb: It contains the data to predict the number of cases 1 to 4 weeks in advance in the 6 biggest provinces in Canada.
- Model_US_Country.ipynb: Similar to the previous one, but for the predictions on US at the country level.
- Model_US_States.ipynb: Similar to the previous one, but for the predictions on US at the state level.
- SIR_Simulations.ipynb: Code to create the simulated SIR, and to show how a simple SIR model with time-varying parameters can describe the complexities of the COVID-19 dynamics.

The provided repository in addition contains the in-house developed python library *MLib*. This library contains custom code for inference in probabilistic graphical models.

Appendix B. Additional Tables

Table A1. Comparison of MAPE between different models across all the states in the US 1 week in advance. The number in parenthesis represents the standard deviation of the MAPE.

	1 Week					
State	**tf-v-SIR**	**SLOW**	**SIMLR**	**LNQ-ens1**	**Best**	**Rank**
Alabama	20(16)	19(16)	20(12)	21(15)	20(12)	1/16
Alaska	16(13)	18(15)	17(15)	18(10)	15(14)	4/15
Arizona	21(18)	25(19)	22(21)	18(16)	18(16)	3/16
Arkansas	20(18)	21(29)	24(29)	19(19)	19(19)	13/16
California	15(11)	20(15)	13(10)	13(10)	13(10)	1/16
Colorado	15(15)	19(11)	16(12)	13(8)	13(8)	2/16
Connecticut	17(12)	19(10)	17(11)	23(17)	17(11)	1/16
Delaware	20(14)	18(14)	19(13)	15(11)	15(11)	4/16
Washington DC	23(15)	19(13)	23(15)	15(10)	15(10)	8/16
Florida	12(11)	13(7)	12(8)	9(7)	9(7)	2/16
Georgia	16(12)	16(13)	16(14)	16(15)	16(15)	3/16
Hawaii	27(22)	23(15)	25(17)	18(13)	18(13)	13/15
Idaho	16(11)	16(10)	14(10)	14(10)	14(10)	2/16
Illinois	13(12)	17(10)	12(9)	12(8)	12(9)	1/17
Indiana	11(10)	17(10)	15(10)	13(11)	13(11)	3/17
Iowa	23(18)	21(15)	22(15)	20(22)	20(14)	5/16
Kansas	16(15)	20(15)	18(12)	21(14)	18(12)	1/16
Kentucky	16(11)	16(8)	15(9)	12(9)	12(9)	2/16
Louisiana	24(17)	23(22)	24(22)	21(19)	21(19)	3/16
Maine	17(15)	19(15)	18(15)	14(11)	14(11)	2/16

Table A1. *Cont.*

	1 Week					
State	tf-v-SIR	SLOW	SIMLR	LNQ-ens1	Best	Rank
Maryland	14(12)	15(12)	13(12)	11(7)	11(7)	2/16
Massachusetts	15(10)	16(11)	13(9)	14(10)	13(9)	1/16
Michigan	15(10)	20(10)	16(11)	19(11)	16(11)	1/16
Minnesota	19(17)	21(16)	20(14)	15(12)	15(12)	4/16
Mississippi	19(16)	17(16)	19(15)	16(12)	16(12)	5/16
Missouri	20(14)	19(13)	21(15)	12(38)	11(36)	14/16
Montana	19(17)	21(12)	19(15)	35(104)	18(13)	2/16
Nebraska	20(18)	20(16)	20(15)	18(13)	18(13)	5/16
Nevada	18(17)	20(15)	20(15)	15(11)	15(11)	5/16
New Hampshire	18(14)	18(13)	16(14)	17(11)	16(14)	1/16
New Jersey	11(10)	13(10)	11(9)	14(10)	11(9)	1/16
New Mexico	15(10)	20(12)	15(11)	15(11)	15(11)	2/16
New York	12(9)	14(10)	13(8)	11(9)	11(9)	2/16
North Carolina	12(10)	14(10)	13(9)	12(9)	12(9)	2/16
North Dakota	22(22)	23(24)	23(23)	16(13)	16(13)	8/16
Ohio	12(9)	16(10)	13(10)	11(8)	11(8)	2/16
Oklahoma	22(23)	24(25)	23(24)	15(11)	15(11)	13/16
Oregon	19(13)	18(13)	18(13)	13(10)	13(10)	4/16
Pennsylvania	13(11)	15(12)	15(11)	11(8)	11(8)	3/17
Rhode Island	14(11)	17(11)	13(11)	23(15)	13(11)	1/16
South Carolina	16(13)	16(11)	16(13)	12(8)	12(8)	7/16
South Dakota	18(12)	17(14)	17(11)	15(10)	15(10)	2/16
Tennessee	18(15)	19(15)	22(16)	18(12)	18(13)	12/16
Texas	24(22)	23(28)	25(29)	20(18)	20(21)	7/16
Utah	14(14)	17(11)	16(13)	11(10)	11(10)	7/16
Vermont	25(20)	20(15)	21(14)	21(15)	21(14)	1/16

Table A2. Comparison of MAPE between different models across all the states in the US 2 weeks in advance. The number in parenthesis represents the standard deviation of the MAPE.

	2 Weeks					
State	tf-v-SIR	SLOW	SIMLR	LNQ-ens1	Best	Rank
Alabama	32(27)	32(30)	32(27)	30(19)	30(24)	3/16
Alaska	30(32)	30(25)	27(25)	28(22)	27(24)	2/15
Arizona	41(32)	46(37)	38(36)	32(28)	32(28)	4/16
Arkansas	39(56)	40(61)	45(61)	32(43)	30(28)	14/16
California	22(20)	41(31)	24(21)	25(19)	24(21)	1/16
Colorado	31(28)	30(19)	33(26)	24(19)	24(19)	11/16
Connecticut	27(25)	29(18)	29(26)	33(18)	29(26)	1/16
Delaware	26(19)	26(19)	26(19)	20(16)	20(16)	5/16
Washington DC	34(22)	26(16)	34(23)	23(13)	23(13)	8/16
Florida	20(14)	22(11)	20(13)	14(10)	14(10)	3/16
Georgia	25(18)	31(19)	27(19)	22(20)	22(20)	4/16
Hawaii	41(38)	32(30)	39(36)	29(23)	28(23)	7/15

Table A2. *Cont.*

	2 Weeks					
State	**tf-v-SIR**	**SLOW**	**SIMLR**	**LNQ-ens1**	**Best**	**Rank**
Idaho	25(24)	27(20)	24(23)	24(16)	24(23)	1/16
Illinois	23(18)	31(19)	27(19)	23(16)	23(16)	3/17
Indiana	27(21)	31(23)	31(23)	24(22)	23(16)	13/17
Iowa	36(45)	33(21)	33(26)	34(32)	31(24)	3/16
Kansas	32(28)	35(29)	33(30)	24(17)	24(17)	5/16
Kentucky	26(22)	28(14)	25(22)	19(15)	19(15)	7/16
Louisiana	31(35)	31(39)	31(39)	29(24)	29(24)	3/16
Maine	34(28)	31(27)	34(30)	23(18)	23(18)	6/16
Maryland	24(18)	26(19)	23(18)	22(16)	22(16)	3/16
Massachusetts	26(18)	28(19)	25(19)	24(16)	24(16)	2/16
Michigan	33(22)	35(19)	33(20)	31(16)	27(16)	4/16
Minnesota	40(34)	39(32)	41(35)	28(23)	28(23)	10/16
Mississippi	26(23)	32(25)	31(24)	22(18)	22(18)	11/16
Missouri	32(30)	29(26)	31(27)	18(41)	13(38)	14/16
Montana	34(29)	35(20)	36(28)	30(25)	26(18)	13/16
Nebraska	29(22)	32(20)	30(20)	27(14)	27(14)	3/16
Nevada	31(22)	37(25)	33(26)	23(18)	23(18)	5/16
New Hampshire	29(23)	32(18)	30(24)	28(16)	28(16)	2/16
New Jersey	19(14)	23(13)	19(14)	25(13)	19(14)	1/16
New Mexico	29(23)	36(20)	30(24)	25(21)	25(21)	4/16
New York	24(18)	24(15)	24(18)	21(13)	21(13)	4/16
North Carolina	22(14)	26(18)	25(18)	17(14)	17(14)	6/16
North Dakota	42(39)	41(42)	48(44)	32(24)	31(20)	13/16
Ohio	25(22)	30(19)	29(24)	20(15)	20(15)	10/16
Oklahoma	34(30)	34(32)	37(31)	25(21)	25(21)	13/16
Oregon	29(24)	28(18)	30(24)	18(15)	18(15)	10/16
Pennsylvania	29(19)	27(16)	31(19)	19(14)	19(14)	9/17
Rhode Island	21(17)	29(19)	24(17)	30(19)	24(17)	1/16
South Carolina	27(19)	26(20)	27(21)	18(13)	18(13)	13/16
South Dakota	30(26)	30(28)	32(25)	27(20)	27(20)	4/16
Tennessee	30(24)	29(26)	34(27)	24(19)	24(19)	12/16
Texas	38(49)	35(52)	38(51)	26(26)	25(34)	8/16
Utah	27(29)	30(20)	32(27)	20(19)	20(19)	10/16
Vermont	29(24)	26(22)	28(25)	29(25)	27(23)	3/16

Table A3. Comparison of MAPE between different models across all the states in the US 3 weeks in advance. The number in parenthesis represents the standard deviation of the MAPE.

	3 Weeks					
State	**tf-v-SIR**	**SLOW**	**SIMLR**	**LNQ-ens1**	**Best**	**Rank**
Alabama	40(43)	42(41)	34(36)	34(27)	34(27)	2/16
Alaska	36(41)	37(35)	32(35)	39(36)	32(35)	1/15
Arizona	49(44)	70(59)	59(60)	42(35)	42(35)	6/16
Arkansas	49(52)	54(69)	53(70)	40(38)	37(26)	12/16
California	41(49)	67(53)	48(50)	34(29)	34(29)	6/16

Table A3. *Cont.*

	3 Weeks					
State	**tf-v-SIR**	**SLOW**	**SIMLR**	**LNQ-ens1**	**Best**	**Rank**
Colorado	50(54)	39(26)	39(27)	31(27)	31(27)	5/16
Connecticut	38(42)	39(24)	40(35)	39(21)	39(21)	2/16
Delaware	39(36)	34(28)	39(35)	30(23)	30(23)	5/16
Washington DC	48(44)	32(23)	35(33)	26(20)	26(20)	5/16
Florida	33(26)	34(20)	29(20)	19(14)	19(14)	3/16
Georgia	41(27)	47(26)	39(27)	29(22)	29(22)	5/16
Hawaii	64(79)	41(38)	54(61)	34(28)	34(28)	6/15
Idaho	38(39)	40(31)	35(35)	34(26)	33(25)	4/16
Illinois	38(29)	40(31)	40(28)	33(26)	32(21)	5/17
Indiana	40(33)	44(38)	42(34)	35(33)	32(23)	11/17
Iowa	45(48)	43(34)	42(33)	47(41)	41(38)	2/16
Kansas	47(47)	51(46)	45(43)	31(20)	31(20)	5/16
Kentucky	38(39)	38(25)	31(23)	25(18)	25(18)	5/16
Louisiana	36(41)	48(58)	46(58)	38(28)	38(28)	4/16
Maine	50(39)	43(41)	46(39)	33(27)	33(27)	5/16
Maryland	34(36)	36(33)	37(37)	32(25)	32(25)	5/16
Massachusetts	38(34)	40(28)	38(30)	33(23)	33(23)	2/16
Michigan	49(35)	48(27)	45(24)	43(22)	39(24)	3/16
Minnesota	55(54)	51(51)	51(50)	40(35)	40(37)	6/16
Mississippi	43(38)	47(41)	46(38)	29(23)	29(23)	12/16
Missouri	36(29)	39(39)	39(39)	23(47)	19(43)	12/16
Montana	51(46)	42(32)	40(34)	40(31)	34(21)	7/16
Nebraska	42(33)	44(33)	43(33)	37(26)	37(26)	4/16
Nevada	41(35)	55(42)	47(44)	34(25)	34(25)	6/16
New Hampshire	43(38)	42(24)	38(22)	34(21)	34(21)	3/16
New Jersey	27(24)	31(20)	25(17)	34(16)	25(17)	1/16
New Mexico	46(47)	52(29)	42(32)	33(32)	33(32)	8/16
New York	37(35)	33(18)	30(28)	29(17)	29(17)	2/16
North Carolina	32(21)	36(28)	32(24)	22(15)	22(15)	4/16
North Dakota	61(67)	61(54)	66(67)	50(36)	45(28)	12/16
Ohio	43(42)	41(31)	38(31)	28(19)	28(19)	5/16
Oklahoma	51(50)	46(47)	49(48)	33(22)	33(22)	12/16
Oregon	47(49)	39(23)	35(26)	28(21)	28(21)	2/16
Pennsylvania	46(40)	37(23)	38(23)	27(17)	27(17)	5/17
Rhode Island	27(26)	37(31)	32(28)	39(21)	32(28)	1/16
South Carolina	36(25)	35(28)	33(28)	23(15)	23(15)	3/16
South Dakota	44(41)	47(40)	43(40)	40(29)	40(29)	3/16
Tennessee	34(29)	40(35)	38(37)	31(25)	31(25)	3/16
Texas	52(54)	48(55)	44(55)	31(27)	31(27)	6/16
Utah	42(44)	43(30)	46(33)	32(24)	32(24)	10/16
Vermont	44(29)	37(21)	41(28)	39(24)	38(24)	3/16

Table A4. Comparison of MAPE between different models across all the states in the US 4 weeks in advance. The number in parenthesis represents the standard deviation of the MAPE.

	4 Weeks					
State	**tf-v-SIR**	**SLOW**	**SIMLR**	**LNQ-ens1**	**Best**	**Rank**
Alabama	54(48)	56(52)	51(46)	40(27)	40(27)	3/16
Alaska	58(66)	50(36)	47(38)	49(32)	46(36)	2/15
Arizona	70(80)	104(103)	93(102)	65(68)	61(64)	7/16
Arkansas	59(56)	68(86)	66(87)	46(53)	45(49)	8/16
California	64(87)	95(83)	81(84)	50(47)	50(47)	7/16
Colorado	73(90)	52(26)	52(33)	41(36)	39(24)	6/16
Connecticut	60(65)	46(34)	58(49)	44(23)	44(23)	6/16
Delaware	44(47)	39(37)	46(41)	34(34)	34(34)	5/16
Washington DC	65(64)	37(30)	46(48)	35(34)	35(34)	5/16
Florida	47(46)	52(42)	47(45)	27(26)	27(26)	5/16
Georgia	48(40)	64(35)	59(33)	38(32)	38(32)	7/16
Hawaii	102(153)	55(43)	77(98)	45(43)	45(43)	6/15
Idaho	55(53)	54(41)	53(44)	41(40)	41(40)	7/16
Illinois	53(38)	51(43)	54(40)	43(37)	39(27)	5/17
Indiana	56(51)	61(55)	56(54)	45(49)	44(33)	6/17
Iowa	61(72)	55(46)	53(46)	55(56)	50(45)	2/16
Kansas	66(68)	68(69)	59(58)	43(26)	43(26)	5/16
Kentucky	50(49)	47(39)	43(36)	34(23)	34(23)	5/16
Louisiana	48(49)	68(66)	64(67)	44(35)	44(35)	7/16
Maine	69(64)	56(55)	62(59)	43(40)	43(40)	6/16
Maryland	51(71)	45(45)	53(61)	42(43)	42(43)	6/16
Massachusetts	49(52)	50(40)	47(45)	45(38)	45(38)	2/16
Michigan	62(67)	56(36)	51(37)	53(32)	51(44)	2/16
Minnesota	74(87)	65(61)	64(62)	55(51)	47(41)	5/16
Mississippi	52(49)	62(52)	59(51)	38(43)	38(43)	6/16
Missouri	48(45)	54(57)	54(57)	32(47)	28(44)	9/16
Montana	70(72)	53(40)	55(39)	52(48)	42(36)	9/16
Nebraska	53(43)	57(46)	56(45)	47(31)	47(35)	5/16
Nevada	67(54)	77(61)	71(65)	41(43)	41(43)	8/16
New Hampshire	52(50)	50(30)	43(33)	40(25)	40(25)	2/16
New Jersey	45(62)	36(24)	40(54)	43(24)	38(23)	3/16
New Mexico	69(80)	73(39)	65(48)	46(48)	45(29)	7/16
New York	48(43)	41(22)	39(31)	37(25)	33(22)	4/16
North Carolina	41(33)	48(41)	45(36)	29(22)	29(22)	6/16
North Dakota	79(112)	83(77)	94(93)	72(63)	60(53)	9/16
Ohio	60(58)	54(44)	52(44)	35(31)	35(31)	6/16
Oklahoma	81(92)	61(67)	70(81)	42(32)	42(40)	9/16
Oregon	63(65)	49(33)	43(35)	39(27)	39(27)	2/16
Pennsylvania	63(54)	47(29)	47(30)	35(27)	35(27)	5/17
Rhode Island	37(39)	46(45)	42(43)	44(27)	42(43)	1/16
South Carolina	45(31)	49(37)	49(36)	28(21)	28(21)	8/16
South Dakota	51(47)	64(46)	62(45)	54(40)	52(30)	9/16
Tennessee	48(48)	58(49)	58(50)	43(31)	43(31)	5/16

Table A4. *Cont.*

	4 Weeks					
State	**tf-v-SIR**	**SLOW**	**SIMLR**	**LNQ-ens1**	**Best**	**Rank**
Texas	63(62)	59(67)	58(67)	37(34)	37(34)	6/16
Utah	55(70)	56(44)	58(50)	40(36)	40(36)	7/16
Vermont	56(67)	41(26)	49(55)	45(27)	41(26)	4/16

References

1. Dong, E.; Du, H.; Gardner, L. An interactive web-based dashboard to track COVID-19 in real time. *Lancet Infect. Dis.* **2020**, *20*, 533–534. [CrossRef]
2. Fauci, A.S.; Lane, H.C.; Redfield, R.R. Covid-19—Navigating the Uncharted *N. Engl. J. Med.* **2020**, *382*, 1268–1269. [CrossRef]
3. Liu, H.; Manzoor, A.; Wang, C.; Zhang, L.; Manzoor, Z. The COVID-19 outbreak and affected countries stock markets response. *Int. J. Environ. Res. Public Health* **2020**, *17*, 2800. [CrossRef]
4. Hale, T.; Angrist, N.; Goldszmidt, R.; Kira, B.; Petherick, A.; Phillips, T.; Webster, S.; Cameron-Blake, E.; Hallas, L.; Majumdar, S.; et al. A global panel database of pandemic policies (Oxford COVID-19 Government Response Tracker). *Nat. Hum. Behav.* **2021**, *5*, 529–538. [CrossRef]
5. Arik, S.; Li, C.L.; Yoon, J.; Sinha, R.; Epshteyn, A.; Le, L.; Menon, V.; Singh, S.; Zhang, L.; Nikoltchev, M. Interpretable Sequence Learning for Covid-19 Forecasting. *arXiv* **2020**, arXiv:2008.00646.
6. Liao, Z.; Lan, P.; Liao, Z.; Zhang, Y.; Liu, S. TW-SIR: Time-window based SIR for COVID-19 forecasts. *Sci. Rep.* **2020**, *10*, 1–15.
7. Watson, G.L.; Xiong, D.; Zhang, L.; Zoller, J.A.; Shamshoian, J.; Sundin, P.; Bufford, T.; Rimoin, A.W.; Suchard, M.A.; Ramirez, C.M. Pandemic velocity: Forecasting COVID-19 in the US with a machine learning & Bayesian time series compartmental model. *PLoS Comput. Biol.* **2021**, *17*, e1008837.
8. Blackwood, J.C.; Childs, L.M. An introduction to compartmental modeling for the budding infectious disease modeler. *Lett. Biomath.* **2018**, *5*, 195–221. [CrossRef]
9. Kermack, W.O.; McKendrick, A.G. A contribution to the mathematical theory of epidemics. *Proc. R. Soc. Lond. Ser. A Contain. Pap. Math. Phys. Character* **1927**, *115*, 700–721.
10. Ramazi, P.; Haratian, A.; Meghdadi, M.; Oriyad, A.M.; Lewis, M.A.; Maleki, Z.; Vega, R.; Wang, H.; Wishart, D.S.; Greiner, R. Accurate long-range forecasting of COVID-19 mortality in the USA. *Sci. Rep.* **2021**, *11*, 1–11.
11. Rudin, C. Stop explaining black box machine learning models for high stakes decisions and use interpretable models instead. *Nat. Mach. Intell.* **2019**, *1*, 206–215. [CrossRef]
12. Holmdahl, I.; Buckee, C. Wrong but useful—What covid-19 epidemiologic models can and cannot tell us. *N. Engl. J. Med.* **2020**, *383*, 303–305. [CrossRef]
13. Jacobs, R.A.; Jordan, M.I.; Nowlan, S.J.; Hinton, G.E. Adaptive mixtures of local experts. *Neural Comput.* **1991**, *3*, 79–87. [CrossRef]
14. Santosh, K. COVID-19 prediction models and unexploited data. *J. Med Syst.* **2020**, *44*, 1–4. [CrossRef]
15. Bhapkar, H.; Mahalle, P.N.; Dey, N.; Santosh, K. Revisited COVID-19 mortality and recovery rates: are we missing recovery time period? *J. Med Syst.* **2020**, *44*, 1–5. [CrossRef]
16. Ioannidis, J.P.; Cripps, S.; Tanner, M.A. Forecasting for COVID-19 has failed. *Int. J. Forecast.* **2020**. [CrossRef] [PubMed]
17. Anastassopoulou, C.; Russo, L.; Tsakris, A.; Siettos, C. Data-based analysis, modelling and forecasting of the COVID-19 outbreak. *PLoS ONE* **2020**, *15*, e0230405. [CrossRef]
18. Chen, Y.C.; Lu, P.E.; Chang, C.S.; Liu, T.H. A time-dependent SIR model for COVID-19 with undetectable infected persons. *IEEE Trans. Netw. Sci. Eng.* **2020**, *7*, 3279–3294. [CrossRef]
19. Liu, X.; Stechlinski, P. Infectious disease models with time-varying parameters and general nonlinear incidence rate. *Appl. Math. Model.* **2012**, *36*, 1974–1994. [CrossRef]
20. Walker, P.G.; Whittaker, C.; Watson, O.J.; Baguelin, M.; Winskill, P.; Hamlet, A.; Djafaara, B.A.; Cucunubá, Z.; Mesa, D.O.; Green, W.; et al. The impact of COVID-19 and strategies for mitigation and suppression in low-and middle-income countries. *Science* **2020**, *369*, 413–422. [CrossRef]
21. Knock, E.S.; Whittles, L.K.; Lees, J.A.; Perez-Guzman, P.N.; Verity, R.; FitzJohn, R.G.; Gaythorpe, K.A.; Imai, N.; Hinsley, W.; Okell, L.C.; et al. Key epidemiological drivers and impact of interventions in the 2020 SARS-CoV-2 epidemic in England. *Sci. Transl. Med.* **2021**. [CrossRef] [PubMed]
22. Jin, X.; Wang, Y.X.; Yan, X. Inter-Series Attention Model for COVID-19 Forecasting. In Proceedings of the SIAM International Conference on Data Mining (SDM), SIAM, Alexandria, VA, USA, 25–27 March 2021; pp. 495–503.
23. Kafieh, R.; Arian, R.; Saeedizadeh, N.; Amini, Z.; Serej, N.D.; Minaee, S.; Yadav, S.K.; Vaezi, A.; Rezaei, N.; Haghjooy Javanmard, S. COVID-19 in Iran: forecasting pandemic using deep learning. *Comput. Math. Methods Med.* **2021**. [CrossRef] [PubMed]
24. Mojjada, R.K.; Yadav, A.; Prabhu, A.; Natarajan, Y. Machine Learning Models for covid-19 future forecasting. *Mater. Today Proc.* **2020**. [CrossRef]
25. Omran, N.F.; Abd-el Ghany, S.F.; Saleh, H.; Ali, A.A.; Gumaei, A.; Al-Rakhami, M. Applying Deep Learning Methods on Time-Series Data for Forecasting COVID-19 in Egypt, Kuwait, and Saudi Arabia. *Complexity* **2021**. [CrossRef]

26. Yeung, A.Y.; Roewer-Despres, F.; Rosella, L.; Rudzicz, F. Machine Learning–Based Prediction of Growth in Confirmed COVID-19 Infection Cases in 114 Countries Using Metrics of Nonpharmaceutical Interventions and Cultural Dimensions: Model Development and Validation. *J. Med Internet Res.* **2021**, *23*, e26628. [CrossRef]
27. Cramer, E.Y.; Huang, Y.; Wang, Y.; Ray, E.L.; Cornell, M.; Bracher, J.; Brennen, A.; Castro Rivadeneira, A.J.; Gerding, A.; House, K.; et al. The United States COVID-19 Forecast Hub dataset. *medRxiv* **2021**. [CrossRef]
28. Murphy, K.P. *Machine Learning: A Probabilistic Perspective*; MIT Press: Cambridge, MA, USA, 2012.
29. Koller, D.; Friedman, N. *Probabilistic Graphical Models: Principles and Techniques*; MIT Press: Cambridge, MA, USA, 2009.
30. Vega, R.; Gorji, P.; Zhang, Z.; Qin, X.; Rakkunedeth, A.; Kapur, J.; Jaremko, J.; Greiner, R. Sample efficient learning of image-based diagnostic classifiers via probabilistic labels. In Proceedings of the International Conference on Artificial Intelligence and Statistics. PMLR, San Diego, CA, USA, 13–15 April 2021, pp. 739–747.
31. Cramer, E.Y.; Lopez, V.K.; Niemi, J.; George, G.E.; Cegan, J.C.; Dettwiller, I.D.; England, W.P.; Farthing, M.W.; Hunter, R.H.; Lafferty, B.; et al. Evaluation of individual and ensemble probabilistic forecasts of COVID-19 mortality in the US. *medRxiv* **2021**. [CrossRef]
32. Alberta, G. COVID-19 Alberta Statistics. 2021. Available online: https://www.alberta.ca/stats/covid-19-alberta-statistics.htm#data-notes (accessed on 30 June 2021).
33. Ontario, G. COVID-19 Case Data: Glossary. 2021. Available online: https://covid-19.ontario.ca/data/covid-19-case-data-glossary (accessed on 30 June 2021).

Article

A Deep Learning Model for Forecasting Velocity Structures of the Loop Current System in the Gulf of Mexico

Ali Muhamed Ali [1,*], Hanqi Zhuang [1], James VanZwieten [1], Ali K. Ibrahim [2] and Laurent Chérubin [2]

[1] EECS Department, Florida Atlantic University, Boca Raton, FL 33431, USA; zhuang@fau.edu (H.Z.); jvanzwi@fau.edu (J.V.)

[2] Harbor Branch Oceanographic Institute, Florida Atlantic University, Boca Raton, FL 33431, USA; aibrahim2014@fau.edu (A.K.I.); lcherubin@fau.edu (L.C.)

* Correspondence: amuhamedali2014@fau.edu

Abstract: Despite the large efforts made by the ocean modeling community, such as the GODAE (Global Ocean Data Assimilation Experiment), which started in 1997 and was renamed as Ocean-Predict in 2019, the prediction of ocean currents has remained a challenge until the present day—particularly in ocean regions that are characterized by rapid changes in their circulation due to changes in atmospheric forcing or due to the release of available potential energy through the development of instabilities. Ocean numerical models' useful forecast window is no longer than two days over a given area with the best initialization possible. Predictions quickly diverge from the observational field throughout the water and become unreliable, despite the fact that they can simulate the observed dynamics through other variables such as temperature, salinity and sea surface height. Numerical methods such as harmonic analysis are used to predict both short- and long-term tidal currents with significant accuracy. However, they are limited to the areas where the tide was measured. In this study, a new approach to ocean current prediction based on deep learning is proposed. This method is evaluated on the measured energetic currents of the Gulf of Mexico circulation dominated by the Loop Current (LC) at multiple spatial and temporal scales. The approach taken herein consists of dividing the velocity tensor into planes perpendicular to each of the three Cartesian coordinate system directions. A Long Short-Term Memory Recurrent Neural Network, which is best suited to handling long-term dependencies in the data, was thus used to predict the evolution of the velocity field in each plane, along each of the three directions. The predicted tensors, made of the planes perpendicular to each Cartesian direction, revealed that the model's prediction skills were best for the flow field in the planes perpendicular to the direction of prediction. Furthermore, the fusion of all three predicted tensors significantly increased the overall skills of the flow prediction over the individual model's predictions. The useful forecast period of this new model was greater than 4 days with a root mean square error less than $0.05 \ \mathrm{cm \cdot s^{-1}}$ and a correlation coefficient of 0.6.

Keywords: deep learning; Loop Current; ocean current forecasting; LSTM; ocean measurements

Citation: Muhamed Ali, A.; Zhuang, H.; VanZwieten, J.; Ibrahim, A.K.; Chérubin, L. A Deep Learning Model for Forecasting Velocity Structures of the Loop Current System in the Gulf of Mexico. *Forecasting* **2021**, *3*, 934–953. https://doi.org/10.3390/forecast3040056

Academic Editor: Roberto Henriques

Received: 8 November 2021
Accepted: 6 December 2021
Published: 14 December 2021

Publisher's Note: MDPI stays neutral with regard to jurisdictional claims in published maps and institutional affiliations.

1. Introduction

Sustained large efforts in the ocean modeling community, such as the GODAE (Global Ocean Data Assimilation Experiment), which started in 1997 [1,2] and was renamed as OceanPredict in 2019 [3], have been made to promote and coordinate the approach to ocean forecasting among the international community. This large effort has seen many achievements in terms of predictive capabilities of ocean features temperature, salinity and sea surface height (SSH) and they are evaluated through a standard set of metrics [4]. However, the prediction of ocean currents has remained a challenge to this day—particularly in ocean regions that are characterized by rapid changes in their circulation due to changes in atmospheric forcing or due to the release of available potential energy through the development of dynamical instabilities. Predictions of ocean currents in the California current system can be found in [5], as well as other studies. This paper shows a correlation coefficient less than 0.3 after two days with a root mean square error (RMSE) of 7 cm·s^{-1} for the vertically integrated velocity component. Using mooring measurements in the same oceanographic region as that studied in Chao et al. [5], Shulman and Paduan [6] showed a significant decrease in the correlation coefficient and RMSE with depth between the observation and the model analyses while assimilating the 33 h filtered high-frequency (HF) radar surface current data. Ocean numerical models' useful forecast window is no longer than two days over a given area with the best initialization possible, as shown by [7] in a dynamically active current system, such as the Loop Current (LC) in the Gulf of Mexico (GoM). The RMSE was 10 cm·s^{-1} and the correlation coefficient was 0.63 for the daily surface averaged predicted current. Ocean numerical model predictions quickly diverge from the observational field throughout the water and become unreliable, despite the fact that they can simulate the observed dynamics through other variables such as temperature, salinity and SSH. Numerical methods such as harmonic analysis are used to predict both short- and long-term tidal currents with significant accuracy. However, they are limited to the areas where the tide was measured.

Today's full-water column predictions primarily rely on the use of finite-difference, finite-volume and finite-element methods to solve the primitive equation of motion in numerical models used to simulate ocean dynamics. The outputs of these models consist of the temporal prediction of three-dimensional fields of ocean state variables including both components of the horizontal velocity field, namely u and v along the x and y axes of the Cartesian coordinate system, respectively. In this study, we evaluate the application of a deep learning (DL—[8]) model to predict the three-dimensional velocity field from in-situ data. We demonstrate that the water column current velocity patterns can be learned by a DL model, which can then be used to predict the layered structure of the flow field. To this end, we show that the DL model is capable of accurately predicting the water column velocities more than four days in advance, doubling the current state of the art prediction window for in-situ currents. In this study, we propose a Recurrent Neural Network (RNN) Long Short-Term Memory (LSTM) model [9] to perform predictions of ocean currents' speed and direction, as described in Section 2. LSTM networks have outperformed fully connected neural networks and other machine learning techniques in natural language processing [10,11] that has many similarities with ocean current predictions, as shown by Immas et al. [12]. RNNs have been the state-of-the-art method in modeling time series data for the last decade. In addition, this type of network has seen an increase in real-life applications, including but not limited to aquaculture [13], wind and solar energy resources management [14], bio science and medical applications [15] and also in industrial applications [16].

In a recent study by Wang et al. [17], an LSTM network was used to demonstrate the feasibility of medium-term (3 months) predictions of the GoM's SSH in the LC region. The LSTM model was trained and tested with 18 years of analyzed daily SSH—"analyzed" indicates that the model calculated SSH was corrected with in-situ and remote sensing observations—from the Hybrid Coordinate Ocean Model (HYCOM)-GoM 1/25° horizontal resolution [18]. The Loop Current (LC) and the mesoscale eddies associated with its nonlinear dynamics are the major drivers of the upper 1000 m water column circulation in the GoM [19]. The nonlinear dynamics of the LC is dominated by the shedding of anticyclonic eddies called Loop Current Eddies (LCE) at irregular time intervals [20–22]. The formation of the latter is primarily caused by the growth of baroclinic instability, which is associated with the formation of deep meanders and eddies [19,23,24]. Using metrics set in the literature for LC forecasting, the deep learning model predicted; overall, the LC system SSH frontal distance from reference points within 40 km nine weeks in advance. Furthermore, the model also predicted the final separation of two consecutive LC eddies through the SSH evolution, namely the eddies Cameron and Darwin 8 and 12 weeks in advance, respectively, an improvement over the 5–6-week useful forecast range of state of the art numerical models for the LC dynamics [25].

In this study, the LSTM model is applied to the prediction of water column velocity three-dimensional tensors. The prediction model is implemented on in-situ full water column current measurements collected in the LC region in the GoM between 2009 and 2011. Section 2 describes the measurements and their four-dimensional structure as well as the metrics used to assess the model's skills. Section 3 presents the LTSM prediction model and its implementation on the velocity data. Section 4 presents the model results and concluding remarks are given in Section 5.

2. Method

2.1. Dataset

Long term times series of 3-dimensional velocity flow fields in the LC region are readily available from various ocean numerical model consortia that provide free online access. Such consortia include HYCOM (https://www.hycom.org (accessed on 12 September 2021) [26]), Navy Coastal Ocean Model (NCOM) (https://www.ncdc.noaa.gov/data-access/model-data/model-datasets/navoceano-ncom-reg (accessed on 12 September 2021) [27]), or ECCO (http://www.ecco.ucsd.edu (accessed on 12 September 2021) [28]) for example. In comparison, long term in-situ measurements of the LC system water column are scarce.

A comprehensive observational study of the LC in the eastern GOM, including 9 tall moorings and 7 short moorings, an array of 25 pressure-equipped inverted echo sounders (PIES), and remote sensing, measured the water column velocity for 2.5 years, beginning in April 2009 [29]. This array was located to cover both the east and west sides of the LC between the West Florida Slope and the Mississippi Fan, and was also centered over the zone where LCEs typically separate from the LC. The horizontal separation between moorings was around 50–80 km and between the PIES sensors was around 40–50 km. These recorded data were used to construct the measurement-based water velocity matrix used in this study.

To create such a matrix, these observations were processed using the optimal interpolation, as described in [30,31]. The horizontal resolution of the resulting data array was roughly based on the correlation length scales of recorded data, with the geostrophic velocity profiles based on the gravest empirical method (GEM) [30,32]. The resulting measurement-based water velocity matrix comprised 50 depth levels down to 3000 m below the surface, and extended horizontally between 88.5° W to 85° W and 24.65° N to 27° N with a horizontal resolution between 30 and 50 km (Figure 1). However, in this study, only the first 500 m was selected, corresponding to 26 vertical layers. The time resolution for the velocity data was 12 h, which corresponds to 1810 data frames for each u and v velocity component. The final matrix dimensions were of $1810 \times 26 \times 29 \times 36$. The current velocity measurements used in this study encompass the period from May 2009 to November 2011, during which three LCEs, namely Ekman, Franklin, and Hadal, were formed.

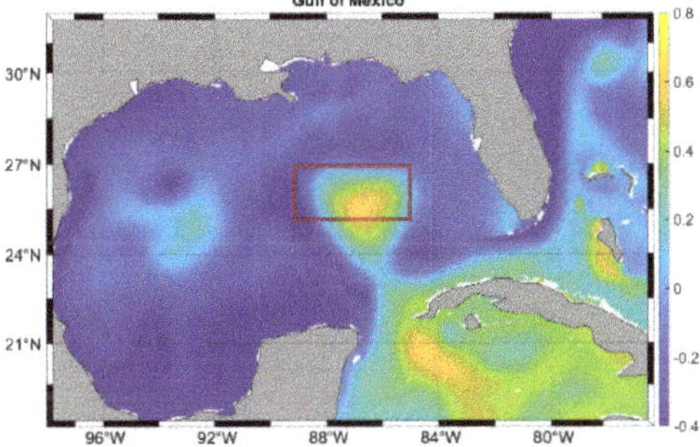

Figure 1. Loop Current SSH from HYCOM [33] (m) during eddy Ekman separation on 1 July 2009. The red rectangle shows the array boundaries.

2.2. 4-Dimensional Tensor Slicing

The time series of the 3-dimensional gridded velocity forms a tensor whose dimensions must be reduced to one so it can be processed by the DL model. At each time step, a gridded velocity cube can be sliced in layers perpendicular to the three Cartesian coordinate axes (Figure 2). Thus, in the vertical direction (z-axis), the volume is split in horizontal layers corresponding to each depth level of the velocity data. Each layer becomes its own time series and can be reduced to a single dimension by EOF decomposition, as was carried out for the SSH field in [17]. For each resulting layer and velocity component, a DL model, trained on its own layer, is used to predict the evolution of that particular layer only. A similar approach can be used for layers perpendicular to the x and y axes and located at each grid point of the respective axis, as shown in Figure 2. As errors are specific to each layer and because the tensor evolves differently in each of the directions, it is expected that the models' skill will vary with the direction of prediction, as explained in the following section.

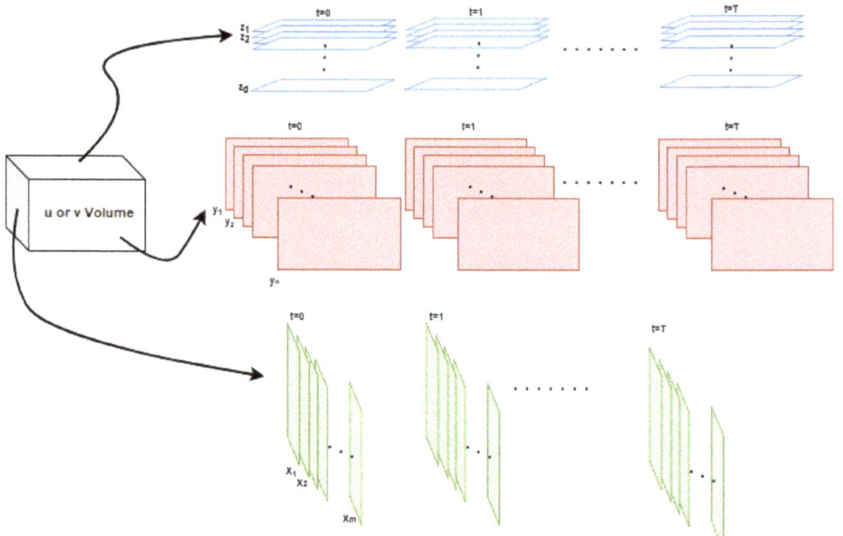

Figure 2. Velocity plane field extractions perpendicular to each of the three Cartesian coordinate directions at each depth (z) in the vertical direction and at each grid point of the x (zonal direction) and y-axis (meridional direction), respectively.

2.3. Volume Slicing Induced Errors

Most numerical model solutions are obtained from discretized partial differential equations solved on one or more embedded volumetric grid, such as the Arakawa C grid [34]. To solve these equations, boundary conditions are provided at the grid boundaries, where virtual grid points are added for computational purposes. Specific boundary conditions allow the radiation of features from within the grid to outside of it without losing the integrity of the signal inside the grid during the outing process. This process can be tracked in all three directions. In the case of a deep learning model, only features contained within the grid are available to the model. There is no influence from the boundaries, which serve to constrain the solution within the model and limit the model solution's drift in numerical models. Therefore, DL model forecast errors in individual layers may grow significantly over time and ultimately change the integrity of the signal, as shown in Figure 3. This is particularly relevant in the case of perturbation simulations, where the phase of the signal in different layers could be changed by the errors in the individual layered predictions. Figure 3 provides an example of what it would look like in each of the planes normal to each direction. Starting with the z-planes (top view), a slight phase shift in the vertical direction will lead to the removal of the red signal in the x-plane in the region outlined by the green shaded area and also in the y-plane the furthest on the outside. As each forecast is sequentially reused for the next, the errors become part of the learning base. Additionally, because horizontal motions are much larger than vertical motions in the ocean, the DL model prediction skills will differ according to the direction of layers used for prediction.

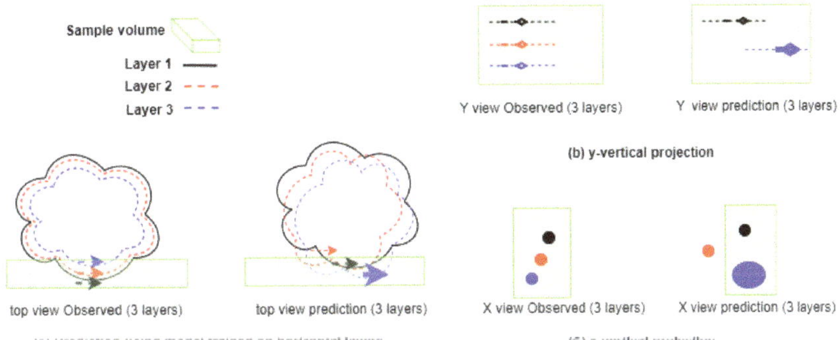

Figure 3. Layered prediction-induced errors. (**a**) Top view, normal to the z-axis. (**b**) Lateral view normal to the y-axis. (**c**) Lateral view normal to the x-axis. The green shaded area highlights the focus area where errors are displayed. In each subplot, the left (right) image shows the observed (predicted) field. Each color corresponds to a different vertical layer as indicated, layer 1(3) being at the top (bottom).

To evaluate the layered prediction errors, the metrics set by GODAE OceanPredict [4] were applied. They identify two types of errors, namely the single point error and the structural error. These errors are quantified by the calculation of the Peak Signal to Noise Ratio (PSNR) including RMSE and correlation coefficient (CC) (see [17] for definitions), and Structural Similarity (SSIM), respectively. The PSNR is based on the mean square error (MSE) [35]. Given an observed plane field Ob of size m, n and its prediction Pr, MSE is defined as:

$$\text{MSE} = \frac{1}{m\,n} \sum_{i=0}^{m-1} \sum_{j=0}^{n-1} [Ob(i,j) - Pr(i,j)]^2 \tag{1}$$

The PSNR (in dB) is defined as:

$$\text{PSNR} = 10 \cdot \log_{10}\left(\frac{Peak^2}{\text{MSE}}\right) \tag{2}$$

where $Peak$ is the maximum value of all data points in both Ob and Pr. In image processing, PSNR is primarily used to assess the quality of an image reconstruction. The PSNR between two images is calculated in decibels. To compare image reconstruction quality, both the mean square error (MSE) and peak signal-to-noise ratio (PSNR) are often utilized.

The SSIM index can be calculated in sub-regions of each layer. It is a measure of similarity between two patterns [35].

$$\text{SSIM}(Ob, Pr) = \frac{(2\mu_{Ob}\mu_{Pr} + c_1)(2\sigma_{ObPr} + c_2)}{(\mu_{Ob}^2 + \mu_{Pr}^2 + c_1)(\sigma_{Ob}^2 + \sigma_{Pr}^2 + c_2)} \tag{3}$$

where:

- μ_{Ob} and σ_{Ob}^2 are the mean and variance of Ob, respectively.
- μ_{Pr} and σ_{Pr}^2 are the mean and the variance of Pr, respectively.
- σ_{ObPr} is the covariance of Ob and Pr.
- $c_1 = (k_1 L)^2, c_2 = (k_2 L)^2$ are used to stabilize the ratio with a weak denominator.

- $L = 2^{\#bits\ per\ gridcell} - 1$ is the dynamic range of the gridded velocity values.
- $k_1 = 0.01$ and $k_2 = 0.03$ are the default values of the two scale factors.

3. Deep Learning Prediction Model

Unlike conventional numerical models which use a set of a dynamical equations to describe a physical system, data-driven deep learning methods rely on neural networks to model physical systems. To achieve this, we first reduced the temporal matrix of each layer to one dimension by applying EOFs and then implemented an LSTM network to model the velocity field in each layer, as shown in Figure 4.

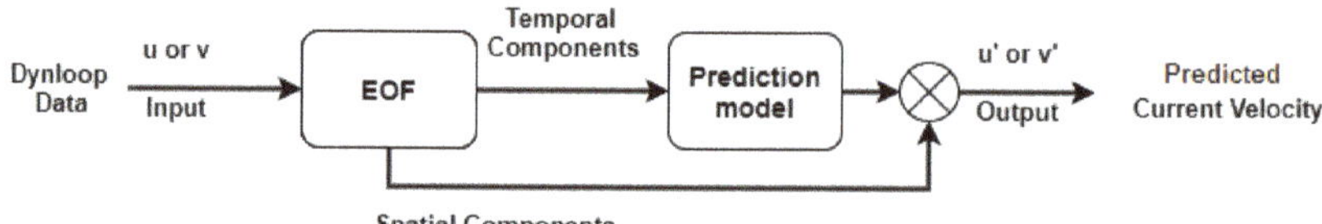

Figure 4. Single layer forecasting model flow chart.

3.1. Empirical Orthogonal Functions

EOF is a major analysis tool in oceanographic, geophysical and meteorological applications [36–38]. EOFs are used to reduce data dimensions by separating spatial components from temporal components. The principle of this decomposition is to extract the most dominant information with fewer dimensions [39]. It provides a dense description of spatial data and temporal variability in terms of an orthogonal basis (eigenvectors). Each associated eigenvalue provides a measure of the fraction of the total variance under the EOF mode. This decomposition provides a statistical description of any dynamical processes by projecting them onto empirical normal modes, rather than the physical or natural modes of the system, which are process specific and therefore unable to encompass all the processes involved in the dynamics of the system being predicted in this study. The projection of the data onto EOF modes is called principal component (PC), which indicates the temporal variations of the variance of it associated spatial pattern [37]. EOF decomposition is carried out by Singular Value Decomposition (SVD), which is written as follows:

$$Q = UDW^T \tag{4}$$

where Q is an $n \times p$ matrix and D is an $n \times p$ rectangular diagonal matrix of non-negative numbers (the singular values of Q). U is an $n \times n$ matrix, the columns of which are orthogonal unit vectors of length n, called the left singular vectors of Q, and W is a $p \times p$ matrix whose columns are orthogonal unit vectors of length p and called the right singular vectors of Q. In addition, UD is the time-dependent principal components (PCs), and W^T is the spatial pattern matrix whose columns are so-called EOF modes.

3.2. Deep Learning Model: Long Short-Term Memory Network

The deep learning model selected for the prediction model is a type of Recurrent Neural Network (RNN). RNNs are well suited for time sequence prediction, and work by feeding the output of each neuron, along with a new input, back into itself, forming loops within its architecture [40]. In an RNN network, a simple RNN neuron or hidden unit's output behavior can be modeled by Equation (5), where x_n and s_n are the input and state at time n, respectively. Furthermore, W_{g_i} and W_{g_s} represent the input and state (recurrence) weights and f an activation function. Note that the output can be obtained from the state whenever it is needed.

$$s_n = f(W_{g_i} x_{n-1} + W_{g_s} s_{n-1}) \tag{5}$$

However, the caveat of the RNN given in Equation (5) is its gradient vanishing problem or memory loss. This occurs because RNNs are typically trained with a stochastic gradient descent algorithm, and gradients may vanish for a multi-layer RNN due to the chain rule in differentiation. LSTM neural networks were designed to solve this problem [41], in which a memory unit m_n was added to avoid the disappearance of gradients. Let α and β be constants and $\oplus$ denote an element-wise multiplication; then, the memory unit is updated by the following rule:

$$m_n = \alpha \oplus m_{n-1} + \beta \oplus f(W_{g_i} x_{n-1} + W_{g_s} s_{n-1}) \tag{6}$$

The state is then related to the memory unit with an activation function. In this way, derivatives will not vanish due to the additive relationship described in Equation (6).

3.3. Prediction Procedure

The LSTM network used in this study was previously adopted for the prediction of SSH time series in [17,42]. In this study, two identical networks were designed to model and predict the velocity components, u and v, respectively. After the EOF decomposition of each velocity component, the PCs were used to train the LSTM model, which in turn was used to predict future PCs. At the beginning of the process, the system was initialized using random weights and then run through all the training data in chronological order, each time adjusting the weights through the gradient descent of the loss function. Each run through the entire training dataset is called an epoch and this step allows the model to optimize its weights (Equation (6)), at which point the model can predict any state learned from the data without the data at any point in time in hindcast mode.

The MATLAB Neural Network Toolbox was used to implement the LSTM network. The Adaptive Moment estimation [43] (Adam) optimization rather than the Stochastic Gradient Descent with Momentum [44] (SGDM) algorithm was used to update network weights iteratively during the training phase due to the improved performance with the former. The hyperparameters of the prediction model were manually tuned to optimize the performance of the prediction model. The resulting hyperparameters were as follows: mini-batch size = 128, initial learning rate = 0.03, number of hidden nodes = 100, and maximum number of Epochs = 500. Only one LSTM layer was used because the overall performance of the model, including the training and prediction processing time, as well as prediction skills, degraded when more layers were added. Training and prediction were carried out on a single NVIDIA GPU, TITAN X (Pascal compatibility) with CUDA toolkit Version 11 with a memory of 12GB. The training times for a single layer and for all layers for each direction are provided in Table 1. Once trained, at each prediction time step, the LSTM updates its state in accordance to its own prediction. This allows the LSTM to continue predicting based on both the training data and future predictions.

Table 1. Training time in seconds for a single and all layers in each directional model.

Training Direction	Single Layer	All Layers
Z direction	25.22 s	649.43 s
Y direction	26.32 s	722.8 s
X direction	25.16 s	652.15 s

The prediction procedure can be summarized as follows: (1) The current velocity components' time series from time t_1 to t_n are reduced to their respective PCs by EOF decomposition. (2) The PCs are used to train the LSTM model sequentially. (3) All the PCs up to time t_n are then used to predict the PCs of the velocity field at time $t_n + 1$. For the next prediction at time $t_n + 2$, the predicted PCs at $t_n + 1$ are used to retrain the LSTM model together with the PCs corresponding to time t_1 to t_n, and this is repeated for all subsequent forecasts. In addition, new data can be added at any time to the training dataset, which will then be used to retrain the LSTM model.

3.4. Layered Prediction Model Approach

As previously described, the water velocity dataset is a time series consisting of two orthogonal components u and v, each of which known as a four-dimensional tensor. To reduce the computational complexity, at each instant, the corresponding velocity cube (u or v) is partitioned into a number of layers (or planes). For each layer, a prediction module consisting of EOF and LSTM is trained and then used to predict the velocity field of that particular layer. Collectively, these prediction modules form a layered prediction model. Layered models are implemented for each spatial direction. As a consequence, they are referred to as prediction models X (29 layers), Y (36 layers), and Z (26 layers), respectively.

4. Layered Prediction Experiments

In all three directions, each layered model was trained using 90% of the available time series and preserving the remaining for prediction validation. Thus, the training used 1629 samples (814.5 days) from 2 March 2009 to 25 May 2011, while the testing period started from 26 May 2011 to 23 August 2011 (90.5 days). The training and testing periods are illustrated in Figure 5. The model prediction period was set to 7 days, which was also the length of the sliding prediction window. This prediction period was chosen in response to the predictive skill goal set for the LC current speed by the United States' National Academies of Sciences, Engineering, and Medicine (NASEM) [45].

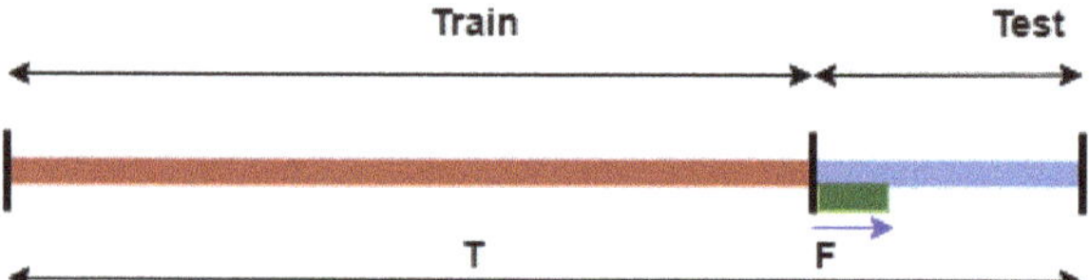

Figure 5. Data partitioning for training and prediction experiments. T is the duration of the dataset (1810 time samples = 905 days), F the length of the prediction window (14 samples = 7 days), and the blue line is the testing period which includes 20 forecast sliding windows, each separated by a 12 h period.

4.1. Model Z Velocity Predictions

Model Z, or the Z directional model, consisted of 26 horizontal layers distributed from the sea surface down to 500 m. The 7-day prediction of the model for the surface layer velocity (layer 1 in the z-direction) is shown in Figure 6. It illustrates that the proposed model was able to predict seven days in advance the formation of a cyclonic eddy in the region highlighted in Figure 6. The model accurately predicted the center of rotation, direction and strength of the velocity vectors at the surface. However, elsewhere, the model prediction differed more significantly from the observations. To assess the overall performance of the model, we computed the average CC, RMSE, SSIM and PSNR, along with their standard deviations, for both u, v on each plane of each tensor and over the twenty sliding windows (Figure 7). These quantities quickly deteriorated over 14 time steps (7 days), which indicates the challenge of predicting LC velocity tensors compared to SSH prediction performed using a similar LSTM structure in [17,42,46], despite the fact that the cyclonic eddy was correctly predicted. Indeed, after 7 days, the anticyclone southwest of the predicted cyclone exhibited a weaker circulation than the observed one, and its northern counterpart was sustained for longer than the observed one.

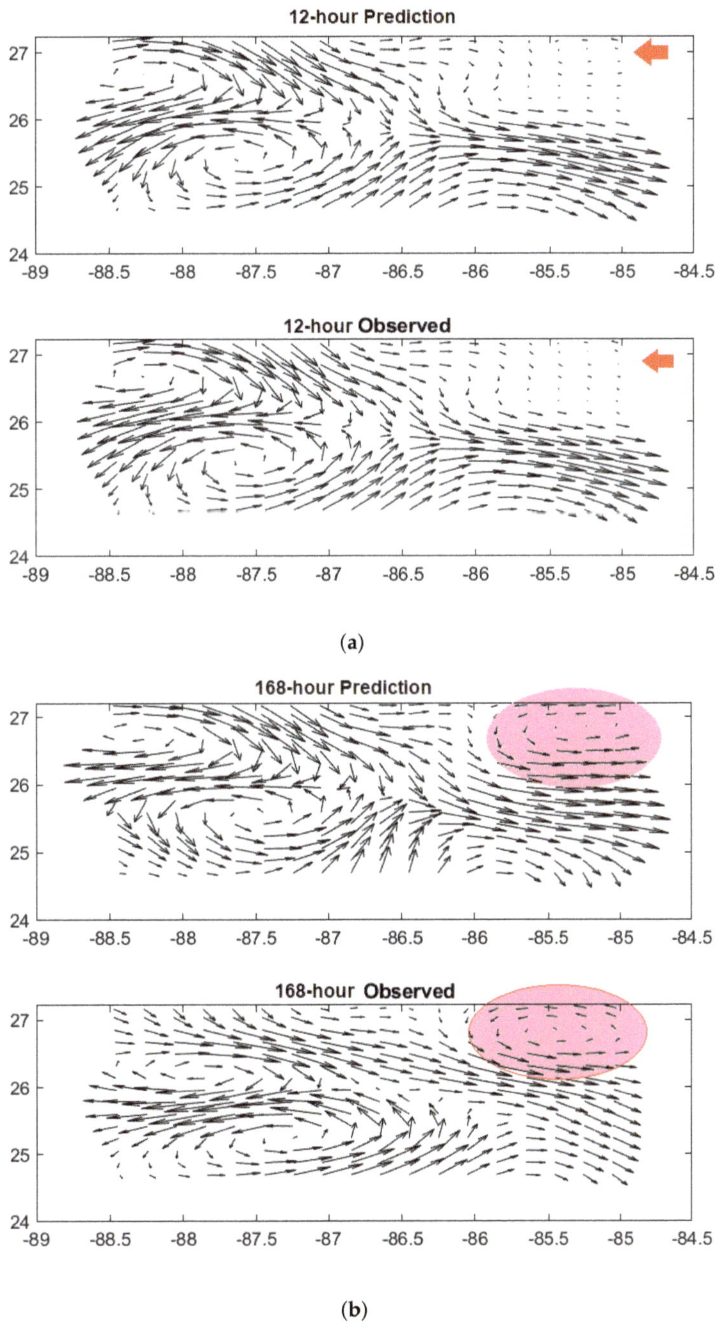

Figure 6. Predicted (top) and corresponding observed (bottom) surface velocity. (**a**) Twelve-hour prediction. (**b**) One hundred sixty-eight-hour (7-day) prediction. The red arrows in (**a**) show the region of formation of the cyclonic eddy predicted in the red highlighted areas in (**b**).

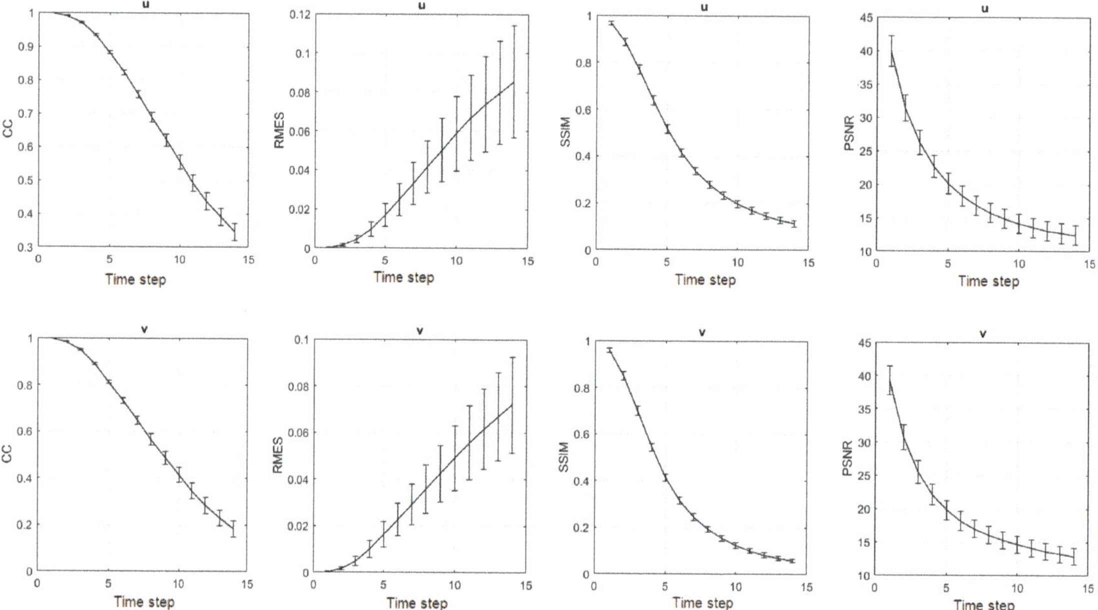

Figure 7. Model Z fourteen time-step (7 days) 20-day sliding window average of CC, SSIM, PSNR, and RMSE of the velocity fields. The unit for the horizontal axis is prediction time steps (one time step is 12 h). The error bar denotes the standard deviation with a 95% confidence interval.

4.2. Directional Velocity Structure Prediction Dependency

We now compare the predictions of all models in each of the three Cartesian directions. In particular, for a given directional model prediction, we compared the other two model predictions along the same direction as the former. Figure 8 shows the comparison in terms of CC, RMSE, SSIM and PSNR for Model X prediction and the other two models (Models Y and Z) in the x-direction. All the metrics were calculated for fourteen time steps and averaged over 20-day sliding windows and over all the layers of the directional model. Model X prediction in the x-direction exhibited a higher CC and similarity index, although the PNSR is very similar between models, especially after the seventh time step. On the other hand, the RMSE is the highest for Model X after the sixth time step. Model Y prediction is also better than Model Z's prediction in the x-direction.

A similar comparison is shown for Model Y in Figure 9. The CC and the similarity index are strikingly much higher for Model Y and than for the other two models. The PSNR is also significantly higher and the RMSE much lower than for the other two models. The prediction of Model Z in the y-direction was also better than the one of Model X. In the x-directions, fewer differences were found between all three models predictions than in the y-direction.

Figure 10 shows the comparison of the three models in the z-direction. As expected, Model Z is better at predicting in the z-direction; however it shows a better CC than for the other two models only after 7 days. The similarity index is much higher while the PSNR is similar to the ones of the other model, showing no significant improvement. The RMSE becomes lower than for the other two models after the seventh time step. Again, as for the x-direction prediction, the differences between the three models are not as different as they were in the y-direction. These results indicate that each model best prediction is associated with its direction of prediction. In addition, in terms of dynamical evolution, the most significant changes were in the y-direction (x-z planes) and better captured by the Model Y.

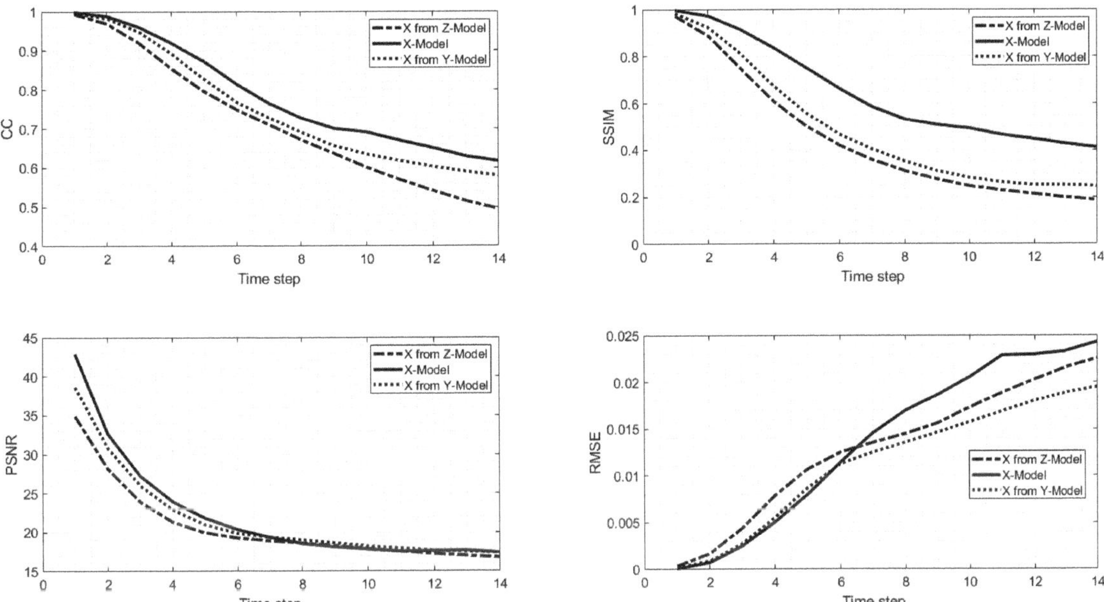

Figure 8. Fourteen time-step (7 days) 20-day sliding window average of CC, SSIM, PSNR, and RMSE of the velocity fields in the x-direction, predicted by Model X (solid line), Model Y (dotted line), and Model Z (dashed line).

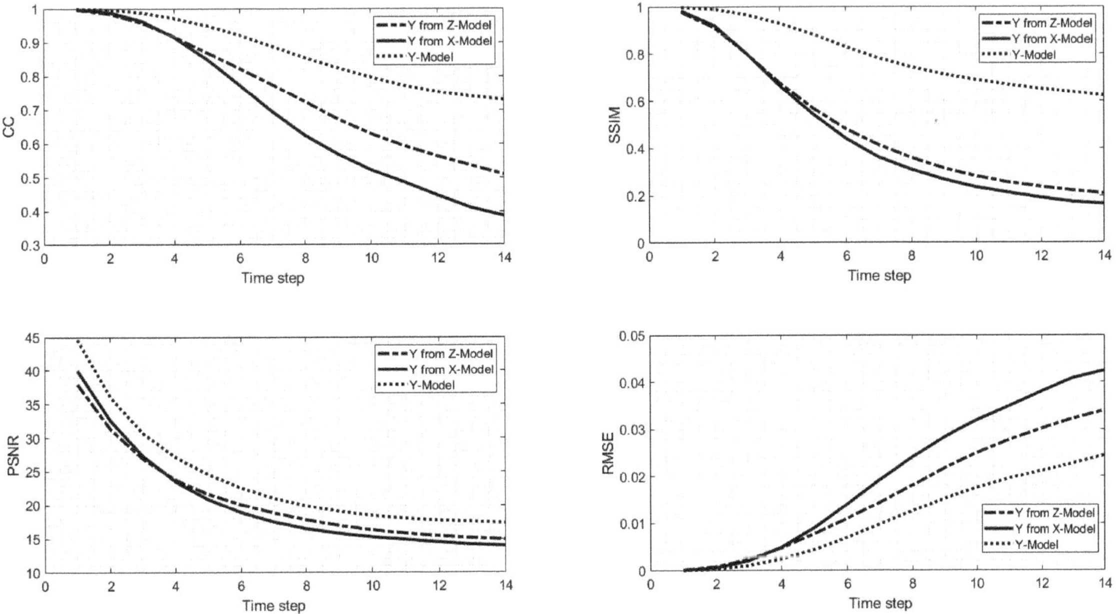

Figure 9. Same as Figure 8 but for the velocity fields in the y-direction.

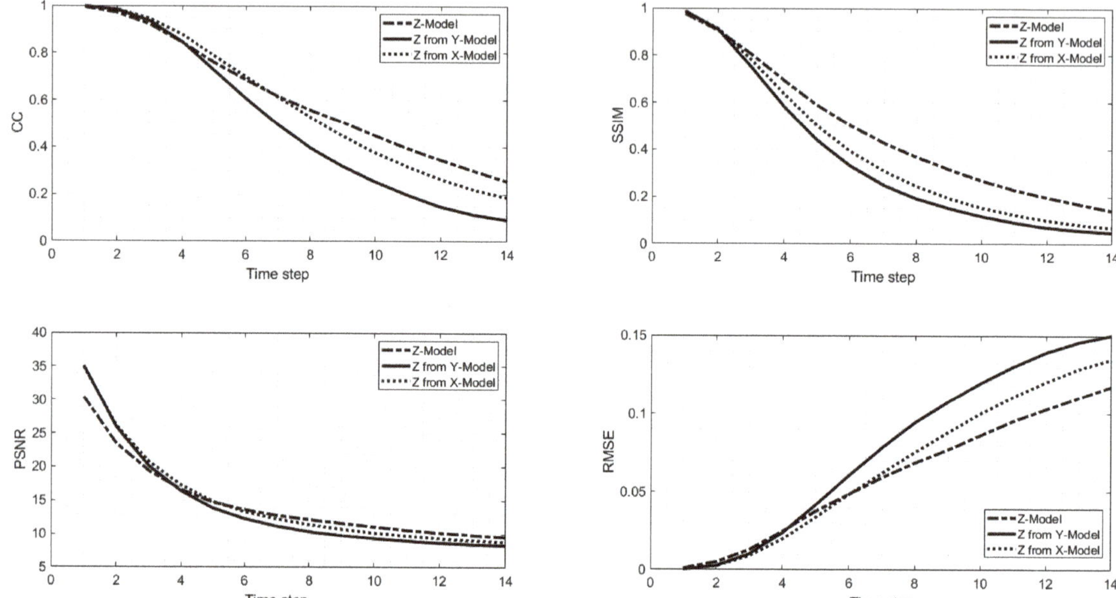

Figure 10. Same as Figure 8 but for the velocity fields in the *z*-direction.

4.3. Vertical Velocity Structure Prediction

Examples of the 7-day predicted flow field are shown in Figures 11–13 for Models X, Y, and Z respectively. Figure 11 shows a vertical section of the velocity magnitude in the *x*-direction for all three models at 87° W. It confirms the metrics results and shows the best agreement between the flow structure of Model X and the observations in the *x*-direction. Similarly, Figure 12 shows the vertical section of the velocity magnitude in the *x*-direction for all three models at 25° N. It confirms the metrics results and shows the best agreement between the flow structure of Model Y and the observations in the *y*-direction. The same consistency between the prediction and observed flow structure is also confirmed for Model Z in the *z*-direction (Figure 13). As each prediction model performs best in its corresponding tensor orientation, we propose fusing the prediction of all three models into one tensor.

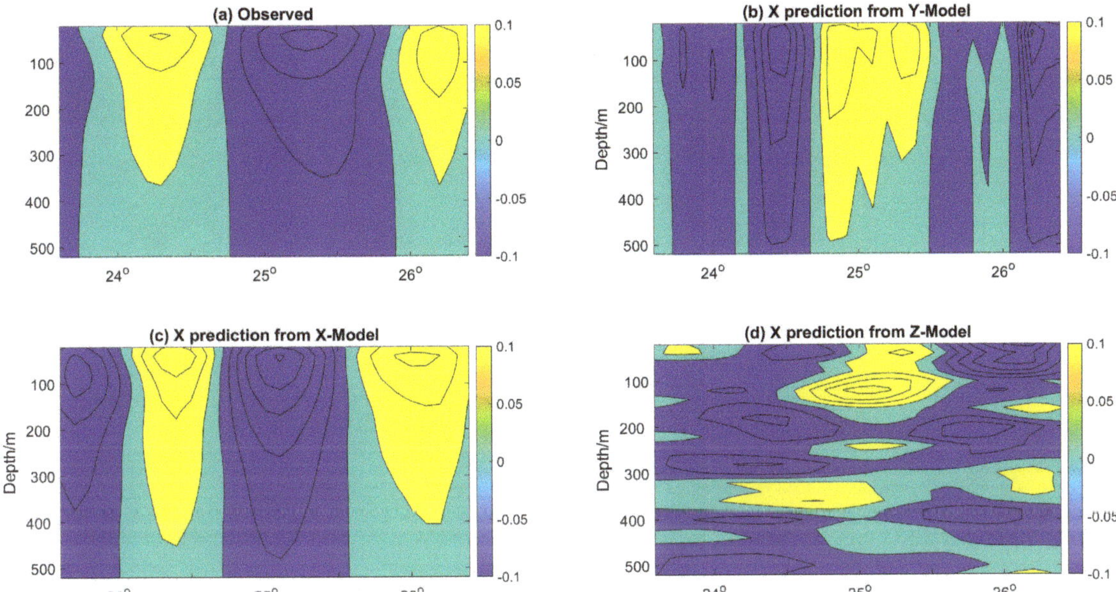

Figure 11. Vertical section of the velocity tensor in the *x*-direction at 87° N on day 7 of the prediction. (**a**) Observations; (**b**) Model *X* prediction; (**c**) Model *Y* prediction; (**d**) Model *Z* prediction.

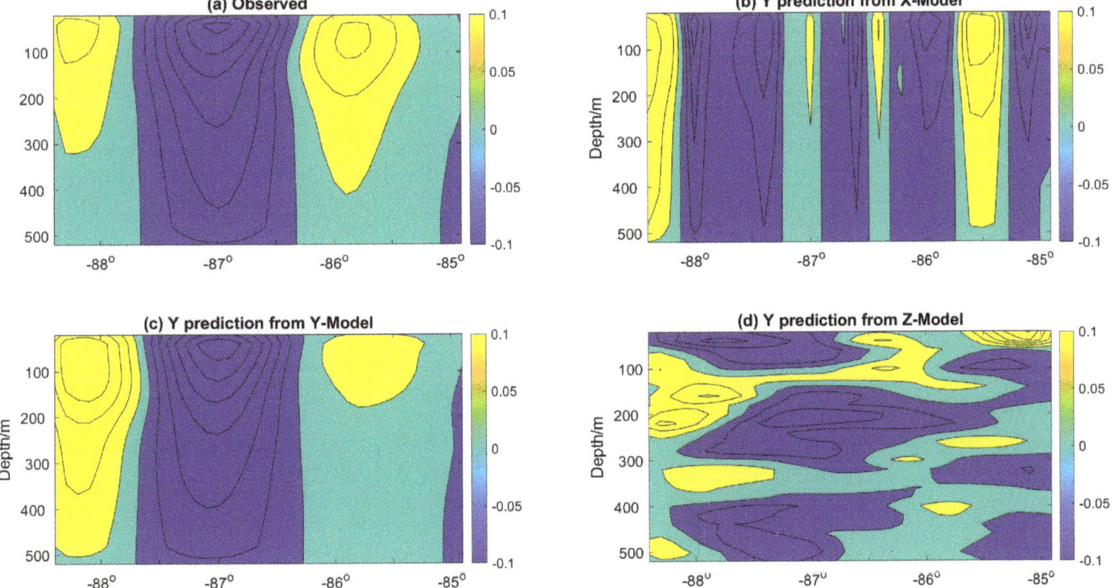

Figure 12. Vertical section of the velocity tensor in the *y*-direction at 25° W.

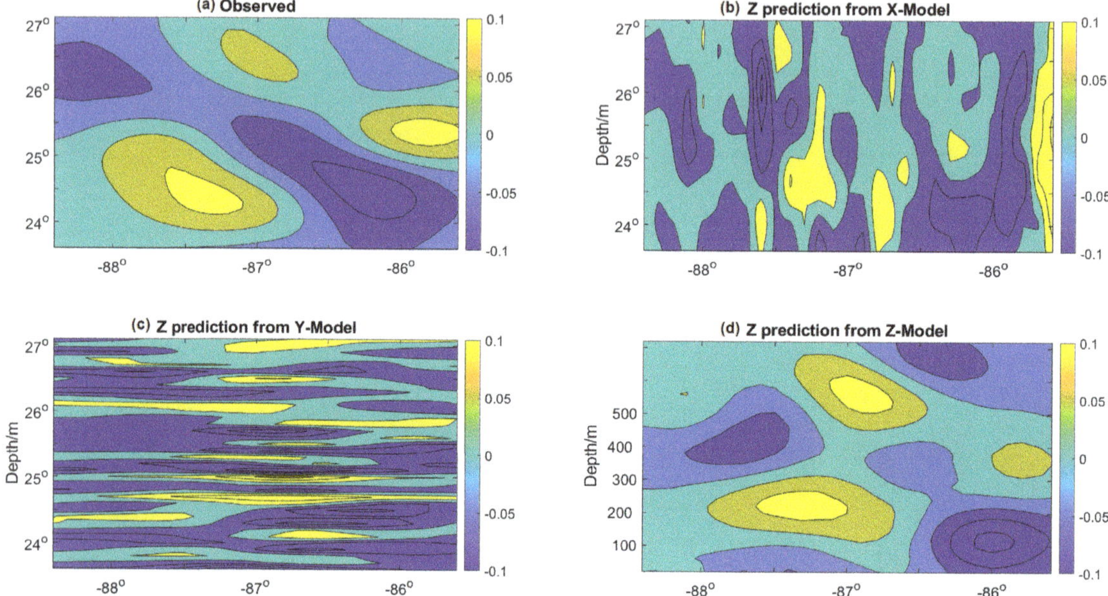

Figure 13. Horizontal section of the velocity tensor at 100 m depth on day 7 of the prediction. (**a**) Observations; (**b**) Model X prediction; (**c**) Model Y prediction; (**d**) Model Z prediction.

4.4. Fusion of the Models' Predictions

As each model can best predict the evolution of the velocity field in its respective layers, we hypothesize that the fusion of the three model predictions would yield an improved prediction of the overall tensor over each individual one. For this purpose, a simple fusion block was added to the prediction system, as shown in Figure 14. Although various methods can be used to fuse all three tensors, such as unweighted or median selection-based average, we chose to apply a three-dimensional Gaussian smoothing procedure [47] as it provides better results than the other two. The results of the fusion process are shown for the 72 h (3-day) and 168 h (7-day) predictions in Figures 15 and 16, respectively. These figures consist of a 3D representation of the normalized relative vorticity of the flow predicted by each of the three individual models and by the fusion method. Despite the noise associated with each model, the fusion approach is able to filter the noise out and deliver a tensor field that is very similar to the observations, even for a 168 h prediction. The significant improvement of the 3D tensor prediction by the fusion process over individual prediction models is further demonstrated by computing the metrics RMSE, PNSR, SSIM and CC of the various predictions (Figure 17). The fusion output showed an overall improvement over individual predictions for all metrics over the 7-day prediction window. In particular, the RMSE was reduced by more than 25% on day 7 of the prediction.

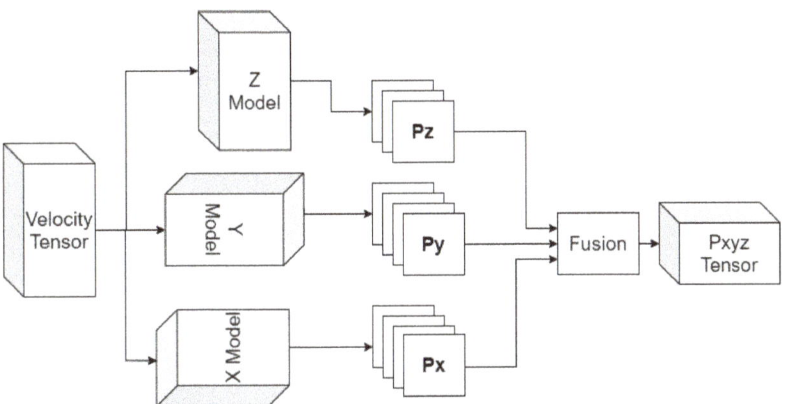

Figure 14. Block diagram of the fusion approach to produce a unified volumetric prediction.

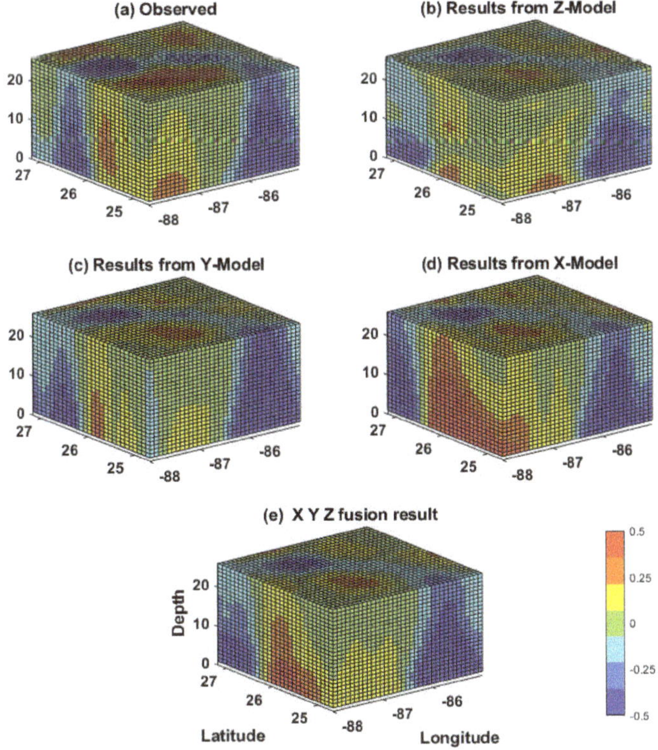

Figure 15. Three-dimensional normalized relative vorticity field of the observed and 72 h predicted tensors. (**a**) Observed, (**b**) Model Z, (**c**) Model X, (**d**) Model Y, (**e**) and fusion result.

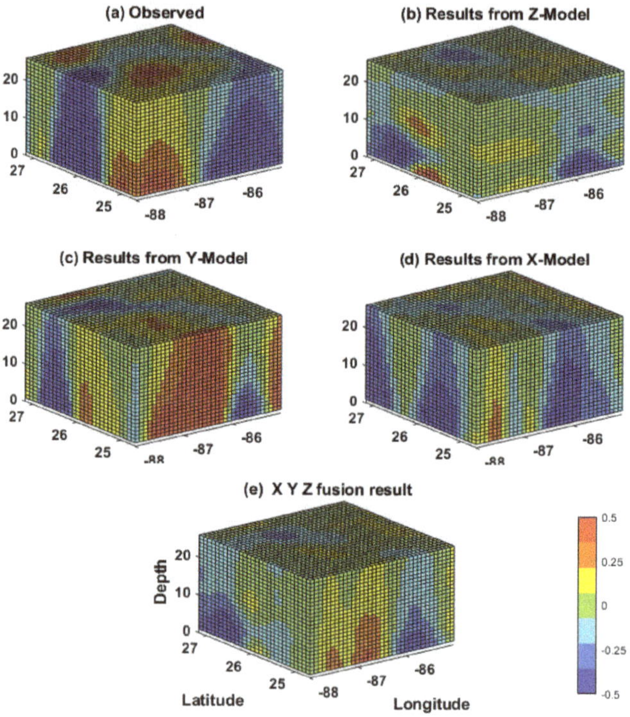

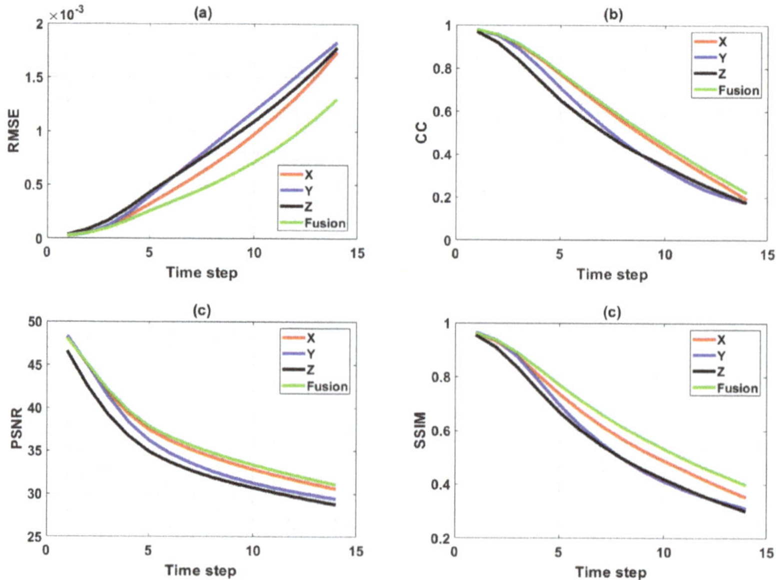

Figure 16. The conditions are same as Figure 15 for the 168 h prediction.

Figure 17. Fusion results. (**a**) Mean Squared Error (MSE), (**b**) Correlation Coefficient (CC), (**c**) Peak Signal-to-Noise Ratio (PNSR) and (**d**) Structural Similarity index (SSIM) between the observed 3-dimensional fields and the predictions from Models Z, X, and Y.

5. Conclusions

Modeling and predicting the LCS subsurface vertical structure in the GoM region is essential to all aspects of life in the region. However, useful forecasts of the flow field by current modeling methods do not exceed two days [5–7]. In this study, we developed a deep learning-based prediction model that was capable of predicting some important features of the 3D velocity fields of the LCS up to seven days in advance in a rectangular region where the LC is most active and commonly sheds eddies (Figure 6). Overall, the fusion model exhibited a CC > 0.5 up to 4.5 days (Figure 17). Subsurface velocity data measured by in-situ sensors for an approximately three-year period [29] were used to train and test a deep learning prediction method. To implement the deep learning model, we reduced the dimensionality of the tensors of each component of the velocity field to one dimension by applying EOF. The obtained PC vectors were used as an input variable to the LSTM model. The prediction model was applied separately to each layer of the tensor. We defined one tensor for each direction of the Cartesian coordinate system, which led to three prediction models associated with each direction, respectively. Each model was composed of one individual LSTM model per layer in each tensor and the final prediction consisted of the final tensor made of all the layered predictions for each velocity component. The results of this approach revealed that the prediction models associated with each of the three directions were the best at predicting the flow field in their respective directions. The errors across layers significantly altered the cross-layer structure of the flow. However, the fusion of all three models' solution with a Gaussian filter delivered an improved prediction field over each individual predictions.

Because the number of layer models necessary to conduct the full three-dimensional prediction is equal to the total number of grid points of the field to be predicted, the implementation of such model for real-time forecast seems unrealistic. However, multithread and parallel computing allows for the simultaneous computation of the predictions in all the layers in an efficient and timely manner. In addition, such dense observation arrays are rare and spatially and temporally limited, which limits the size and number of the layers to be predicted as well. In any case, when compared to ocean numerical model operations, even though numerical models are much cheaper to operate, they are unable to reliably predict the evolution of the ocean state without being constrained by ocean observations. It is true that observing arrays are ephemeral, but when they do exist they can be used to make forecasts that do not require numerical models, which simplifies the data processing and streamlines the forecasting process since only one variable is used versus the multitude of state and atmospheric variables required by ocean numerical models. Table 1 shows that the computational time for training is less than 800 s for all layers in a given direction. Assuming that each direction can be computed by one thread, then the overall prediction time would be less than 800 s, which makes this approach adequate for real-time forecasting, even at a hourly rate. HF radar ocean surface current measurements provide a good test-bed for the application of our method, where in this case, only one layer is predicted. The latter are now increasingly used for monitoring coastal circulation in many areas of the coast around the world [48]. The other limitation of the deep learning method is the duration of the measurements. Such methods' accuracy strongly depends on the diversity of events captured by the measurements and therefore their prediction skills can be limited by the duration of the measurements used to create the deep learning models. Ideally, a times series that captures the full extent of the variability in the natural system would yield the best forecast by such methods. However, it is not explicitly clear how prediction improvement is correlated to the duration of the measurements in this tensor prediction method. In a point-wise prediction exercise of ocean current velocity for unmanned underwater vehicle navigation, Immas et al. [12] showed that they could predict with an LSTM model one month of current with one month of training data.

The layered prediction method applied in this study was originally developed by Wang et al. [17] to predict the evolution of the SSH, a two-dimensional field. Predicting the three-dimensional velocity fields with this two-dimensional method has revealed the importance of the relative changes between layers in the accuracy of the predicted tensor. Future work will be focused on the inclusion of the relationship between individual nodes and their surrounding nodes in the domain, in order to account for the relative evolution between nodes. This node's spatio-temporal connectivity could be learned through another DL model ultimately coupled with the prediction model. We anticipate that such multi-model approach could provide longer reliable three-dimensional forecasts than the approach herein.

Author Contributions: A.M.A., H.Z. and L.C. conceived the project idea, and designed the experiments; A.M.A. and A.K.I. designed the model; A.M.A. performed the experiments; A.M.A. and H.Z.; A.M.A., H.Z. and L.C. wrote the paper. J.V. edited the paper. All authors have read and agreed to the published version of the manuscript.

Funding: This work was partially supported by an Understanding Gulf Ocean Systems Phase II grant (#NAS/GRP 2000011052) from the National Academies of Sciences, Engineering, and Medicine (NASEM), and by an MRI grant (#1828181) from National Science Foundation (NSF).

Institutional Review Board Statement: Not applicable.

Informed Consent Statement: Not applicable.

Data Availability Statement: The dataset used in this study is provided by Kathleen Donohue and her team.

Acknowledgments: We would like to thank Kathleen Donohue and her team for providing the Dynloop dataset, which made this study possible.

Conflicts of Interest: The authors declare no conflict of interest.

References

1. Smith, N. The global ocean data assimilation experiment (GODAE). In Proceedings of the Monitoring the Oceans in the 2000s: An Integrated Approach, Biarritz, France, 15–17 October 1997.
2. Schiller, A.; Davidson, F.; DiGiacomo, P.M.; Wilmer-Becker, K. Better Informed Marine Operations and Management: Multi-disciplinary Efforts in Ocean Forecasting Research for Socioeconomic Benefit. *Bull. Am. Meteorol. Soc.* **2016**, *97*, 1553–1559. [CrossRef]
3. Vinayachandran, P.N.; Davidson, F.; Chassignet, E.P. Toward Joint Assessments, Modern Capabilities, and New Links for Ocean Prediction Systems. *Bull. Am. Meteorol. Soc.* **2020**, *101*, E485–E487. [CrossRef]
4. MERSEA IP. *List of Internal Metrics for the MERSEA-GODAE Global Ocean: Specification for Implementation*; Mercator Ocean: Ramonville Saint-Agne, France, 14 March 2006. Available online: https://www.clivar.org/sites/default/files/documents/wgomd/GODAE_MERSEA-report.pdf (accessed on 12 September 2021).
5. Chao, Y.; Li, Z.; Farrara, J.; McWilliams, J.C.; Bellingham, J.; Capet, X.; Chavez, F.; Choi, J.K.; Davis, R.; Doyle, J.; et al. Development, implementation and evaluation of a data-assimilative ocean forecasting system off the central California coast. *Deep Sea Res. Part II Top. Stud. Oceanogr.* **2009**, *56*, 100–126. [CrossRef]
6. Shulman, I.; Paduan, J.D. Assimilation of HF radar-derived radials and total currents in the Monterey Bay area. *Deep Sea Res. Part II Top. Stud. Oceanogr.* **2009**, *56*, 149–160. [CrossRef]
7. Cooper, C.; Danmeier, D.; Frolov, S.; Stuart, G.; Zuckerman, S.; Anderson, S.; Sharma, N. Real Time Observing and Forecasting of Loop Currents in 2015. In Proceedings of the Offshore Technology Conference, Houston, TX, USA, 2–5 May 2016. [CrossRef]
8. LeCun, Y.; Bengio, Y.; Hinton, G. Deep learning. *Nature* **2015**, *521*, 436–444. [CrossRef]
9. Hochreiter, S.; Schmidhuber, J. Long short-term memory. *Neural Comput.* **1997**, *9*, 1735–1780. [CrossRef]
10. Salehinejad, H.; Sankar, S.; Barfett, J.; Colak, E.; Valaee, S. Recent advances in recurrent neural networks. *arXiv* **2017**, arXiv:1801.01078.
11. Al-Rfou, R.; Choe, D.; Constant, N.; Guo, M.; Jones, L. Character-Level Language Modeling with Deeper Self-Attention. *Proc. AAAI Conf. Artif. Intell.* **2019**, *33*, 3159–3166. [CrossRef]
12. Immas, A.; Do, N.; Alam, M.R. Real-time in situ prediction of ocean currents. *Ocean Eng.* **2021**, *228*, 108922. [CrossRef]
13. Banan, A.; Nasiri, A.; Taheri-Garavand, A. Deep learning-based appearance features extraction for automated carp species identification. *Aquac. Eng.* **2020**, *89*, 102053. [CrossRef]
14. Shamshirband, S.; Rabczuk, T.; Chau, K.W. A Survey of Deep Learning Techniques: Application in Wind and Solar Energy Resources. *IEEE Access* **2019**, *7*, 164650–164666. [CrossRef]
15. Muhamed Ali, A.; Zhuang, H.; Ibrahim, A.; Oneeb, R.; Huang, M.; Wu, A. A machine learning approach for the classification of kidney cancer subtypes using mirna genome data. *Appl. Sci.* **2018**, *8*, 2422. [CrossRef]

16. Fan, Y.; Xu, K.; Wu, H.; Zheng, Y.; Tao, B. Spatiotemporal modeling for nonlinear distributed thermal processes based on KL decomposition, MLP and LSTM network. *IEEE Access* **2020**, *8*, 25111–25121. [CrossRef]

17. Wang, J.L.; Zhuang, H.; Chérubin, L.M.; Ibrahim, A.K.; Muhamed Ali, A. Medium-Term Forecasting of Loop Current eddy Cameron and eddy Darwin formation in the Gulf of Mexico with a Divide-and-Conquer Machine Learning Approach. *J. Geophys. Res. Oceans* **2019**, *124*, 5586–5606. [CrossRef]

18. Dukhovskoy,D.; Leben, R.; Chassignet, E.; Hall, C.; Morey, S.;Nedbor-Gross, R. Characterization of the uncertainty of loop current metrics using a multidecadal numerical simulation and altimeter observations. *Elsevier* **2015** *100*, 140–158 [CrossRef]

19. Donohue, K.A.; Watts, D.; Hamilton, P.; Leben, R.; Kennelly, M. Loop Current Eddy formation and baroclinic instability. *Dyn. Atmos. Ocean.* **2016**, *76*, 195–216. [CrossRef]

20. Vukovich, F.M.; Maul, G.A. Cyclonic eddies in the eastern Gulf of Mexico. *J. Phys. Oceanogr.* **1985**, *15*, 105–117. [CrossRef]

21. Sturges, W.; Leben, R. Frequency of Ring Separations from the Loop Current in the Gulf of Mexico: A Revised Estimate. *J. Phys. Oceanogr.* **2000**, *30*, 1814–1819. [CrossRef]

22. Leben, R.R. Altimeter-derived loop current metrics. *Geophys. Monogr. Am. Geophys. Union* **2005**, *161*, 181.

23. Chérubin, L.M.; Sturges, W.; Chassignet, E.P. Deep flow variability in the vicinity of the Yucatan Straits from a high-resolution numerical simulation. *J. Geophys. Res. Ocean.* **2005**, *110*. [CrossRef]

24. Chérubin, L.M.; Morel, Y.; Chassignet, E.P. Loop Current Ring Shedding: The Formation of Cyclones and the Effect of Topography. *J. Phys. Oceanogr.* **2006**, *36*, 569–591. [CrossRef]

25. Oey, L.; Ezer, T.; Lee, H. Loop Current, rings and related circulation in the Gulf of Mexico: A review of numerical models and future challenges. *Geophys. Monogr. Am. Geophys. Union* **2005**, *161*, 31.

26. Chassignet, E.; Hurlburt, H.; Metzger, E.; Smedstad, O.; Cummings, A.; Halliwell, G.; Bleck, R.; Baraille, R.; Wallcraft, A.; Lozano, C.; et al. US GODAE: Global Ocean Prediction with the HYbrid Coordinate Ocean Model (HYCOM). *Oceanography* **2009**, *22*, 64–75. [CrossRef]

27. Rowley, C.; Mask, A. Regional and coastal prediction with the Relocatable Ocean Nowcast/Forecast System. *Oceanography* **2014**, *27*, 44–55. [CrossRef]

28. Gopalakrishnan, G.; Cornuelle, B.D.; Hoteit, I.; Rudnick, D.L.; Owens, W.B. State estimates and forecasts of the loop current in the Gulf of Mexico using the MITgcm and its adjoint. *J. Geophys. Res. Ocean.* **2013**, *118*, 3292–3314. [CrossRef]

29. Hamilton, P.; Lugo-Fernández, A.; Sheinbaum, J. A Loop Current experiment: Field and remote measurements. *Dyn. Atmos. Ocean.* **2016**, *76*, 156–173. [CrossRef]

30. Donohue, K.A.; Watts, D.R.; Tracey, K.L.; Greene, A.D.; Kennelly, M. Mapping circulation in the Kuroshio Extension with an array of current and pressure recording inverted echo sounders. *J. Atmos. Ocean. Technol.* **2010**, *27*, 507–527. [CrossRef]

31. Daley, R. *Atmospheric Data Analysis*; Number 2; Cambridge University Press: Cambridge, UK, 1993.

32. Watts, D.R.; Sun, C.; Rintoul, S. A two-dimensional gravest empirical mode determined from hydrographic observations in the Subantarctic Front. *J. Phys. Oceanogr.* **2001**, *31*, 2186–2209. [CrossRef]

33. Suranjana, S.; Moorthi, S.; Pan, H.; Wu, X.; Wang, J.; Nadiga, S.; Tripp, P.; Kistler, R.; Woollen, J.; Behringer, D.; et al. The NCEP climate forecast system reanalysis. *Bull. Am. Meteorol. Soc.* **2010**, *91*, 1015–1057.

34. Arakawa, A.; Lamb, V.R. Computational Design of the Basic Dynamical Processes of the UCLA General Circulation Model. Methods in Computational Physics. *Adv. Res. Appl.* **1977**, *17*, 173–265. [CrossRef]

35. Wang, Z.; Bovik, A.C.; Sheikh, H.R.; Simoncelli, E.P. Image quality assessment: From error visibility to structural similarity. *IEEE Trans. Image Process.* **2004**, *13*, 600–612. [CrossRef]

36. Zeng, X.; Li, Y.; He, R. Predictability of the loop current variation and eddy shedding process in the Gulf of Mexico using an artificial neural network approach. *J. Atmos. Ocean. Technol.* **2015**, *32*, 1098–1111. [CrossRef]

37. Casagrande, G.; Stephan, Y.; Varnas, A.C.W.; Folegot, T. A novel empirical orthogonal function (EOF)-based methodology to study the internal wave effects on acoustic propagation. *IEEE J. Ocean. Eng.* **2011**, *36*, 745–759. [CrossRef]

38. Beckers, J.M.; Rixen, M. EOF calculations and data filling from incomplete oceanographic datasets. *J. Atmos. Ocean. Technol.* **2003**, *20*, 1839–1856. [CrossRef]

39. Thomson, R.E.; Emery, W.J. *Data Analysis Methods in Physical Oceanography*; Newnes: Amsterdam, The Netherlands, 2014.

40. Yin, W.; Kann, K.; Yu, M.; Schütze, H. Comparative Study of CNN and RNN for Natural Language Processing. *arXiv* **2017**, arXiv:1702.01923.

41. Bengio, Y.; Simard, P.; Frasconi, P. Learning long-term dependencies with gradient descent is difficult. *IEEE Trans. Neural Netw.* **1994**, *5*, 157–166. [CrossRef] [PubMed]

42. Ali, A.M.; Zhuang, H.; Ibrahim, A.K.; Wang, J.L. Preliminary results of forecasting of the loop current system in Gulf of Mexico using robust principal component analysis. In Proceedings of the 2018 IEEE International Symposium on Signal Processing and Information Technology (ISSPIT), Louisville, KY, USA, 6–8 December 2018; pp. 1–5. [CrossRef]

43. Kingma, D.P.; Ba, J. Adam: A method for stochastic optimization. *arXiv* **2014**, arXiv:1412.6980.

44. Murphy, K.P. *Machine Learning: A Probabilistic Perspective*; MIT Press: Cambridge, MA, USA, 2012.

45. National Academies of Sciences Engineering and Medicine. *Understanding and Predicting the Gulf of Mexico Loop Current: Critical Gaps and Recommendations*; National Academies Press: Washington, DC, USA, 2018.

46. Wang, J.L.; Zhuang, H.; Chérubin, L.; Muhamed Ali, A.; Ibrahim, A. Loop Current SSH Forecasting: A New Domain Partitioning Approach for a Machine Learning Model. *Forecasting* **2021**, *3*, 570–579. [CrossRef]

47. Haddad, R.A.; Akansu, A.N. A class of fast Gaussian binomial filters for speech and image processing. *IEEE Trans. Signal Process.* **1991**, *39*, 723–727. [CrossRef]
48. Harlan, J.; Terrill, E.; Hazard, L.; Keen, C.; Barrick, D.; Whelan, C.; Howden, S.; Kohut, J. The Integrated Ocean Observing System high-frequency radar network: Status and local, regional, and national applications. *Mar. Technol. Soc. J.* **2010**, *44*, 122–132. [CrossRef]

Article

Model-Free Time-Aggregated Predictions for Econometric Datasets

Kejin Wu [1,†] **and Sayar Karmakar** [2,*,†]

1 Department of Mathematics, University of California San Diego, La Jolla, CA 92093, USA; kwu@ucsd.edu
2 Department of Statistics, University of Florida, Gainesville, FL 32611, USA
* Correspondence: sayarkarmakar@ufl.edu
† These authors contributed equally to this work.

Abstract: Forecasting volatility from econometric datasets is a crucial task in finance. To acquire meaningful volatility predictions, various methods were built upon GARCH-type models, but these classical techniques suffer from instability of short and volatile data. Recently, a novel existing normalizing and variance-stabilizing (NoVaS) method for predicting squared log-returns of financial data was proposed. This model-free method has been shown to possess more accurate and stable prediction performance than GARCH-type methods. However, whether this method can sustain this high performance for long-term prediction is still in doubt. In this article, we firstly explore the robustness of the existing NoVaS method for long-term time-aggregated predictions. Then, we develop a more parsimonious variant of the existing method. With systematic justification and extensive data analysis, our new method shows better performance than current NoVaS and standard GARCH(1,1) methods on both short- and long-term time-aggregated predictions. The success of our new method is remarkable since efficient predictions with short and volatile data always carry great importance. Additionally, this article opens potential avenues where one can design a model-free prediction structure to meet specific needs.

Keywords: ARCH-GARCH; model-free; aggregated forecasting

Citation: Wu, K.; Karmakar, S. Model-Free Time-Aggregated Predictions for Econometric Datasets. *Forecasting* **2021**, *3*, 920–933. https://doi.org/10.3390/forecast3040055

Academic Editor: Alessia Paccagnini

Received: 3 November 2021
Accepted: 6 December 2021
Published: 8 December 2021

Publisher's Note: MDPI stays neutral with regard to jurisdictional claims in published maps and institutional affiliations.

1. Introduction

Accurate and robust volatility forecasting is a central focus in financial econometrics. This type of forecasting is crucial for practitioners and traders to make decisions in risk management, asset allocation, pricing of derivative instruments and strategic decisions regarding fiscal policies, etc. Standard methods to perform volatility forecasting are typically built upon applying GARCH-type models to predict squared financial log-returns. With the model-free prediction principle, first proposed by Politis [1], a model-free volatility prediction method—NoVaS—has been proposed recently for efficient forecasting without the assumption of normality. Some previous studies have shown that the NoVaS method possesses better predictive performance than GARCH-type models when forecasting squared log-returns, e.g., Gulay and Emec [2] showed that the NoVaS method could overcome GARCH-type models (GARCH, EGARCH and GJR-GARCH) with generalized error distributions by comparing the pseudo-out-of-sample (POOS) forecasting performance on S&P500 and BIST 100 return series (here the pseudo-out-of-sample forecasting analysis means using data up to and including the current time to predict future values). Chen and Politis [3] showed that the "time-varying" NoVaS method is robust against possible non-stationarities in the data. Furthermore, Chen and Politis [4] extended this NoVaS approach to perform multi-step-ahead predictions of squared log-returns.

However, to the best of our knowledge, such methods have not been evaluated for time-aggregated prediction. Time-aggregated prediction here stands for the prediction of $Y_{n+1} + \cdots + Y_{n+h}$ after observing $\{Y_t\}_{t=1}^n$. Such predictions remain crucial for strategic decisions implemented by commodity or service providers, ([5,6]), trust funds, pension management, insurance companies, portfolio management of specific derivatives ([7]) and assets ([8]). Time-aggregated forecasting is also able to provide some degree of confidence in understanding the general trend in the near future, potentially for the entire following week or months ahead, which is definitely more meaningful than merely understanding what might happen for any single step ahead (predicting Y_{n+h} for one value of h) in the time horizon. In fact, the quality of forecasts for econometric data has been evaluated through such time-aggregated metrics in [9,10]. In this article, we continue utilizing these time-aggregated metrics to challenge the ability of the NoVaS method for short- and long-term time-aggregated predictions on squared log-returns series. For exploring such capabilities of the existing NoVaS method, we set up comprehensive data analyses to substantiate the efficiency of the NoVaS method and also address the lack of data experiments in NoVaS studies. Apart from this, we also attempt to improve the existing one further by proposing a more parsimonious model. Based on extensive data analysis, our new method shows more stable performance than the state-of-the-art NoVaS method regardless of whether simulation or real-world data are used. We also find that the state-of-the-art NoVaS method is even surpassed by the standard GARCH(1,1) model sometimes. On the other hand, our new method returns consistently excellent forecasting. Notably, our method achieves a remarkable improvement when the dataset at hand is short and volatile.

The rest of this article is organized as follows. In Section 2, we firstly introduce the theoretical background and structure of the existing NoVaS method. Then, our new method is proposed and a simple comparison is made to show the stability of our new method. In Section 3, we substantiate our proposal by extensive simulations and data analysis. Moreover, we utilize the CW test to support our parsimonious model. Finally, a summary and discussion are given in Sections 4 and 5, respectively.

2. Method

2.1. The Existing NoVaS Method

The NoVaS method is a model-free prediction principle. The main idea lies in applying an invertible transformation H, which can map the non-*i.i.d.* vector $\{Y_i\}_{i=1}^{t}$ to a vector $\{\epsilon_i\}_{i=1}^{t}$ that has *i.i.d.* components. This leads to the prediction of Y_{t+1} by inversely transforming the prediction of ϵ_{t+1} [11]. The starting point to build the transformation of the existing NoVaS method is the ARCH model [12]. Then, Politis [1] made some adjustments to determine the final form of H as:

$$W_t = \frac{Y_t}{\sqrt{\alpha s_{t-1}^2 + \tilde{a}_0 Y_t^2 + \sum_{i=1}^{p} a_i Y_{t-i}^2}} \quad \text{for } t = p+1, \cdots, n. \tag{1}$$

In Equation (1), $\{Y_t\}_{t=1}^{n}$ is the log-returns vector in this article; $\{W_t\}_{t=p+1}^{n}$ is the transformed vector, which we hope to transform to *i.i.d.*; α is a fixed-scale invariant constant; s_{t-1}^2 is calculated by $(t-1)^{-1} \sum_{i=1}^{t-1} (Y_i - \mu)^2$, with μ being the mean of $\{Y_i\}_{i=1}^{t-1}$; $\tilde{a}_0$ is the coefficient corresponding with the currently observed value Y_t^2. For reaching a qualified transformation function, Equation (2) is required to stabilize the variance.

$$\alpha \in (0,1), \tilde{a}_0 \geq 0, a_i \geq 0 \text{ for all } i \geq 1, \alpha + \tilde{a}_0 + \sum_{i=1}^{p} a_i = 1 \tag{2}$$

Then, α and $\tilde{a}_0, a_1, \cdots, a_p$ are finally determined by minimizing $|Kurtosis(W_t) - 3|$. In practice, the transformed $\{W_t\}$ is usually uncorrelated; see [11] for additional processes for correlated $\{W_t\}$. This method is model-free in the sense that we do not assume any particular distribution for the innovation $\{W_t\}$ except for matching its kurtosis to 3. Once

H is found, H^{-1} can be obtained immediately. For example, H^{-1} corresponding with Equation (1) is:

$$Y_t = \sqrt{\frac{W_t^2}{1 - \tilde{a}_0 W_t^2}\left(\alpha s_{t-1}^2 + \sum_{i=1}^{p} a_i Y_{t-i}^2\right)} \text{ for } t = p+1, \cdots, n. \tag{3}$$

To obtain the prediction of Y_{n+1}^2, Politis [11] defined two types of optimal predictors under L_1 (Mean Absolute Deviation) and L_2 (Mean Squared Error) criteria after observing historical information set $\mathscr{F}_n = \{Y_t, 1 \leq t \leq n\}$:

L_1-optimal predictor of Y_{n+1}^2 :

$$\text{Median}\left\{Y_{n+1,m}^2 : m = 1, \cdots, M \middle| \mathscr{F}_n\right\}$$

$$= \text{Median}\left\{\frac{W_{n+1,m}^2}{1 - \tilde{a}_0 W_{n+1,m}^2}\left(\alpha s_n^2 + \sum_{i=1}^{p} a_i Y_{n+1-i}^2\right) : m = 1, \cdots, M \middle| \mathscr{F}_n\right\}$$

$$= \left(\alpha s_n^2 + \sum_{i=1}^{p} a_i Y_{n+1-i}^2\right)\text{Median}\left\{\frac{W_{n+1,m}^2}{1 - \tilde{a}_0 W_{n+1,m}^2} : m = 1, \cdots, M\right\} \tag{4}$$

L_2-optimal predictor of Y_{n+1}^2 :

$$\text{Mean}\left\{Y_{n+1,m}^2 : m = 1, \cdots, M \middle| \mathscr{F}_n\right\}$$

$$= \text{Mean}\left\{\frac{W_{n+1,m}^2}{1 - \tilde{a}_0 W_{n+1,m}^2}\left(\alpha s_n^2 + \sum_{i=1}^{p} a_i Y_{n+1-i}^2\right) : m = 1, \cdots, M \middle| \mathscr{F}_n\right\}$$

$$= \left(\alpha s_n^2 + \sum_{i=1}^{p} a_i Y_{n+1-i}^2\right)\text{Mean}\left\{\frac{W_{n+1,m}^2}{1 - \tilde{a}_0 W_{n+1,m}^2} : m = 1, \cdots, M\right\}$$

where $\{W_{n+1,m}\}_{m=1}^{M}$ are generated M times from its empirical distribution or a normal distribution. Here, the normal distribution is an asymptotic limit of the empirical distribution of $\{W_{n+1}\}$. More details about this procedure and multi-step prediction are presented in Section 2.2. $\{Y_{n+1,m}^2\}_{m=1}^{M}$ are given by plugging $\{W_{n+1,m}\}_{m=1}^{M}$ into Equation (3) and setting t as $n+1$. During the optimization process, different forms of unknown parameters in Equation (2) are applied so that various NoVaS methods are established. Chen [13] pointed out that the Generalized Exponential NoVaS (GE-NoVaS) method with exponentially decayed unknown parameters presented in Equation (5) is superior to other NoVaS-type methods.

$$\alpha \neq 0, \tilde{a}_0 = c', a_i = c'e^{-ci} \text{ for all } 1 \leq i \leq p, c' = \frac{1 - \alpha}{\sum_{i=0}^{p} e^{-ci}} \tag{5}$$

2.2. A New Method with Less Parameters

However, during our investigation, we found that the GE-NoVaS method returns extremely large predictions under the L_2 criterion sometimes. The reason for this phenomenon is that the denominator of Equation (3) will be quite small when the generated $\{W^*\}$ (from empirical or normal distribution) is very close to $1/\tilde{a}_0$. In this situation, the prediction error will be amplified. Moreover, when the long-term ahead prediction is desired, this amplification will be accumulated and the final prediction will be dampened. Therefore, a removing-$\tilde{a}_0$ idea is proposed to avoid such issues in this article. H and H^{-1} of the GE-NoVaS-without-$\tilde{a}_0$ method can be rewritten as below:

$$W_t = \frac{Y_t}{\sqrt{\alpha s_{t-1}^2 + \sum_{i=1}^{p} a_i Y_{t-i}^2}} \; ; \; Y_t = \sqrt{W_t^2\left(\alpha s_{t-1}^2 + \sum_{i=1}^{p} a_i Y_{t-i}^2\right)} \; ; \text{ for } t = p+1, \cdots, n. \tag{6}$$

We should notice that even without the $\tilde{a}_0$ term, the causal prediction rule is still satisfied. It is easy to obtain the analytical form of the first-step-ahead Y_{n+1}, which can be expressed as below:

$$Y_{n+1} = \sqrt{W_{n+1}^2 \left(\alpha s_n^2 + \sum_{i=1}^{p} a_i Y_{n+1-i}^2 \right)} \tag{7}$$

More specifically, when the first-step GE-NoVaS-without-$\tilde{a}_0$ prediction is performed, $\{W_{n+1}^*\}$ are generated M (i.e., 5000 in this article) times from a standard normal distribution by the Monte Carlo method or bootstrapped from its empirical distribution $\hat{F}_w$ which is calculated from Equation (1). Then, plugging these $\{W_{n+1,m}^*\}_{m=1}^{M}$ into Equation (7), M pseudo-predictions $\{\hat{Y}_{n+1,m}^*\}_{m=1}^{M}$ are obtained. According to the strategy implied by Equation (4), we choose L_1 and L_2 risk optimal predictors $\hat{Y}_{n+1}^2$ as the sample median and mean of $\{\hat{Y}_{n+1,1}^*, \cdots, \hat{Y}_{n+1,M}^*\}$, respectively. We can even predict the general form of Y_{n+h}, such as $g(Y_{n+h})$, by adopting the sample mean or median of $\{g(\hat{Y}_{n+1,1}^*), \cdots, g(\hat{Y}_{n+1,M}^*)\}$. Similarly, the two-steps-ahead Y_{n+2} can be expressed as:

$$Y_{n+2} = \sqrt{W_{n+2}^2 \left(\alpha s_{n+1}^2 + a_1 Y_{n+1}^2 + \sum_{i=2}^{p} a_i Y_{n+2-i}^2 \right)} \tag{8}$$

When the prediction of Y_{n+2} is required, M pairs of $\{W_{n+1}^*, W_{n+2}^*\}$ are still generated by bootstrapping or Monte Carlo method from empirically or standard normal distributions, respectively. Y_{n+1}^2 is replaced by the predicted value $\hat{Y}_{n+1}^2$ which is derived from running the first-step GE-NoVaS-without-$\tilde{a}_0$ prediction with simulated $\{W_{n+1,m}^*\}_{m=1}^{M}$ under the L_1 or L_2 criterion. Subsequently, we choose L_1 and L_2 risk optimal predictors of Y_{n+2} as the sample median and mean of $\{\hat{Y}_{n+2,1}^*, \cdots, \hat{Y}_{n+2,M}^*\}$.

Finally, iterating the process described above, we can accomplish multi-step-ahead NoVaS predictions. $Y_{n+h}, h \geq 3$ can be expressed as:

$$Y_{n+h} = \sqrt{W_{n+h}^2 \left(\alpha s_{n+h-1}^2 + \sum_{i=1}^{p} a_i Y_{n+h-i}^2 \right)} \tag{9}$$

To obtain the prediction of Y_{n+h}, we generate M number of $\{W_{n+1}^*, \cdots, W_{n+h}^*\}$ and plug $\{Y_{n+k}\}_{k=1}^{h-1}$ with NoVaS predicted values $\{\hat{Y}_{n+k}\}_{k=1}^{h-1}$, which are computed iteratively. L_1 and L_2 risk optimal predictors of Y_{n+h} are computed by the sample median and mean of $\{\hat{Y}_{n+h,1}^*, \cdots, \hat{Y}_{n+h,M}^*\}$. In short, we can summarize that Y_{n+h} is determined by:

$$Y_{n+h} = f_{\text{GE-NoVaS-without}-\tilde{a}_0}(W_{n+1}, \cdots, W_{n+h}, \mathscr{F}_n) \tag{10}$$

Since $\mathscr{F}_n$ is the observed information set, we can simplify the expression of Y_{n+h} as:

$$Y_{n+h} = f_{\text{GE-NoVaS-without}-\tilde{a}_0}(W_{n+1}, \cdots, W_{n+h}) \tag{11}$$

For applying the GE-NoVaS method, we can still build the relationship between Y_{n+h} and $\{W_{n+1}, \cdots, W_{n+h}\}$ as:

$$Y_{n+h} = f_{\text{GE-NoVaS}}(W_{n+1}, \cdots, W_{n+h}) \tag{12}$$

We should notice that simulated $\{W_{n+1,m}^*, \cdots, W_{n+h,m}^*\}_{m=1}^{M}$ for obtaining GE-NoVaS method prediction of Y_{n+h} should be generated by the bootstrapping or Monte Carlo method from an empirically or trimmed standard normal distribution. The reason for using the trimmed distribution is $|W_t| \leq 1/\sqrt{\tilde{a}_0}$ from Equation (1). Here, we summarize Algorithm 1 to perform h-step-ahead time-aggregated prediction using the GE-NoVaS-without-$\tilde{a}_0$ method. The algorithm of GE-NoVaS can be written out similarly.

Remark (The advantage of removing the $\tilde{a}_0$ term): First, after removing the $\tilde{a}_0$ term, the prediction of the NoVaS method under the L_2 criterion is more stable. More details will be shown in Section 2.3. Second, the suggestion of removing $\tilde{a}_0$ can also lead to less time complexity of our new method. The reason for this phenomenon is simple. If we consider the limiting distribution of $\{W_t\}$ series, $1/\sqrt{\bar{a}_0}$ is required to be larger than or equal to 3 to ensure that $\{W_t\}$ has a sufficiently large range, i.e., $\tilde{a}_0$ is required to be less than or equal to 0.111 (recall that the mass of standard normal data is within $[-3,3]$). However, the optimal combination of NoVaS coefficients may not render a suitable $\tilde{a}_0$. For this situation, we need to increase the NoVaS transformation order p and repeat the normalizing and variance-stabilizing process till $\tilde{a}_0$ in the optimal combination of coefficients is suitable. This repeating process definitely increases the computation workload.

Algorithm 1: The h-step ahead prediction for the GE-NoVaS-without-$\tilde{a}_0$ method.

Step 1	Define a grid of possible α values, $\{\alpha_k;\ k=1,\cdots,K\}$. Fix $\alpha = \alpha_k$, then calculate the optimal combination of $\alpha_k, a_1, \cdots, a_p$ of the GE-NoVaS-without-$\tilde{a}_0$ method, which minimizes $\lvert Kurtosis(W_t) - 3\rvert$.
Step 2	Derive the analytic form of Equation (11) using $\alpha_k, a_1, \cdots, a_p$ from the first step.
Step 3	Generate $\{W^*_{n+1}, \cdots, W^*_{n+h}\}$ M times from a standard normal distribution or the empirical distribution $\hat{F}_w$. Plug $\{W^*_{n+1}, \cdots, W^*_{n+h}\}$ into the analytic form of Equation (11) to obtain M pseudo-values $\{\hat{Y}^*_{n+h,1}, \cdots, \hat{Y}^*_{n+h,M}\}$
Step 4	Calculate the optimal predictor of $g(Y_{n+h})$ by taking the sample mean (under L_2 risk criterion) or sample median (under L_1 risk criterion) of the set $\{g(\hat{Y}^*_{n+h,1}), \cdots, g(\hat{Y}^*_{n+h,M})\}$.
Step 5	Repeat above steps with different α values from $\{\alpha_k;\ k=1,\cdots,K\}$ to get K prediction results.

2.3. The Potential Instability of the GE-NoVaS Method

Next, we provide an illustration to compare the GE-NoVaS and GE-NoVaS-without-$\tilde{a}_0$ methods in predicting the volatility of the Microsoft Corporation (MSFT) daily closing price from 8 January 1998 to 31 December 1999 and show an interesting finding that the long-term time-aggregated predictions of the GE-NoVaS method are unstable under the L_2 criterion. Based on the finding of Awartani and Corradi [14], squared log-returns can be used as a proxy for volatility to render a correct ranking of different GARCH models in terms of a quadratic loss function. Log-return series $\{Y_t\}$ can be computed by the equation shown below:

$$Y_t = 100 \times log(X_{t+1}/X_t) \tag{13}$$

where $\{X_t\}$ is the corresponding MSFT daily closing price series. For achieving a comprehensive comparison, we use 250 financial log-returns as a sliding window to perform POOS 1-step, 5-step and 30-step (long-term) ahead time-aggregated predictions under the L_2 criterion. Then, we roll this window through the whole dataset, i.e., we use $\{Y_1, \cdots, Y_{250}\}$ to predict $Y^2_{251}, \{Y^2_{251}, \cdots, Y^2_{255}\}$ and $\{Y^2_{251}, \cdots, Y^2_{280}\}$; then, we use $\{Y_2, \cdots, Y_{251}\}$ to predict $Y^2_{252}, \{Y^2_{252}, \cdots, Y^2_{256}\}$ and $\{Y^2_{252}, \cdots, Y^2_{281}\}$, for 1-step, 5-step and 30-step aggregated predictions, respectively, and so on. We can define all 1-step, 5-step and 30-step-ahead time-aggregated predictions as $\{\hat{Y}^2_{k,1}\}$, $\{\hat{Y}^2_{i,5}\}$ and $\{\hat{Y}^2_{j,30}\}$, which are presented as below:

Assume that there are a total of N log-return data points:

$$\hat{Y}^2_{k,1} = \hat{Y}^2_{k+1}, \ k = 250, 251, \cdots, N-1$$

$$\hat{Y}^2_{i,5} = \sum_{m=1}^{5} \hat{Y}^2_{i+m}, \ i = 250, 251, \cdots, N-5 \tag{14}$$

$$\hat{Y}^2_{j,30} = \sum_{m=1}^{30} \hat{Y}^2_{j+m}, \ j = 250, 251, \cdots, N-30$$

In Equation (14), $\hat{Y}^2_{k+1}, \hat{Y}^2_{i+m}, \hat{Y}^2_{j+m}$ are single-step predictions of squared log-returns by the two NoVaS-type methods. To obtain the "Prediction Errors" for the two methods, we can calculate the "loss" by comparing the aggregated prediction results with the realized aggregated values based on Equation (15):

$$L_{p,h} = \sum_{p} (\hat{Y}^2_{p,h} - \sum_{m=1}^{h} (Y^2_{p+m}))^2, \ p \in \{k, i, j\}; \ h \in \{1, 5, 30\} \tag{15}$$

where $\{Y^2_{p+m}\}$ are realized squared log-returns. To show the potential instability of the GE-NoVaS method under the L_2 criterion, we take α to be 0.5 to build a toy example. In the algorithm when performing the GE-NoVaS method, α could take an optimal value from a discrete set $\{0.1, \cdots, 0.8\}$ based on the prediction performance.

From Figure 1, we can clearly see that the GE-NoVaS-without-$\tilde{a}_0$ method can better capture different steps' true time-aggregated features. On the other hand, the GE-NoVaS method returns unstable results for 30-step-ahead time-aggregated predictions. Besides, we can see that the 1-step-ahead POOS prediction returned by the GE-NoVaS method is almost a flat curve, which is actually meaningless. Similarly, for the 5-step-ahead time-aggregated prediction case, the POOS prediction of the GE-NoVaS method fails to match the true time-aggregated values.

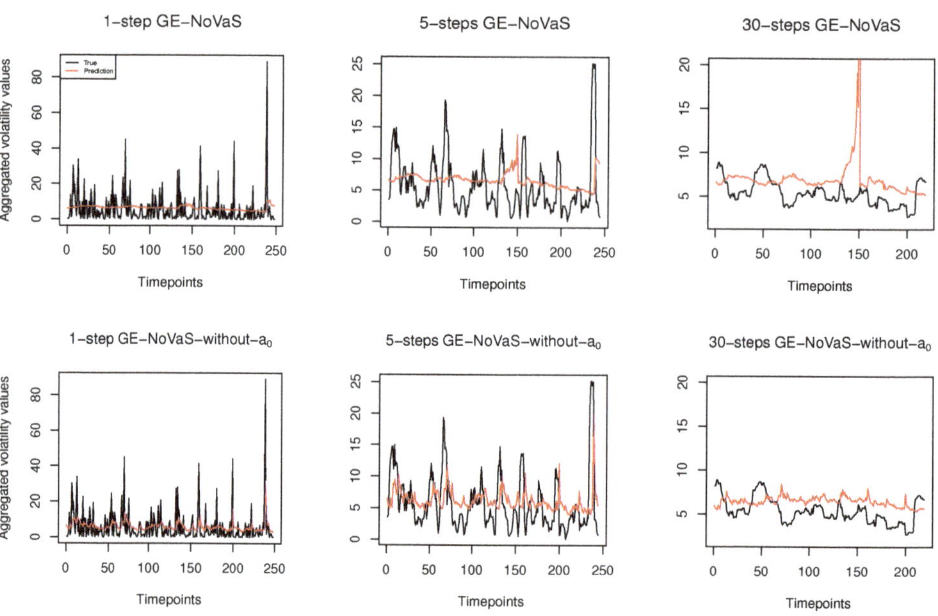

Figure 1. Curves of the true and predicted time-aggregated squared log-returns from GE-NoVaS and GE-NoVaS-without-$\tilde{a}_0$ methods.

3. Data Analysis and Results

To perform extensive data analysis in a bid to validate our method, we deploy POOS predictions using two NoVaS and standard GARCH(1,1) methods with simulated and real-world data. All results are collated in Table 1. The optimal results for each data cases are highlighted in bold. For controlling the dependence of the prediction performance on the length of the dataset, we build datasets with two fixed lengths—250 or 500—to mimic 1-year or 2-year data, respectively. At the same time, we choose the window size for our rollover forecasting analysis to be 100 or 250 for the 1-year or 2-year datasets.

3.1. Simulation Study

We use the same simulation Models 1–4 from [4], shown below, to mimic four 1-year datasets. Recall that one NoVaS method can generate the L_1 or L_2 predictor and $\{W^*\}$ can be chosen from a normal distribution or empirical distribution; thus, there are four variants of one specific NoVaS method. We take the best-performing result among four variants of a specific NoVaS method to be its final prediction. Finally, we continue applying the formula in Equation (15) to measure the performance of the different methods, as described in Section 2.3.

Model 1: Time-varying GARCH(1,1) with Gaussian errors
$X_t = \sigma_t \epsilon_t$, $\sigma_t^2 = 0.00001 + \beta_{1,t}\sigma_{t-1}^2 + \alpha_{1,t}X_{t-1}^2$, $\{\epsilon_t\} \sim i.i.d.\ N(0,1)$
$\alpha_{1,t} = 0.1 - 0.05t/n$; $\beta_{1,t} = 0.73 + 0.2t/n$, $n = 250$
Model 2: Standard GARCH(1,1) with Gaussian errors
$X_t = \sigma_t \epsilon_t$, $\sigma_t^2 = 0.00001 + 0.73\sigma_{t-1}^2 + 0.1X_{t-1}^2$, $\{\epsilon_t\} \sim i.i.d.\ N(0,1)$
Model 3: (Another) Standard GARCH(1,1) with Gaussian errors
$X_t = \sigma_t \epsilon_t$, $\sigma_t^2 = 0.00001 + 0.8895\sigma_{t-1}^2 + 0.1X_{t-1}^2$, $\{\epsilon_t\} \sim i.i.d.\ N(0,1)$
Model 4: Standard GARCH(1,1) with Student-t errors
$X_t = \sigma_t \epsilon_t$, $\sigma_t^2 = 0.00001 + 0.73\sigma_{t-1}^2 + 0.1X_{t-1}^2$,
$\{\epsilon_t\} \sim i.i.d.\ t$ distribution with five degrees of freedom

Result analysis: From the first block of Table 1, we can read that both NoVaS methods are superior to the GARCH(1,1) model. Although these simulated datasets are generated from GARCH(1,1)-type models, the GE-NoVaS-without-$\tilde{a}_0$ method can bring around 66% and 48% improvements compared to the GARCH(1,1) model for 5-step-ahead time-aggregated predictions of Model-4 and Model-1 data, respectively. Notably, GARCH(1,1) brings poor results for the 30-step-ahead time-aggregated predictions of Model-4 simulated data, which implies that such a classical method is impaired by error accumulation problems when long-term predictions are required. On the other hand, the model-free NoVaS method can avoid this issue. Taking a closer look at these results, we can observe that almost all optimal results come from applying the GE-NoVaS-without-$\tilde{a}_0$ method. Moreover, the GE-NoVaS method is surpassed by GARCH(1,1) when forecasting 30-step-ahead time-aggregated Model-2 data. On the other hand, the GE-NoVaS-without-$\tilde{a}_0$ method provides consistently stable results. These results imply that the GE-NoVaS-without-$\tilde{a}_0$ method dominates the GE-NoVaS method when predicting long-term or short-term time-aggregated predictions. Besides, using the same generated models from the previous study of the NoVaS method [4] ensures fairness. Additionally, with simulation implementations, the ability against model misspecification of NoVaS methods is verified in Appendix A.

3.2. A Few Real Datasets

We also present a variety of real-world datasets of different size and intrinsic behavior:

- 2-year period data: 2018~2019 stock price data.
- 1-year period data: 2019 stock price and index data.
- 1-year period volatile data due to pandemic: 11.2019~10.2020 stock price, currency and index data.

Taking into account three types of real-world data is necessary to challenge our new method and explore the existing method in different regimes. We also tactically pay more attention to short and volatile data since this is a more challenging task to handle. Equation (13) is continually used to obtain the log-return series of different datasets.

Before comparing in depth the forecasting performance of the NoVaS-type and GARCH methods, we first investigate the properties of the used datasets. From Figure 2, we can see that there were huge variations in the four datasets during 11.2019~10.2020, which implies the extreme fluctuations in global economics due to the COVID-19 pandemic. We wished to apply such datasets to test whether the NoVaS-type methods can achieve good forecasting performance for such volatile data.

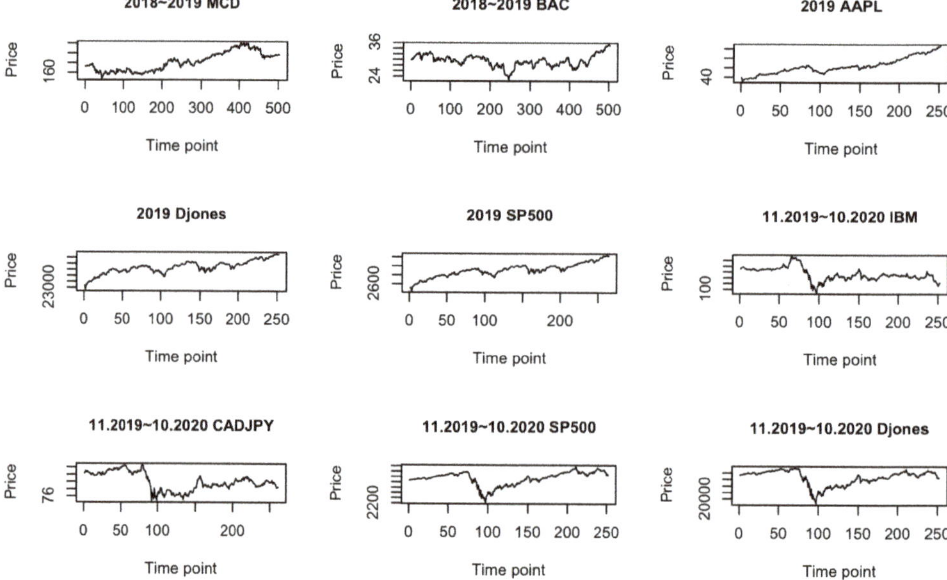

Figure 2. Price series of selected 9 datasets.

Besides, it is natural to question whether these datasets are stationary. In a comprehensive manner, we choose three statistical tests—Augmented Dickey–Fuller (ADF) Test [15], Phillips–Perron (PP) Unit Root Test [16] and Kwiatkowski–Phillips–Schmidt–Shin (KPSS) Test [17]—to check the stationarity of the squared log-returns series of each selected dataset. One aspect that should be noticed is that the number of lags is crucial for the ADF test. If the included lag is too small, then the remaining serial correlation in the errors will bias the test. If this number is larger, the power of the test will suffer. Here, we consider taking the longest lag that is statistically significant. More specifically, we determine this longest lag by observing the last lag that crosses through the confidence interval lines of the auto-correlation plot. Besides, we apply a long version of the truncation lag parameter on both the PP and KPSS tests. The results of the three tests are tabulated in Table A4. Combining these results, we can argue that most of the squared log-return series in the normal time period are stationary. However, during the volatile time period, the squared log-returns of IBM, SP500 and Dow Jones are thought to be non-stationary by the ADF test. The KPSS test also returns small p-values for these three datasets. These results are consistent with our conjecture that data tend to show non-stationarity during volatile periods. Again, it will be interesting to see if the NoVaS-type methods can offer good forecasting performance for non-stationary data. Recall that Chen and Politis [3] found that the NoVaS methodology generally outperforms the GARCH benchmark on the one-step-ahead point prediction of non-stationary data (involving local stationarity and/or structural breaks). However, they only considered two real-world time series. Here, we extend such empirical study to short- and long-term time-aggregated predictions with sufficient data examples.

Remark (One ARCH-type model for non-stationary data): Since our stationarity tests suggest that some series may not be stationary, we can consider applying ARCH-without-intercept, which is a variant of the ARCH model. This variant is non-stationary but stable in the sense that the observed process has non-degenerated distribution. Moreover, it appears to be an alternative to common stationary but highly persistent GARCH models [18]. Inspired by this ARCH-type model, the NoVaS method may be further improved by removing the corresponding intercept term αs_{t-1}^2 in Equations (1) and (6). More empirical experiments could be conducted along this direction.

Result analysis: From the last three blocks of Table 1, there is no optimal result that comes from the GARCH(1,1) method. When the target data are short and volatile, GARCH(1,1) gives poor results for 30-step-ahead time-aggregated predictions, such as the volatile Djones, CADJPY and IBM cases. Among the two NoVaS methods, the GE-NoVaS-without-$\tilde{a}_0$ method outperforms the GE-NoVaS method for the three types of real-world data. More specifically, around 70% and 30% improvements are created by our new method compared to the existing GE-NoVaS method when forecasting 30-step-ahead time-aggregated volatile Djones and CADJPY data, respectively. We should also notice that the GE-NoVaS method is again surpassed by the GARCH(1,1) model on 30-step-ahead aggregated predictions of 2018~2019 BAC data. On the other hand, the GE-NoVaS-without-$\tilde{a}_0$ method performs stably. These comprehensive prediction comparisons cover the shortage of empirical analyses of NoVaS methods, and imply that NoVaS-type methods are indeed valid and efficient for real-world short- or long-term predictions of three main types of econometric data. See Appendix A for more results.

3.3. Statistical Significance

However, one may suggest that the victory of our new methods is only specific to these samples. Therefore, we challenge this superiority by testing the statistical significance. Noting that the GE-NoVaS-without-$\tilde{a}_0$ method is a nested method (taking $\tilde{a}_0 = 0$ in the larger model) compared with the GE-NoVaS method, we deploy the CW test [19] to ensure that the removing-$\tilde{a}_0$ idea is also statistically reasonable; see the p-value column in Table 1 for the tests' results. The reason for not performing CW tests on the simulation cases is that the prediction performance of each simulation is the average value of 5 replications. These CW test results imply that the null hypothesis should not be rejected for almost all cases under a 5% level of significance, which confirms the equivalence of the new method to the existing one.

Table 1. Comparisons of different methods' forecasting performance.

		GE-NoVaS	GE-NoVaS-without-$\tilde{a}_0$	GARCH(1,1)	*p*-Value(CW Test)
Simulated-1-year-data	Model-1-1step	0.91369	**0.88781**	1.00000	
	Model-1-5steps	0.61001	**0.52872**	1.00000	
	Model-1-30steps	0.77250	**0.73604**	1.00000	
	Model-2-1step	0.97796	**0.94635**	1.00000	
	Model-2-5steps	0.98127	**0.96361**	1.00000	
	Model-2-30steps	1.38353	**0.98872**	1.00000	
	Model-3-1step	0.99183	**0.92829**	1.00000	
	Model-3-5steps	0.77088	**0.67482**	1.00000	
	Model-3-30steps	0.79672	**0.71003**	1.00000	
	Model-4-1step	0.83631	**0.78087**	1.00000	
	Model-4-5steps	0.38296	**0.34396**	1.00000	
	Model-4-30steps	**0.00199**	0.00201	1.00000	
2-years-data	2018~2019-MCD-1step	0.99631	**0.99614**	1.00000	0.00053
	2018~2019-MCD-5steps	0.95403	**0.92120**	1.00000	0.03386
	2018~2019-MCD-30steps	0.75730	**0.62618**	1.00000	0.19691
	2018~2019-BAC-1step	0.98393	**0.97966**	1.00000	0.09568
	2018~2019-BAC-5steps	0.98885	**0.95124**	1.00000	0.07437
	2018~2019-BAC-30steps	1.14111	**0.87414**	1.00000	0.03643
1-year-data	2019-AAPL-1step	0.84533	**0.80948**	1.00000	0.25096
	2019-AAPL-5steps	0.85401	**0.68191**	1.00000	0.06387
	2019-AAPL-30steps	0.99043	**0.73823**	1.00000	0.17726
	2019-Djones-1step	0.96752	**0.96365**	1.00000	0.34514
	2019-Djones-5steps	0.98725	**0.89542**	1.00000	0.24529
	2019-Djones-30steps	0.86333	**0.80304**	1.00000	0.23766
	2019-SP500-1step	0.96978	**0.92183**	1.00000	0.45693
	2019-SP500-5steps	0.96704	**0.75579**	1.00000	0.24402
	2019-SP500-30steps	0.34389	**0.29796**	1.00000	0.08148
Volatile-1-year-data	11.2019~10.2020-IBM-1step	**0.80222**	0.80744	1.00000	0.16568
	11.2019~10.2020-IBM-5steps	**0.38933**	0.40743	1.00000	0.03664
	11.2019~10.2020-IBM-30steps	0.01143	**0.00918**	1.00000	0.15364
	11.2019~10.2020-CADJPY-1step	**0.46940**	0.48712	1.00000	0.16230
	11.2019~10.2020-CADJPY-5steps	**0.11678**	0.13549	1.00000	0.06828
	11.2019~10.2020-CADJPY-30steps	0.00584	**0.00394**	1.00000	0.15174
	11.2019~10.2020-SP500-1step	0.97294	**0.92349**	1.00000	0.05536
	11.2019~10.2020-SP500-5steps	0.96590	**0.75183**	1.00000	0.17380
	11.2019~10.2020-SP500-30steps	0.34357	**0.29793**	1.00000	0.16022
	11.2019~10.2020-Djones-1step	**0.56357**	0.57550	1.00000	0.11099
	11.2019~10.2020-Djones-5steps	**0.09810**	0.11554	1.00000	0.45057
	11.2019~10.2020-Djones-30steps	4.32×10^{-5}	**1.24×10^{-5}**	1.00000	0.68487

Note: The values presented in the GE-NoVaS and GE-NoVaS-without-$\tilde{a}_0$ columns reflect the relative performance compared with the 'standard' GARCH(1,1) method. The null hypothesis of the CW test is that parsimonious and larger models have equal mean squared prediction error (MSPE). The alternative is that the larger model has a smaller MSPE.

4. Summary

In previous studies of NoVaS methods, only a few real-word data analyses were performed [2–4]. Here, we provide extensive data analyses to address the lack of real-world data experiments. Our results are consistent with previous findings and substantiate the effectiveness of the NoVaS method again, i.e., the NoVaS method is more efficient and stable than the classical GARCH method for short-term predictions. Further, we reveal the ability of NoVaS-type methods to perform long-term time-aggregated forecasting. Beyond this, we propose a new NoVaS method that outperforms the state-of-the-art GE-NoVaS method. Our findings in this article are summarized as follows:

- Existing GE-NoVaS and new GE-NoVaS-without-$\tilde{a}_0$ methods provide substantial improvements for time-aggregated prediction, which hints towards the stability of NoVaS-type methods for providing long-horizon inferences.
- Our new method has superior performance to the GE-NoVaS method, especially for shorter sample sizes or more volatile data. This is significant given that GARCH-type models are difficult to estimate in shorter samples.
- We provide a statistical hypothesis test that shows that our model provides a more parsimonious fit, especially for long-term time-aggregated predictions.

5. Discussion

In this article, we explored the GE-NoVaS method toward short and long time-aggregated predictions and proposed a new variant that is based on a parsimonious model, has better empirical performance and yet is statistically reasonable. Although our new method is in a parsimonious form, it still obeys the autoregressive prediction rule and it is more stable for performing predictions under L_2 risk criterion than current the GE-NoVaS method. We should note that the unknown coefficients of both the GE-NoVaS $(\tilde{a}_0, a_1, \cdots, a_p)$ and GE-NoVaS-without-$\tilde{a}_0$ $(a_1, \cdots, a_p)$ methods are in exponential form, which implies that the correlations within series data are decreasing in exponential speed with the increasing time order. However, this specific form is not suitable for use for predicting all datasets. In other words, we anticipate performing NoVaS prediction without fixing the unknown coefficients in an invariant form to satisfy the variety of real-world econometric datasets. Therefore, building a NoVaS method with a more arbitrary coefficient form can be a future research direction. In addition, we should also note that there is a high demand to perform efficient forecasting for integer time series data. For example, a relevant topic regarding such integer-value prediction is forecasting COVID-19 cases. It will be beneficial to develop a variant of NoVaS for integer-value data. Moreover, in the financial market, the stock data move together. Thus, it would be exciting to see if one can perform model-free predictions in a multiple time series scenario. We hope that this article will open up avenues where one can explore other specific transformation structures to improve the existing forecasting frameworks and aid in specific tasks.

From a statistical inference point of view, one can also construct prediction intervals for these predictions using bootstrap. Such prediction intervals are well sought after in the econometrics literature and some results on the asymptotic validity of these can be provided. Additionally, we can also explore dividing the dataset into test and training in some optimal way and see if this can improve the performance of these methods.

In addition, there are some model-free methods based on machine learning to perform prediction tasks. These modern techniques enjoy high accuracy, but are time-consuming and lack of statistical inference. On the other hand, our new method and existing NoVaS methods are time-efficient and outperform classical GARCH-type methods significantly. More importantly, NoVaS-type methods can provide concrete statistical inference. Thus, it will be interesting to challenge NoVaS-type methods' forecasting accuracy with machine-learning-based methods.

Author Contributions: Data curation, K.W. and S.K.; Formal analysis, S.K.; Investigation, K.W. and S.K.; Methodology, S.K.; Software, K.W. and S.K.; Visualization, K.W.; Writing—original draft, K.W.; Writing—review and editing, S.K. All authors have read and agreed to the published version of the manuscript.

Funding: The second author's research is partially funded by NSF-DMS 2124222.

Data Availability Statement: We collected all data presented here from www.investing.com (accessed on 3 November 2021) manually. Then, we transformed the closing price data into financial log-returns based on Equation (13).

Acknowledgments: We are thankful to the two anonymous referees who helped us to improve the paper significantly. The first author is thankful to Professor Politis for the introduction to the topic and useful discussions.

Conflicts of Interest: The authors declare no conflicts of interest.

Appendix A. Additional Simulation Study and Data Analysis Results

Appendix A.1. Additional Simulation Study: Model Misspecification

In the real world, it is difficult to convincingly state whether the data obey one particular type of GARCH model, so we wish to provide four more GARCH-type models to simulate one-year datasets to see if our methods are satisfactory regardless of the underlying distribution and GARCH-type model. The simulation study results are presented in

Table A1, which implies that the NoVaS-type methods are more robust against model misspecification and GE-NoVaS-without-$\tilde{a}_0$ is the best method.

Model 5: Another time-varying GARCH(1,1) with Gaussian errors
$X_t = \sigma_t \epsilon_t,\ \sigma_t^2 = \omega_{0,t} + \beta_{1,t}\sigma_{t-1}^2 + \alpha_{1,t}X_{t-1}^2,\ \{\epsilon_t\} \sim i.i.d.\ N(0,1)$
$g_t = t/n; \omega_{0,t} = -4sin(0.5\pi g_t) + 5; \alpha_{1,t} = -1(g_t - 0.3)^2 + 0.5; \beta_{1,t} = 0.2sin(0.5\pi g_t) + 0.2,\ n = 250$
Model 6: Exponential GARCH(1,1) with Gaussian errors
$X_t = \sigma_t \epsilon_t,\ \log(\sigma_t^2) = 0.00001 + 0.8895 \log(\sigma_{t-1}^2) + 0.1\epsilon_{t-1} + 0.3(|\epsilon_{t-1}| - E|\epsilon_{t-1}|),$
$\{\epsilon_t\} \sim i.i.d.\ N(0,1)$
Model 7: GJR-GARCH(1,1) with Gaussian errors
$X_t = \sigma_t \epsilon_t,\ \sigma_t^2 = 0.00001 + 0.5\sigma_{t-1}^2 + 0.5X_{t-1}^2 - 0.5I_{t-1}X_{t-1}^2,\ \{\epsilon_t\} \sim i.i.d.\ N(0,1)$
$I_t = 1$ if $X_t \leq 0; I_t = 0$ otherwise
Model 8: Another GJR-GARCH(1,1) with Gaussian errors
$X_t = \sigma_t \epsilon_t,\ \sigma_t^2 = 0.00001 + 0.73\sigma_{t-1}^2 + 0.1X_{t-1}^2 + 0.3I_{t-1}X_{t-1}^2,\ \{\epsilon_t\} \sim i.i.d.\ N(0,1)$
$I_t = 1$ if $X_t \leq 0; I_t = 0$ otherwise

Table A1. Comparisons of different methods' forecasting performance on simulated 1-year data.

	GE-NoVaS	GE-NoVaS-without-$\tilde{a}_0$	GARCH(1,1)
M5-1step	0.91538	0.83168	1.00000
M5-5steps	0.49169	0.43772	1.00000
M5-30steps	0.25009	0.22659	1.00000
M6-1step	0.95939	0.94661	1.00000
M6-5steps	0.93594	0.84719	1.00000
M6-30steps	0.84401	0.70301	1.00000
M7-1step	0.84813	0.73553	1.00000
M7-5steps	0.50849	0.46618	1.00000
M7-30steps	0.06832	0.06479	1.00000
M8-1step	0.79561	0.76586	1.00000
M8-5steps	0.48028	0.38107	1.00000
M8-30steps	0.00977	0.00918	1.00000

Appendix A.2. Additional Data Analysis: 1-Year Datasets

To make our data analysis more comprehensive, we present more results of predictions on 1-year real-world datasets in Table A2. One interesting finding is that the GE-NoVaS method is significantly overcome by using the GARCH(1,1) model for some cases, such as the BAC, TSLA and Smallcap datasets. The GE-NoVaS-without-$\tilde{a}_0$ method still maintains great forecasting performance.

Table A2. Comparisons of different methods' forecasting performance on real-world 1-year data.

	GE-NoVaS	GE-NoVaS-without-$\tilde{a}_0$	GARCH(1,1)
2019-MCD-1step	0.95959	0.93141	1.00000
2019-MCD-5steps	1.00723	0.90061	1.00000
2019-MCD-30steps	1.05239	0.80805	1.00000
2019-BAC-1step	1.04272	0.97757	1.00000
2019-BAC-5steps	1.22761	0.89571	1.00000
2019-BAC-30steps	1.45020	1.01175	**1.00000**
2019-MSFT-1step	1.03308	0.98469	1.00000
2019-MSFT-5steps	1.22340	1.02387	**1.00000**
2019-MSFT-30steps	1.23020	0.97585	1.00000
2019-TSLA-1step	1.00428	0.98646	1.00000
2019-TSLA-5steps	1.06610	0.97523	1.00000
2019-TSLA-30steps	2.00623	0.87158	1.00000
2019-Bitcoin-1step	0.89929	0.86795	1.00000
2019-Bitcoin-5steps	0.62312	0.55620	1.00000
2019-Bitcoin-30steps	0.00733	0.00624	1.00000
2019-Nasdaq-1step	0.99960	0.93558	1.00000
2019-Nasdaq-5steps	1.15282	0.84459	1.00000
2019-Nasdaq-30steps	0.68994	0.58924	1.00000
2019-NYSE-1step	0.92486	0.90407	1.00000

Table A2. *Cont.*

	GE-NoVaS	GE-NoVaS-without-$\tilde{a}_0$	GARCH(1,1)
2019-NYSE-5steps	0.86249	**0.69822**	1.00000
2019-NYSE-30steps	0.22122	**0.18173**	1.00000
2019-Smallcap-1step	1.02041	**0.98731**	1.00000
2019-Smallcap-5steps	1.15868	**0.87700**	1.00000
2019-Samllcap-30steps	1.30467	**0.88825**	1.00000
2019-BSE-1step	0.70667	**0.67694**	1.00000
2019-BSE-5steps	0.25675	**0.23665**	1.00000
2019-BSE-30steps	0.03764	**0.02890**	1.00000
2019-BIST-1step	0.96807	**0.95467**	1.00000
2019-BIST-5steps	0.98944	**0.82898**	1.00000
2019-BIST-30steps	2.21996	**0.88511**	1.00000

Appendix A.3. Additional Data Analysis: Volatile 1-Year Datasets

Similarly, we consider more volatile 1-year datasets. All prediction results are tabulated in Table A3. It is clear that both NoVaS-type methods still outperform the GARCH(1,1) model for short- and long-term time-aggregated forecasting. Although the GE-NoVaS method yields optimal performance in some cases, we should note that the GE-NoVaS-without-$\tilde{a}_0$ method still gives almost the same but slightly worse results. Interestingly, the GE-NoVaS-without-$\tilde{a}_0$ method can introduce a significant improvement compared with the GE-NoVaS method for 30-step-ahead predictions. This again hints towards the superior robustness of our new method specifically for long-term aggregated predictions.

Table A3. Comparisons of different methods' forecasting performance on volatile 1-year data.

	GE-NoVaS	GE-NoVaS-without-$\tilde{a}_0$	GARCH(1,1)
11.2019~10.2020-MCD-1step	**0.51755**	0.58018	1.00000
11.2019~10.2020-MCD-5steps	**0.10725**	0.17887	1.00000
11.2019~10.2020-MCD-30steps	3.32×10^{-5}	**7.48×10^{-6}**	1.00000
11.2019~10.2020-AMZN-1step	0.97099	**0.90200**	1.00000
11.2019~10.2020-AMZN-5steps	0.88705	**0.71789**	1.00000
11.2019~10.2020-AMZN-30steps	0.58124	**0.53460**	1.00000
11.2019~10.2020-SBUX-1step	**0.68206**	0.69943	1.00000
11.2019~10.2020-SBUX-5steps	**0.24255**	0.30528	1.00000
11.2019~10.2020-SBUX-30steps	0.00499	**0.00289**	1.00000
11.2019~10.2020-MSFT-1step	**0.80133**	0.84502	1.00000
11.2019~10.2020-MSFT-5steps	**0.35567**	0.37528	1.00000
11.2019~10.2020-MSFT-30steps	0.01342	**0.00732**	1.00000
11.2019~10.2020-EURJPY-1step	0.95093	**0.94206**	1.00000
11.2019~10.2020-EURJPY-5steps	**0.76182**	0.76727	1.00000
11.2019~10.2020-EURJPY-30steps	0.16202	**0.15350**	1.00000
11.2019~10.2020-CNYJPY-1step	**0.77812**	0.79877	1.00000
11.2019~10.2020-CNYJPY-5steps	**0.38875**	0.40569	1.00000
11.2019~10.2020-CNYJPY-30steps	0.08398	**0.06270**	1.00000
11.2019~10.2020-Smallcap-1step	**0.58170**	0.60931	1.00000
11.2019~10.2020-Smallcap-5steps	**0.10270**	0.10337	1.00000
11.2019~10.2020-Smallcap-30steps	7.00×10^{-5}	**5.96×10^{-5}**	1.00000
11.2019~10.2020-BSE-1step	**0.39493**	0.39745	1.00000
11.2019~10.2020-BSE-5steps	**0.03320**	0.04109	1.00000
11.2019~10.2020-BSE-30steps	2.45×10^{-5}	**1.82×10^{-5}**	1.00000
11.2019~10.2020-NYSE-1step	**0.55741**	0.57174	1.00000
11.2019~10.2020-NYSE-5steps	**0.08994**	0.10182	1.00000
11.2019~10.2020-NYSE-30steps	1.36×10^{-5}	**6.64×10^{-6}**	1.00000
11.2019~10.2020-USDXfuture-1step	1.14621	**0.99640**	1.00000
11.2019~10.2020-USDXfuture-5steps	0.61075	**0.54834**	1.00000
11.2019~10.2020-USDXfuture-30steps	0.10723	**0.10278**	1.00000
11.2019~10.2020-Nasdaq-1step	**0.71380**	0.75350	1.00000
11.2019~10.2020-Nasdaq-5steps	**0.29332**	0.33519	1.00000
11.2019~10.2020-Nasdaq-30steps	0.01223	**0.00599**	1.00000
11.2019~10.2020-Bovespa-1step	0.60031	**0.57558**	1.00000
11.2019~10.2020-Bovespa-5steps	0.08603	**0.07447**	1.00000
11.2019~10.2020-Bovespa-30steps	6.87×10^{-6}	**2.04×10^{-6}**	1.00000

Appendix B. Stationarity Test Results of Some Real-World Datasets

Table A4. *p*-values of three stationarity tests.

	ADF	KPSS	PP
2018~2019 MCD	0.01	0.10	0.01
2018~2019 BAC	0.01	0.10	0.01
2019 AAPL	0.01	0.10	0.01
2019 Djones	0.10	0.10	0.01
2019 SP500	0.18	0.10	0.01
11.2019~10.2020 IBM	0.31	0.05	0.01
11.2019~10.2020 CADJPY	0.01	0.10	0.01
11.2019~10.2020 SP500	0.23	0.08	0.01
11.2019~10.2020 Djones	0.22	0.08	0.01

Note: The null hypothesis of the ADF and PP tests is that the tested series is non-stationary. Therefore, if the ADF and PP tests are rejected, it means that this tested series is stationary. On the other hand, the null hypothesis of KPSS is that the series is stationary.

References

1. Politis, D.N. *A Normalizing and Variance-Stabilizing Transformation for Financial Time Series*; Elsevier Inc.: Amsterdam, The Netherlands, 2003.
2. Gulay, E.; Emec, H. Comparison of forecasting performances: Does normalization and variance stabilization method beat GARCH (1, 1)-type models? Empirical evidence from the stock markets. *J. Forecast.* **2018**, *37*, 133–150. [CrossRef]
3. Chen, J.; Politis, D.N. Time-varying NoVaS Versus GARCH: Point Prediction, Volatility Estimation and Prediction Intervals. *J. Time Ser. Econom.* **2020**, *1*. [CrossRef]
4. Chen, J.; Politis, D.N. Optimal multi-step-ahead prediction of ARCH/GARCH models and NoVaS transformation. *Econometrics* **2019**, *7*, 34. [CrossRef]
5. Chudý, M.; Karmakar, S.; Wu, W.B. Long-term prediction intervals of economic time series. *Empir. Econ.* **2020**, *58*, 191–222. [CrossRef]
6. Karmakar, S.; Chudy, M.; Wu, W.B. Long-term prediction intervals with many covariates. *arXiv* **2020**, arXiv:2012.08223.
7. Kitsul, Y.; Wright, J.H. The economics of options-implied inflation probability density functions. *J. Financ. Econ.* **2013**, *110*, 696–711. [CrossRef]
8. Bansal, R.; Kiku, D.; Yaron, A. Risks for the long run: Estimation with time aggregation. *J. Monet. Econ.* **2016**, *82*, 52–69. [CrossRef]
9. Starica, C. *Is GARCH (1, 1) as Good a Model as the Accolades of the Nobel Prize Would Imply?* University Library of Munich: Munich, Germany, 2003; SSRN 637322.
10. Fryzlewicz, P.; Sapatinas, T.; Rao, S.S. Normalized least-squares estimation in time-varying ARCH models. *Ann. Stat.* **2008**, *36*, 742–786. [CrossRef]
11. Politis, D.N. The Model-Free Prediction Principle. In *Model-Free Prediction and Regression*; Springer: Berlin/Heidelberg, Germany, 2015; pp. 13–30.
12. Engle, R.F. Autoregressive conditional heteroscedasticity with estimates of the variance of United Kingdom inflation. *Econom. J. Econom. Soc.* **1982**, *50*, 987–1007. [CrossRef]
13. Chen, J. Prediction in Time Series Models and Model-free Inference with a Specialization in Financial Return Data. Ph.D. Thesis, University of California, San Diego, CA, USA, 2018.
14. Awartani, B.M.; Corradi, V. Predicting the volatility of the S&P-500 stock index via GARCH models: The role of asymmetries. *Int. J. Forecast.* **2005**, *21*, 167–183.
15. Said, S.E.; Dickey, D.A. Testing for unit roots in autoregressive-moving average models of unknown order. *Biometrika* **1984**, *71*, 599–607. [CrossRef]
16. Perron, P. Trends and random walks in macroeconomic time series: Further evidence from a new approach. *J. Econ. Dyn. Control* **1988**, *12*, 297–332. [CrossRef]
17. Kwiatkowski, D.; Phillips, P.C.; Schmidt, P.; Shin, Y. Testing the null hypothesis of stationarity against the alternative of a unit root: How sure are we that economic time series have a unit root? *J. Econom.* **1992**, *54*, 159–178. [CrossRef]
18. Hafner, C.M.; Preminger, A. An ARCH model without intercept. *Econ. Lett.* **2015**, *129*, 13–17. [CrossRef]
19. Clark, T.E.; West, K.D. Approximately normal tests for equal predictive accuracy in nested models. *J. Econom.* **2007**, *138*, 291–311. [CrossRef]

forecasting

MDPI

Article

Bootstrapped Holt Method with Autoregressive Coefficients Based on Harmony Search Algorithm

Eren Bas [1], Erol Egrioglu [1,*] and Ufuk Yolcu [2]

[1] Department of Statistics, Faculty of Arts and Science, Giresun University, Giresun 28200, Turkey; eren.bas@giresun.edu.tr

[2] Department of Statistics, Faculty of Arts and Science, Marmara University, Istanbul 34722, Turkey; ufuk.yolcu@giresun.edu.tr

* Correspondence: erol.egrioglu@giresun.edu.tr; Tel.: +90-454-3101400

Abstract: Exponential smoothing methods are one of the classical time series forecasting methods. It is well known that exponential smoothing methods are powerful forecasting methods. In these methods, exponential smoothing parameters are fixed on time, and they should be estimated with efficient optimization algorithms. According to the time series component, a suitable exponential smoothing method should be preferred. The Holt method can produce successful forecasting results for time series that have a trend. In this study, the Holt method is modified by using time-varying smoothing parameters instead of fixed on time. Smoothing parameters are obtained for each observation from first-order autoregressive models. The parameters of the autoregressive models are estimated by using a harmony search algorithm, and the forecasts are obtained with a subsampling bootstrap approach. The main contribution of the paper is to consider the time-varying smoothing parameters with autoregressive equations and use the bootstrap method in an exponential smoothing method. The real-world time series are used to show the forecasting performance of the proposed method.

Keywords: Holt method; subsampling bootstrapped; harmony search algorithm; forecasting

Citation: Bas, E.; Egrioglu, E.; Yolcu, U. Bootstrapped Holt Method with Autoregressive Coefficients Based on Harmony Search Algorithm. *Forecasting* **2021**, *3*, 839–849. https://doi.org/10.3390/forecast3040050

Academic Editor: Kuo-Ping Lin

Received: 20 September 2021
Accepted: 3 November 2021
Published: 4 November 2021

Publisher's Note: MDPI stays neutral with regard to jurisdictional claims in published maps and institutional affiliations.

1. Introduction

Exponential smoothing methods were published in the late 1950s [1–3], and they are known as some of the most successful forecasting methods in the literature. There are many exponential smoothing methods in the literature, such as the single exponential smoothing method, Holt method, Holt-Winters method, etc. Each exponential smoothing method is used in different situations. If data has no trend and no seasonality, a simple exponential smoothing method is used for forecasting. If data has a linear trend and no seasonality, the Holt method is used for forecasting. If data has both trend and seasonality, the Holt-Winters method is used for forecasting. In the coming years, the damped trend model was proposed by [4] if data has an over-trend. The reason why exponential smoothing methods are popular in the literature is that the forecasting success of exponential smoothing methods is superior to complicated approaches such as [5–7]. In addition to these methods, [8] proposed a simple modification of the exponential smoothing method named the ATA method, which is an effective and simple method to use compared with complex approaches in recent years.

Moreover, ref. [9,10] developed state-of-the-art guidelines for the application of the exponential smoothing methodology. Ref. [11] proposed a uniformly-sampled-autoregressive-moving-average model for a second-order linear stochastic system. Ref. [12] introduced the optimal procedure of the Boolean Kalman filter over a finite horizon. Ref. [13] presented a general benchmarking framework applicable to computational intelligence algorithms for solving forecasting problems. Ref. [14] proposed a new enhanced optimization model based on the bagged echo state network and improved by a differential evolution algorithm to estimate energy consumption. Ref. [15] introduced a two-stage Bayesian optimization framework for scalable and efficient inference in state-space models.

The method proposed by [2] is one of the effective exponential smoothing methods for forecasting data with trend. The Holt method has a forecasting equation and two smoothing equations, which are for the level of the series and slope of the trend as given in Equations (1)–(3).

$$\hat{x}_{n+1} = \hat{l}_n + \hat{b}_n \tag{1}$$

$$\hat{l}_n = \lambda_1 x_n + (1 - \lambda_1) x_n \tag{2}$$

$$\hat{b}_n = \lambda_2 \left(\hat{l}_n - \hat{l}_{n-1} \right) + (1 - \lambda_2) \hat{b}_{n-1} \tag{3}$$

In Equations (1)–(3), λ_1 and λ_2 are the smoothing parameters of mean level and slope, respectively, and these parameters get values between zero and one. In these equations, the initial values are obtained by applying simple linear regression to the series. In addition, in these equations, trend and level update formulas are only based on a lag.

In this study, the Holt method is modified by using time-varying smoothing parameters instead of fixed on time, and the smoothing parameters of mean level and slope are obtained for each observation with first-order autoregressive models. The parameters of the autoregressive models are estimated by using the harmony search algorithm (HSA). With these contributions, the proposed method eliminates the initial parameter determination problem. Moreover, the forecasts for the proposed method are obtained from sampling distributions of forecasts.

The proposed method is applied to Istanbul Stock Exchange data sets between the years 2000 and 2017 with different test lengths. The obtained results are compared with many methods in the literature. The brief information for HSA is given in Section 2. The proposed method is introduced, and the implementation results are given in Sections 3 and 4 respectively. The final section is for conclusion and discussion.

2. Harmony Search Algorithm

HSA algorithm was proposed by [16]. HSA is a heuristic algorithm that simulates the notes of musicians. The principle of HSA is that the musicians in an orchestra play the best melody harmonically with the notes they play. Just as a chromosome in the genetic algorithm or a particle in particle swarm optimization represents a solution, a harmony in a harmony memory represents a solution in the harmony search algorithm. In HSA, each musician has a decision variable and each note in the memory of each musician corresponds to a different solution of that decision variable. Each harmony consists of different notes and each note corresponds to the decision variable. HSA aims to investigate whether the obtained solution vector is better than the worst solution in memory. The HSA is given below in steps in Algorithm 1.

Algorithm 1 The algorithm of HSA

Step 1. Determination of parameters to be used in HSA:
- XHM: Harmony memory;
- HMS: Harmony memory search;
- HMCR: Harmony memory considering rate;
- PAR: Pitch adjusting rate;
- n: the number of variables.

Algorithm 1 Cont.

Step 2. Creating of the harmony memory.

HM for HSA is generated as in Equation (4).

$$
HM = \begin{bmatrix} x_{11} & x_{12} & \cdots & x_{1n} \\ x_{21} & x_{22} & \cdots & x_{2n} \\ \vdots & \vdots & \vdots & \vdots \\ x_{HMS1} & x_{HMS2} & x_{HMS3} & x_{HMSn} \end{bmatrix} = \begin{bmatrix} x'_1 \\ x'_2 \\ \vdots \\ x'_{HMS} \end{bmatrix} \tag{4}
$$

Here, $x_{ij}, i = 1, 2, \ldots HMS\,; j = 1, 2, \cdots, n$ is expressed as a note value and is generated randomly.

In HSA, each solution vector is denoted by $x'_i, i = 1, 2, \cdots, HMS$. In HSA, there are HMS solution vectors. The representation of the first solution vector is given in Equation (5).

$$
x'_1 = [x_{11}, x_{12}, \cdots, x_{1n}] \tag{5}
$$

Step 3. Calculation of objective function values.

The objective function values are calculated for each solution vector generated randomly as given in Equation (6).

$$
\begin{bmatrix} x_{11} & x_{12} & \cdots & x_{1n} \\ x_{21} & x_{22} & \cdots & x_{2n} \\ \vdots & \vdots & \vdots & \vdots \\ x_{HMS1} & x_{HMS2} & x_{HMS3} & x_{HMSn} \end{bmatrix} - \begin{bmatrix} x'_1 \\ x'_2 \\ \vdots \\ x'_{HMS} \end{bmatrix} - \begin{bmatrix} f(x'_1) \\ f(x'_2) \\ \vdots \\ f(x'_{HMS}) \end{bmatrix} \tag{6}
$$

Step 4. Improvement of a new harmony.

While the probability of $HMCR$ with a value between 0 and 1 is to select a value from the existing values in the HM, (1-HMCR) value is the ratio of a random value selected from the possible value ranges. The new harmony is obtained with the help of Equation (7).

$$
x_{ijnew} = \begin{cases} x_{ijnew} \in \left\{ x_{ij}; \mathrm{i} = 1, , 2, \cdots, HMS \right\} & if rnd < HMCR \\ x_{ijnew} \in \left\{ \left[\min(x_{ij}), \max(x_{ij}) \right]; i = 1, 2, \ldots HMS \right\} & otherwise \end{cases} \tag{7}
$$

It is decided by the PAR parameter whether the toning process can be applied to each selected decision variable with the possibility of $HMCR$ or not as given in Equation (8).

$$
x_{ijnewpitch} = \begin{cases} Yes & rnd < PAR \\ No & otherwise \end{cases} \tag{8}
$$

In Equation (8), rnd is generated randomly between $U(0, 1)$. If this random number is smaller than the PAR value, this value is changed to the closest value to it. If the tonalization will be made for each x_{ijnew} decision variable and the value of x_{ijnew} is assumed to be the kth value within the vector of the value variable, the new value of $x_{ijnew}(k)$ is $x_{ij} \leftarrow x_{ij}(k + m)$, and $m \in \{ \cdots, -2, -1, 1, 2, \cdots \}$ is the neighboring index.

Step 5. Updating the harmony memory.

If the new harmony vector is better than the worst vector in the HM, the worst vector is removed from the memory, and the new harmony vector is included in the HM instead of the removed vector.

Step 6. Stop condition check.

Steps 4–6 are repeated until the termination criteria are met. Possible values for HMCR and PAR in literature are between 0.7–0.95 and 0.05–0.7, respectively [17].

---

3. Proposed Method

Although the Holt method is used as an efficient forecasting method, it has many problems that are obvious and need to be resolved. The first of these problems is the determination of initial trend and level values. The second problem of the Holt method is that the trend and level update formulas are only based on a lag. To avoid these problems and increase the forecasting performance of the Holt method, the advantages and innovations of the proposed method are given step by step as below:

- The smoothing parameters are varied from observation to observation using first-order autoregressive equations;
- The optimal parameters of the Holt method are determined with HSA;
- The forecasts are obtained by the Sub-sampling Bootstrap method.

The algorithm of the proposed method is also given in Algorithm 2.

Algorithm 2 The algorithm of the proposed method

Step 1. Determine the parameters of the training process:
- # observation of test set: *ntest*;
- HMS;
- HMCR;
- PAR;
- # bootstrap samples: *nbst*;
- bootstrap sample size: *bss*.

Step 2. Select bootstrap samples from the training set randomly.

Steps from 2.1. to 2.2 are repeated *nbst* times. $x_{t,j}^*$ presents *j*th bootstrap time series.

Step 2.1. Select a starting point of the block (spb) as an integer from a discrete uniform distribution with parameters $[1, ntrain\text{-}bss+1]$.

Step 2.2. Create bootstrap time series as given in Equation (9).

$$x_{t,j}^* = \left\{ x_{spb}, x_{spb+1}, \ldots, x_{spb+bss-1} \right\}, \qquad j = 1, 2, \ldots nbst \tag{9}$$

Step 3. Apply regression analysis to determine the initial bounds for level $(L(0))$ and trend $(B(0))$ parameters by using $x_{t,j}^*$ bootstrap time series as the training set by using Equations (10)–(12).

$$X = [1\ 1\ \cdots 1; 1\ 2\ \cdots bss]'_{bss*2} \tag{10}$$

$$Y = x_{t,j}^* = \left[x_{spb}, x_{spb+1}, \ldots, x_{spb+bss-1} \right]' \tag{11}$$

$$\hat{\beta} = \left[\begin{array}{c} \hat{\beta}_0 \\ \hat{\beta}_1 \end{array} \right] = (X\prime X)^{-1} X \prime Y \tag{12}$$

$(L(0) \in [\hat{\beta}_0/2, 2\hat{\beta}_0])$ and trend $(B(0) \in [\hat{\beta}_1/2, 2\hat{\beta}_1])$

Step 4. HSA is used to obtain the optimal parameters of the Holt method with autoregressive coefficients for each bootstrap time series. Steps 4.1 and 4.4 are repeated for each bootstrap time series.

Step 4.1. Generate the initial positions of HSA. The positions of harmony are $L(0)$, $B(0), \lambda_1(0), \lambda_2(0), \phi_{11}, \phi_{12}, \phi_{21}$ and ϕ_{22}.

$L(0)$ and $B(0)$ are generated from $U(\hat{\beta}_0/2, 2\hat{\beta}_0)$ and $U(\hat{\beta}_1/2, 2\hat{\beta}_1)$, respectively. $\lambda_1(0)$, $\lambda_2(0), \phi_{11}$ and ϕ_{21} are generated from $U(0, 1)$. ϕ_{12} and ϕ_{22} are generated from $U(-1, 1)$. The creation of the harmony memory for the proposed method is given in Equation (13), and the parameters that correspond to *k*th harmony are given in Table 2.

$$HM = \left[\begin{array}{ccccc} x_1^{\ 1} & x_2^{\ 1} & x_3^{\ 1} & \cdots & x_8^{\ 1} \\ x_1^{\ 2} & x_2^{\ 2} & x_3^{\ 2} & \cdots & x_8^{\ 2} \\ \vdots & \vdots & \vdots & \vdots & \vdots \\ x_1^{\ HMS} & x_2^{\ HMS} & x_3^{\ HMS} & \cdots & x_8^{\ HMS} \end{array} \right] \tag{13}$$

Algorithm 2 Cont.

Step 4.2. According to the initial positions of each harmony, fitness functions are calculated. The root of mean square error (RMSE) is preferred to use as a fitness function and is calculated as given in Equation (14).

$$f_i = RMSE_i = \sqrt{\frac{1}{bss}\sum_{t=1}^{bss}\left(x_{t,j}^* - \hat{x}_{t,j}^*\right)^2}, \, i = 1,2,\ldots,HMS \tag{14}$$

In Equation (14), $\hat{x}_{t,j}^{[?]}$ is the output for j^{th} bootstrap time series data and k^{th} harmony. $\hat{x}_{t,j}^*$ is obtained by using Equations (15)–(19).

$$\lambda_1(t) = \phi_{11} + \phi_{12}\lambda_1(t-1) \tag{15}$$

$$\lambda_2(t) = \phi_{21} + \phi_{22}\lambda_2(t-1) \tag{16}$$

$$L(t) = \lambda_1(t)x_{t,j}^* + (1-\lambda_1(t))(L(t-1) + B(t-1)) \tag{17}$$

$$B(t) = \lambda_2(t)(L(t) - L(t-1)) + (1-\lambda_2(t))(B(t-1)) \tag{18}$$

$$\hat{x}_{t+1,j}^* = L(t) + B(t) \tag{19}$$

Obtain RMSE values for each harmony, and save the best harmony which has the smallest RMSE.

Step 4.3. Improve new harmony.

$HMCR$ shows the probability that the value of a decision variable is selected from the current harmony memory. (1-$HMCR$) represents the random selection of the new decision variable from the existing solution space. x_i' shows the new harmony, obtained as in Equation (20).

$$x_i' = \begin{cases} x_i' \in \{x_i^1, x_i^2, \cdots, x_i^{HMS}\} & if \text{ rand} < HMCR \\ x_i' \in X, & otherwise \end{cases} \tag{20}$$

After this step, each decision variable is evaluated to determine whether a tonal adjustment is necessary. This is determined by the PAR parameter, which is the tone adjustment ratio. The new harmony vector is produced according to the randomly selected tones in the memory of harmony as given in Equation (21). Whether the variables are selected from the harmonic memory is determined by the HMCR ratio, which is between 0 and 1.

$$x_i' = \begin{cases} x_i' + rnd(0,1) * bw & if \text{ rnd} < PAR \\ x_i' & otherwise \end{cases} \tag{21}$$

bw is a bandwidth selected randomly; rnd (0; 1) represents a random number generated between 0 and 1.

Step 4.4. Harmony memory update.

In this step, the comparison between the newly created harmonies and the worst harmonies in the memory is made in terms of the values of the objective functions. If the newly created harmony vector is better than the worst harmony, the worst harmony vector is removed from the memory, and the new harmony vector is substituted for it.

Calculate RMSE values for j^{th} bootstrap time series data and k^{th} harmony. Find the best harmony which has the minimum RMSE value for j^{th} bootstrap time series data.

Step 5. Calculate the forecasts for test data by using the best harmony for each bootstrap sample and their statistics.

The obtained forecasts from the updated Equations for j^{th} bootstrap time series at t time is represented by F_t^i. Forecasts and their statistics are calculated just as in Table 1. In addition, the flowchart of the proposed method is given in Figure 1.

Table 1. Forecasts for bootstrap samples.

Time (*t*)/Bootstrap Sample	1	2	...	*nbst*	Median	Standard Deviation
1	F_1^1	F_1^2	...	F_1^{nbst}	$\hat{F}_1$	SE $(\hat{F}_1)$
2	F_2^1	F_2^2	...	F_2^{nbst}	$\hat{F}_2$	SE $(\hat{F}_2)$
$\vdots$	$\vdots$	$\vdots$	...	$\vdots$	$\vdots$	$\vdots$
ntest	F_{ntest}^1	F_{ntest}^2	...	F_{ntest}^{nbst}	$\hat{F}_{ntest}$	SE $(\hat{F}_{ntest})$

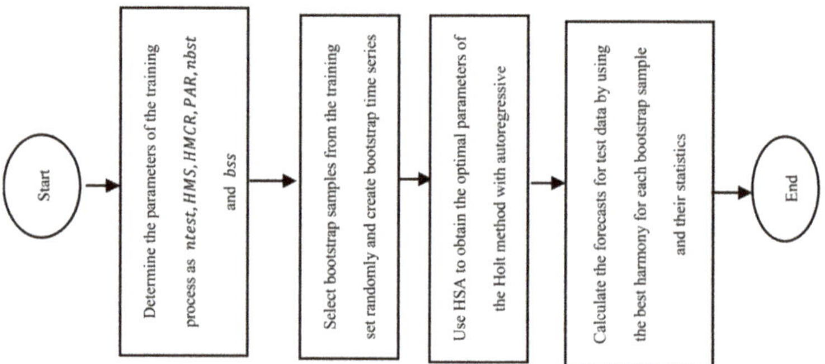

Figure 1. The flowchart of the proposed method.

Table 2. The parameters corresponding to *k*th harmony.

x_1^k	x_2^k	x_3^k	x_4^k	x_5^k	x_6^k	x_7^k	x_8^k
$L(0)$	$B(0)$	$\lambda_1(0)$	$\lambda_2(0)$	ϕ_{11}	ϕ_{12}	ϕ_{21}	ϕ_{22}

4. Applications

To evaluate the performance of the proposed method, the proposed method is applied to the Istanbul Stock Exchange (BIST) data sets observed daily between the years 2000 and 2017 with different test lengths as 10 and 20. To evaluate the performance of the proposed method, the proposed method is compared with the ATA method proposed by [8], Holt method, fuzzy regression functions approach (FF) proposed by [18], random walk (RW), multilayer perceptron artificial neural networks (MLP-ANN) and adaptive neural-fuzzy inference systems (ANFIS) method proposed by [19]. For a fair comparison of the methods, we used both statistical and computational intelligence forecasting methods. While the random walk was used as a simple forecasting method, the Holt and ATA methods were used as statistical forecasting methods. Moreover, MLP-ANN, ANFIS, and FF methods were used as computational intelligence forecasting methods. In the analysis process, the number of bootstrap samples and the bootstrap sample size is given as 100 for each data set. The RMSE and MAPE criteria were used for the comparison of the methods. The mean absolute percentage error (MAPE) is one of the most widely used measures of forecast accuracy, due to its advantages of scale-independency and interpretability [20]. The use of RMSE is very common, and it is considered an excellent general-purpose error metric for numerical predictions [21]. Table 3 gives the all-analysis results for each data set for the RMSE criterion when the length of the test set is 10.

Table 3. All analysis results for each data set for RMSE criterion when the length of the test set is 10.

Data	ATA	Holt	FF	RW	MLP-ANN	ANFIS	PP
BIST2000	279.79	296.17	310.42	286.15	343.9	619.21	278.82
BIST2001	204.84	237.69	272.31	206.5	1106.89	710.82	189.75
BIST2002	325.08	319.78	357	331.87	620.78	399.13	332.13
BIST2003	354.79	355.55	380.82	349.79	1859.21	420.75	328.25
BIST2004	315.62	315.79	390.15	325.69	1807.8	641.43	313.7
BIST2005	316.75	315.36	328.84	342.69	2071.98	559.2	304.98
BIST2006	354.03	348.58	352.07	356.81	423.98	389.3	346.98
BIST2007	768.29	734.55	673.14	734.14	897.02	550.97	728.92
BIST2008	283.99	277.2	256.98	253.67	444.74	340.41	260.52
BIST2009	505.05	483.8	558.06	551.97	3117.96	736.78	473.4
BIST2010	577.68	594.9	583.52	591.88	725.15	588.36	576.4
BIST2011	697.64	710.04	849.68	726.5	1733.88	1037.87	737.83
BIST2012	355.5	350.46	368.26	358.17	3237.45	406.68	358.15
BIST2013	1905.64	1898.61	2105.05	1922.14	4369.35	2104.39	1871.45
BIST2014	1068.36	1025.18	1177.56	1059.97	2631.25	1435.01	1036.6
BIST2015	772.84	767.71	758.89	779.07	1080.69	714.69	751.41
BIST2016	431.86	433.52	450.67	434.01	520.1	424.34	652.25
BIST2017	861.26	869.23	1113.74	911.21	3777.62	1283.75	827.15

In Table 3, the proposed method has 59% success compared with the other methods in terms of the RMSE criterion when the test set is 10. To see the actual comparison results of the proposed method with other methods, we compare the rank values of each method and obtain the average rank values. For this purpose, we rank each method according to their success status for each time series analyzed. In such a ranking, the method with the lowest RMSE value will be named as the best method, and the rank value of it will be taken as 1. For this purpose, all methods were calculated according to rank order considering the RMSE criterion when the length of the test set is 10, and average rank values were obtained as in Figure 2.

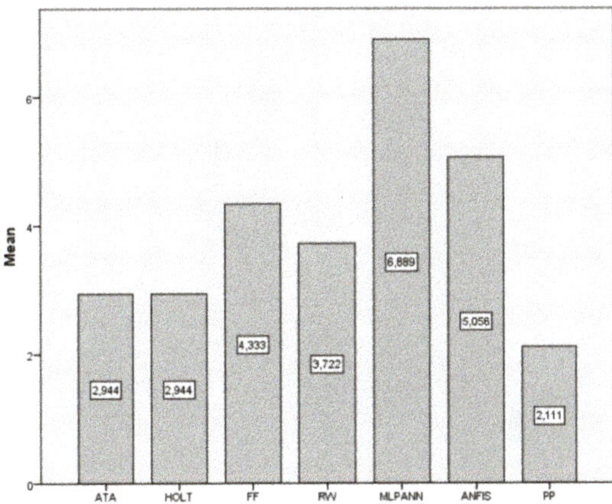

Figure 2. The average rank values of each method for RMSE criterion when the length of the test set is 10.

From Figure 2, it is seen that the proposed method has a minimum average rank value compared with other methods, and the proposed method is the best method for RMSE criterion when the length of the test set is 10. In addition, Table 4 gives the all-analysis results for each data set for the MAPE criterion given in Equation (22) when the length of the test set is 10.

$$MAPE = \frac{1}{ntest} \sum_{t=1}^{ntest} \left| \frac{X_t - \hat{X}_t}{X_t} \right| \tag{22}$$

Table 4. All-analysis results for each data set for MAPE criterion when the length of the test set is 10.

Data	ATA	Holt	FF	RW	MLP-ANN	ANFIS	PP
BIST2000	0.0222	0.0233	0.0268	0.0236	0.0293	0.0507	0.0223
BIST2001	0.011	0.0124	0.0161	0.0112	0.0818	0.0506	0.0103
BIST2002	0.0253	0.0241	0.0267	0.0256	0.043	0.0287	0.0256
BIST2003	0.0163	0.0163	0.0178	0.0162	0.1008	0.0208	0.0154
BIST2004	0.0099	0.01	0.0129	0.0103	0.0735	0.0241	0.0099
BIST2005	0.0068	0.0069	0.0069	0.0074	0.0519	0.0126	0.0066
BIST2006	0.0068	0.0066	0.007	0.0067	0.0082	0.0076	0.0073
BIST2007	0.0098	0.01	0.0087	0.0095	0.0138	0.008	0.0095
BIST2008	0.0082	0.0075	0.0073	0.0071	0.0154	0.0093	0.0075
BIST2009	0.0067	0.0066	0.0077	0.0076	0.0595	0.0114	0.0071
BIST2010	0.006	0.0064	0.0058	0.0063	0.0085	0.0068	0.0061
BIST2011	0.0113	0.0116	0.0137	0.0118	0.0316	0.0165	0.0119
BIST2012	0.004	0.0039	0.004	0.0039	0.041	0.0042	0.004
BIST2013	0.022	0.0219	0.0254	0.0223	0.0608	0.0258	0.0216
BIST2014	0.0092	0.009	0.0101	0.0094	0.0304	0.0118	0.0089
BIST2015	0.0083	0.0082	0.0078	0.0082	0.0109	0.0087	0.0082
BIST2016	0.0049	0.0048	0.0047	0.0048	0.0051	0.0046	0.0062
BIST2017	0.0052	0.0053	0.0076	0.0058	0.0318	0.0092	0.0049

In Table 4, the proposed method has 39% success compared with the other methods in terms of the MAPE criterion when the test set is 10. Looking at the rank evaluation results for the MAPE criterion when the test set length is 10 given in Figure 3, it is seen that the proposed method is in third place among all methods.

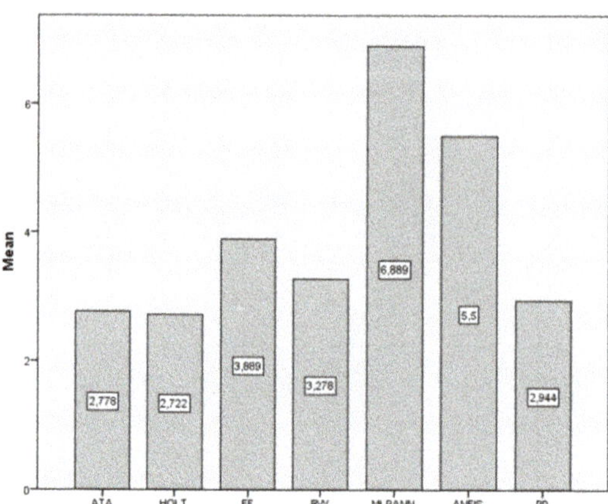

Figure 3. The average rank values of each method for MAPE criterion when the length of the test set is 10.

Table 5 also gives the all-analysis results for each data set for the RMSE criterion when the length of the test set is 20. In Table 5, the proposed method has a 61% success rate. Considering the situations where the proposed method is not the best, it stands out as the second-best method in many time-series analyses. Moreover, the rank evaluation results for all methods for the RMSE criterion when the length of the test set is 20 are given in

Figure 4. In addition, Table 6 gives the all-analysis results for each data set for the MAPE criterion when the length of the test set is 20.

Table 5. All-analysis results for each data set for RMSE criterion when the length of the test set is 20.

Data	ATA	Holt	FF	RW	MLP-ANN	ANFIS	PP
BIST2000	680.61	680.33	713.87	682.74	2868.94	825.58	681.58
BIST2001	315.19	326.20	372.36	312.96	1030.82	540.36	296.32
BIST2002	388.51	389.17	390.47	393.48	392.16	432.21	383.70
BIST2003	313.25	339.08	456.83	311.18	2201.77	558.18	288.38
BIST2004	329.12	329.30	366.48	335.16	1479.79	554.62	319.35
BIST2005	426.84	415.74	496.57	433.66	2940.74	632.79	463.17
BIST2006	539.71	551.20	581.55	547.72	742.07	625.98	556.77
BIST2007	814.90	783.40	789.45	774.91	854.08	660.30	762.16
BIST2008	575.72	571.80	589.64	542.31	766.02	624.59	541.21
BIST2009	492.91	510.09	518.55	516.25	2794.96	623.04	492.17
BIST2010	867.04	921.85	885.97	850.14	1193.33	965.97	864.93
BIST2011	757.81	728.63	849.50	790.69	1141.08	772.13	774.14
BIST2012	592.96	564.85	605.32	544.81	5641.93	1224.80	517.44
BIST2013	1687.26	1680.69	1888.99	1709.07	2453.56	1821.80	1669.36
BIST2014	1318.63	1315.91	1323.78	1315.91	1936.51	1610.11	1318.91
BIST2015	1242.98	1263.71	1223.85	1225.07	2322.70	1189.75	1213.33
BIST2016	650.22	662.26	648.96	599.52	699.81	728.46	604.62
BIST2017	1010.73	1011.04	1165.70	1031.37	2981.64	1134.55	833.03

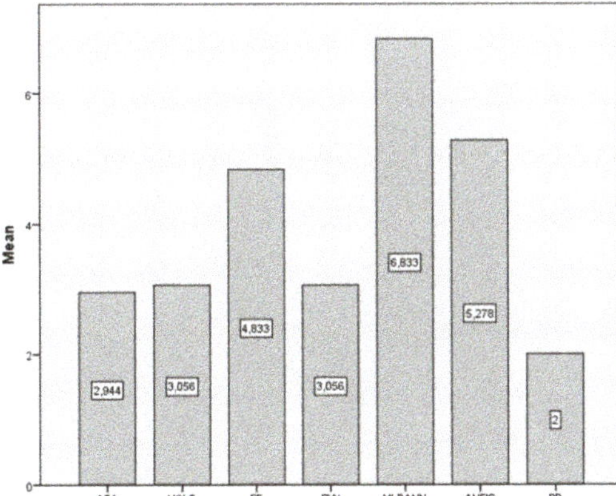

Figure 4. The average rank values of each method for RMSE criterion when the length of the test set is 20.

When the analysis results given in Table 6 are examined, even in the analyses in which the proposed method is not the best method, the proposed method often appears to be either the second-best or third-best method. We examine rank values to verify and highlight these results given in Figure 5.

Considering the average rank obtained from all methods, it can be said that the proposed method for the MAPE criterion has more successful results than other methods. As a final comment, when all analysis results are examined, it can be said from both average rank results and analysis results that the proposed method is a more successful method than other methods used in the comparison.

Table 6. All-analysis results for each data set for MAPE criterion when the length of the test set is 20.

Data	ATA	Holt	FF	RW	MLP-ANN	ANFIS	PP
BIST2000	0.0540	0.0547	0.0615	0.0557	0.3091	0.0748	0.0546
BIST2001	0.0176	0.0182	0.0212	0.0175	0.0746	0.0355	0.0178
BIST2002	0.0261	0.0263	0.0275	0.0272	0.0260	0.0311	0.0269
BIST2003	0.0145	0.0147	0.0219	0.0146	0.1216	0.0242	0.0144
BIST2004	0.0104	0.0101	0.0121	0.0108	0.0587	0.0184	0.0107
BIST2005	0.0087	0.0082	0.0097	0.0091	0.0744	0.0134	0.0096
BIST2006	0.0098	0.0102	0.0109	0.0102	0.0143	0.0124	0.0105
BIST2007	0.0113	0.0108	0.0106	0.0104	0.0127	0.0095	0.0105
BIST2008	0.0180	0.0175	0.0193	0.0164	0.0223	0.0183	0.0167
BIST2009	0.0076	0.0079	0.0080	0.0080	0.0529	0.0097	0.0074
BIST2010	0.0101	0.0112	0.0103	0.0099	0.0129	0.0125	0.0100
BIST2011	0.0118	0.0110	0.0131	0.0124	0.0188	0.0117	0.0121
BIST2012	0.0063	0.0061	0.0064	0.0058	0.0728	0.0142	0.0056
BIST2013	0.0180	0.0179	0.0206	0.0182	0.0278	0.0202	0.0180
BIST2014	0.0121	0.0121	0.0122	0.0122	0.0203	0.0151	0.0120
BIST2015	0.0140	0.0145	0.0133	0.0135	0.0285	0.0129	0.0134
BIST2016	0.0065	0.0065	0.0059	0.0058	0.0068	0.0067	0.0058
BIST2017	0.0073	0.0073	0.0088	0.0077	0.0246	0.0087	0.0065

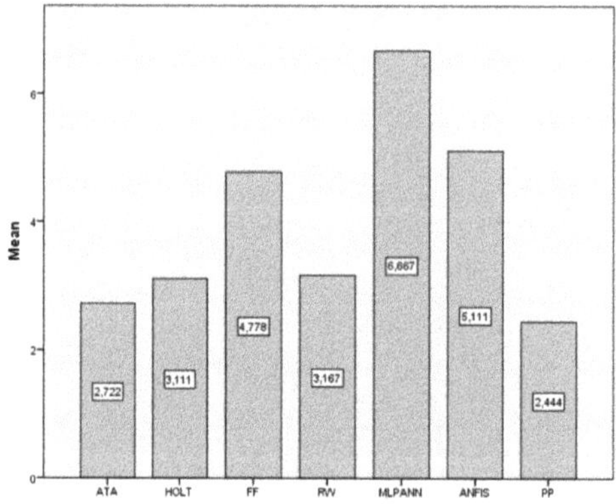

Figure 5. The average rank values of each method for MAPE criterion when the length of the test set is 20.

5. Conclusions and Discussion

Although the Holt method is used as a traditional time series forecasting method, it is known that it has some problems, such as the determination of the initial trend and level values and determining the trend and level update formulas. In this study, to overcome these problems, the parameters of the Holt method are optimized by using HSA, the smoothing parameters are varied by using first-order autoregressive equations, and the forecasting performance is improved by using the subsample bootstrap method.

When comparing the classical Holt method and the proposed method, it is clear that time-varying smoothing parameters and HSA provide important improvements in the forecasting results. The proposed method produces smaller RMSE values than the classical Holt method by about 70% in all analyses. If we compare the computation time of the proposed method with the classical Holt method, the proposed method needs more computation time because of using bootstrap and HSA algorithms, as expected. However, the computation time of the proposed method is very close to computational intelligence forecasting methods, and the computation time is not a problem for today's personal computers. For the BIST series, the computation time is about three minutes.

In future studies, different artificial intelligence optimization techniques can be used to determine the optimal parameters of the Holt method, or the forecasts can be obtained by different bootstrap methods.

Author Contributions: Conceptualisation, E.E. and E.B.; methodology, E.E., U.Y. and E.B.; software, E.E. and E.B.; validation, E.E., U.Y. and E.B.; formal analysis, E.B.; investigation, E.E., U.Y. and E.B.; writing—original draft preparation, E.E. and E.B. All authors have read and agreed to the published version of the manuscript.

Funding: This research received no external funding.

Data Availability Statement: The data set is available at https://datastore.borsaistanbul.com/. The access date is 1 November 2020.

Conflicts of Interest: The authors declare no conflict of interest.

References

1. Brown, R.G. *Statistical Forecasting for Inventory Control*; McGraw-Hill: New York, NY, USA, 1959.
2. Holt, C.E. *Forecasting Seasonals and Trends by Exponentially Weighted Averages (O.N.R. Memorandum No. 52)*; Carnegie Institute of Technology Pittsburgh: Pittsburgh, PA, USA, 1957.
3. Winters, P.R. Forecasting sales by exponentially weighted moving averages. *Manag. Sci.* **1960**, *6*, 324–342. [CrossRef]
4. Gardner, E.S., Jr., McKenzie, E. Forecasting trends in time series. *Manag. Sci.* **1985**, *31*, 1237–1246. [CrossRef]
5. Makridakis, S.G.; Fildes, R.; Hibon, M.; Parzen, E. The forecasting accuracy of major time series methods. *J. R. Stat. Soc. Ser. D Stat.* **1985**, *34*, 261–262.
6. Makridakis, S.; Hibon, M. The M3-competition: Results, conclusions and implications. *Int. J. Forecast.* **2000**, *16*, 451–476. [CrossRef]
7. Koning, A.J.; Franses, P.H.; Hibon, M.; Stekler, H.O. The M3 competition: Statistical tests of the results. *Int. J. Forecast.* **2005**, *21*, 397–409. [CrossRef]
8. Yapar, G.; Selamlar, H.T.; Capar, S.; Yavuz, I. ATA method. *Hacet. J. Math. Stat.* **2019**, *48*, 1838–1844. [CrossRef]
9. Gardner, E.S., Jr. Exponential smoothing. The state of the art. *J. Forecast.* **1985**, *4*, 1–28. [CrossRef]
10. Gardner, E.S., Jr. Exponential smoothing: The state of the art—Part II. *Int. J. Forecast.* **2006**, *22*, 637–666. [CrossRef]
11. Pandit, S.M.; Wu, S.M. Exponential smoothing as a special case of a linear stochastic system. *Oper. Res.* **1974**, *22*, 868–879. [CrossRef]
12. Imani, M.; Braga-Neto, U.M. Optimal finite-horizon sensor selection for Boolean Kalman Filter. In Proceedings of the 2017 51st Asilomar Conference on Signals, Systems, and Computers, IEEE, Pacific Grove, CA, USA, 29 October–1 November 2017; pp. 1481–1485.
13. Oprea, M. A general framework and guidelines for benchmarking computational intelligence algorithms applied to forecasting problems derived from an application domain-oriented survey. *Appl. Soft Comput.* **2020**, *89*, 106103. [CrossRef]
14. Hu, H.; Wang, L.; Peng, L.; Zeng, Y.R. Effective energy consumption forecasting using enhanced bagged echo state network. *Energy* **2020**, *193*, 116778. [CrossRef]
15. Imani, M.; Ghoreishi, S.F. Two-Stage Bayesian Optimization for Scalable Inference in State-Space Models. *IEEE Trans. Neural Netw. Learn. Syst.* **2021**, 1–12. [CrossRef]
16. Geem, Z.W.; Kim, J.H.; Loganathan, G.V. A new heuristic optimization algorithm: Harmony search. *Simulation* **2001**, *76*, 60–68. [CrossRef]
17. Geem, Z.W. Optimal cost design of water distribution networks using harmony search. *Eng. Optim.* **2006**, *38*, 259–277. [CrossRef]
18. Turkşen, I.B. Fuzzy functions with LSE. *Appl. Soft Comput.* **2008**, *8*, 1178–1188. [CrossRef]
19. Jang, J.S. Anfis: Adaptive-network-based fuzzy inference system. *IEEE Trans. Syst. Man Cybern.* **1993**, *23*, 665–685. [CrossRef]
20. Kim, S.; Kim, H. A new metric of absolute percentage error for intermittent demand forecasts. *Int. J. Forecast.* **2016**, *32*, 669–679. [CrossRef]
21. Neill, S.P.; Hashemi, M.R. Ocean Modelling for Resource Characterization. *Fundam. Ocean Renew. Energy* **2018**, 193–235. [CrossRef]

forecasting

MDPI

Article

A Real-Time Data Analysis Platform for Short-Term Water Consumption Forecasting with Machine Learning

Aida Boudhaouia and Patrice Wira *

IRIMAS Laboratory, University of Haute Alsace, 61 rue Albert Camus, 68093 Mulhouse, France; aida.miled@uha.fr
* Correspondence: patrice.wira@uha.fr

Abstract: This article presents a real-time data analysis platform to forecast water consumption with Machine-Learning (ML) techniques. The strategy fully relies on a web-oriented architecture to ensure better management and optimized monitoring of water consumption. This monitoring is carried out through a communicating system for collecting data in the form of unevenly spaced time series. The platform is completed by learning capabilities to analyze and forecast water consumption. The analysis consists of checking the data integrity and inconsistency, in looking for missing data, and in detecting abnormal consumption. Forecasting is based on the Long Short-Term Memory (LSTM) and the Back-Propagation Neural Network (BPNN). After evaluation, results show that the ML approaches can predict water consumption without having prior knowledge about the data and the users. The LSTM approach, by being able to grab the long-term dependencies between time steps of water consumption, allows the prediction of the amount of consumed water in the next hour with an error of some liters and the instants of the 5 next consumed liters in some milliseconds.

Keywords: load curve; unevenly spaced time series; long short-term memory (LSTM); back-propagation neural network (BPNN); machine learning; water consumption

Citation: Boudhaouia, A.; Wira, P. A Real-Time Data Analysis Platform for Short-Term Water Consumption Forecasting with Machine Learning. *Forecasting* **2021**, *3*, 682–694. https://doi.org/10.3390/forecast3040042

Academic Editor: Sonia Leva

Received: 3 August 2021
Accepted: 22 September 2021
Published: 26 September 2021

1. Introduction

Water consumption analysis is crucial as it assists building managers and operators to adopt better strategies to plan usages [1]. Forecasting is an important part for continuous monitoring and efficient management of consumption [2]. Furthermore, an accurate forecasting of consumption is essential for efficiently detecting and avoiding water leakages and wastes in distribution networks and installations [3]. Various methods to predict near-real-time water consumption and demand have been investigated. A complete literature review has been proposed in [4]. Among then, statistical methods, filtering and signal processing techniques, fuzzy logic, intelligent techniques and combinations of several models have shown more or less success. More recently, innovative models such as Machine-Learning (ML) techniques showed superior results when compared with classical models. Specifically, deep neural networks have emerged as efficient forecasting approaches. Regardless of the method, the robustness of the forecasting performance mainly depends on not only on the past water demand data but on contextual and environmental information (weather conditions, well-identified user profiles, knowledge about the architecture of the water distribution system, etc.), on redundancy of measurements and on the short, medium and long-term planning decisions to be addressed. Water demand forecasting remains a major research problem when no information is available behind the consumption of a single water meter.

This article presents a real-time data analysis platform to forecast water consumption with ML techniques only based on past water consumption, i.e., with no prior and contextual information. The strategy fully relies on a web-oriented architecture to ensure better management and optimized monitoring of water consumption [5]. It is a complete

Advanced Metering Infrastructure (AMI) based on integrated Internet of Things (IoT) technologies [6] that offers the possibility of collecting, analyzing and monitoring daily water consumption [7]. To predict water consumption, we also propose a framework based on ML algorithms such as the Long Short-Term Memory (LSTM) [8] and the Back-Propagation Neural Network (BPNN). The water consumption data are stored as unevenly spaced time series constructed from the collected data issued from distributed smart meters. Then, time series are handled in two different ways, with an explicitly and an implicitly sampling [9]. With explicitly sampled time series, the ML approaches predict the quantity of water consumed in the next coming hours [10]. With implicitly sampled time series, the ML approaches predict the instants when the next liters will be consumed. Both cases are achieved using the LSTM [8,11] and the BPNN [12]. The accuracy and usability of the forecast are evaluated and compared. This study can be generalized for any other type of consumption such as electricity and gas for example.

The rest of this article is organized as follows: Section 2 briefly presents appropriate ML approaches for analyzing consumption data with different forecasting horizons. Section 3 details the architecture of the AMI for collecting data. Consumption data are presented in terms of water volumes, indexes and dates of events. In other words, these data are considered to be unevenly spaced time series or Load Curves (LC). A preprocessing strategy is also developed in this section to handle and to compensate for missing and abnormal water consumptions. The two ML strategy, for forecasting the number of consumed liters in the next hour and the instants of the future consumed liters, are presented in Section 4. This section also includes some experimental tests and evaluations. Finally, concluding remarks are provided in Section 5.

2. Machine-Learning Algorithms for Water Consumption Forecasting

2.1. Forecasting with Machine-Learning Algorithms

Short-term forecasts, whether in water [2,13], in electricity [14,15] or even in gas [16], have been reported in the literature with a variety of approaches and with different horizons. However, very few of them have treated individual customers in domestic buildings [8] with high resolution. In fact, the approach proposed in [17] is based on a model of non-homogeneous Markov chains allowing knowledge of the dynamics of water consumption. This model can predict behaviors of daily consumption based on other parameters such as exogenous factors represented by the climate [18], the day type, etc. Another study [19] deals with the water demand forecasting on weekly and hourly scales with an autoregressive model based on a periodic component on time series data to refine daily demand values and hours. This prediction uses a multitude of period models. Most of these studies focus on forecasting consumption by introducing other parameters using different predictive models depending on the nature of the input data and the sought objectives. Indeed, we note that the provided forecast horizon mainly depends on the input databases of the models. These database generally have annual, seasonal, monthly, weekly, daily or hourly resolutions. Most of the work, even based on intelligent techniques, are based on additional information. For example, the study in [20] uses support vector machines with monthly water demands, number of users, and total water consumption bills. Ref. [21] discusses residential water demand management based on pricing, restriction policies, climate, weather and demographic characteristics. For now, there is no study based on learning architectures such as direct or recurrent BPNN, Hopfield networks or LSTM to predict the water demand based on historical data from only one single measurement point.

On the other side, we propose more precise forecasts with data issued from smart meters with high resolution and no additional contextual information. In this paper, we focus on forecasting water consumption from a private building without any knowledge about appliances using water and the number of inhabitants.

2.2. Forecasting Framework Based on LSTM

The LSTM [8] is a special type of recurrent neural network [8]. It is a sequential learning model which can establish temporal correlations between a previous instant $t - 1$ and a current instant t. Consequently, the LSTM seems the most suitable model for forecasting consumption processes, given its ability to deduce the intrinsic daily consumption resident routines. The LSTM is based on the Back-Propagation Through Time (BPTT) learning algorithm [8] to calculate the weights. It is made up of units called memory blocks. Each memory block contains an "input gate", an "output gate" and a "forget gate", as shown in Figure 1.

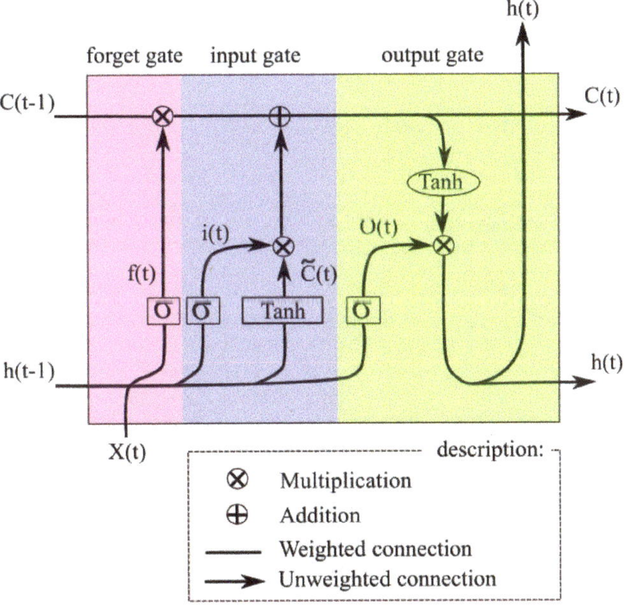

Figure 1. The LSTM unit architecture.

The behavior of each gate is represented by an equation. The input gate $i(t)$ given in (1) consists of transmitting the output h at the previous instant $t - 1$ and the input x at instant t through a sigmoid function $\sigma(x) = \frac{1}{1+e^{-x}}$:

$$i(t) = \sigma(W_i.[h(t-1), x(t)] + b_i) \tag{1}$$

A hyperbolic tangent function is applied to the input and the output data from the previous step to create a vector of a new value $\tilde{C}(t)$ to be an internal state. The update of the internal state is carried out through:

$$\tilde{C}(t) = tanh(W_c.[h(t-1), x(t)] + b_c) \tag{2}$$

The forget gate $f(t)$ is calculated with another sigmoid function that takes for its input the output $h(t-1)$ and the input $x(t)$:

$$f(t) = \sigma\left(W_f.[h(t-1), x(t)] + b_f\right) \tag{3}$$

Finally, the output gate $O(t)$ described by (5) is based on the state $C(t)$. This state is updated with a hyperbolic tangent multiplied with the output of a sigmoid:

$$C(t) = f(t) \times C(t-1) + i(t) \times \tilde{C}(t) \tag{4}$$

$$O(t) = \sigma(W_o.[h(t-1), x(t)] + b_o) \tag{5}$$

W_i, W_c, W_f, W_o, and b_i, b_c, b_f, b_o represent respectively the weights and the biases at the different levels in the LSTM memory block. They are adjusted iteratively with the BPTT learning algorithm [8] until convergence. At each step of the learning process, the performance of the LSTM can be evaluated by an error such as the Root Mean Square Error (*RMSE*) [22] where y_i, $\tilde{y}_i$ and n are respectively the reference, the estimated value and the number of data:

$$RMSE = \sqrt{\frac{1}{n}\sum_{i=1}^{n}(y_i - \tilde{y}_i)^2} \tag{6}$$

This learning approach will be used in the following to forecast short-term water consumption.

3. Proposed Architecture and ML Framework to Collect and Analyze Water Consumption Data

3.1. Data Collecting with Smart Meters

All the data used in this study are collected in an online database from smart water meters. Smart meters are IoT devices that are appropriate to build a sustainable and advanced consumption data system [23]. Most water distributors collect data from smart meters with a resolution of several minutes, for example every 15, 30 or 60 min, or even once a day [5]. This implies that the capacities of smart water meters are clearly not fully exploited [7]. This also means that the resolution of the consumption data is low. We use smart meters with the communication strategy proposed and developed in [24] to compress and to transmit the data with a very high resolution [7] according to industrial specifications. This strategy allows the dating on the server side of each liter consumed and reduces the energy consumption on the meter side. Indeed, emission duration that consumes a lot of energy for the smart meters have been greatly reduced. This strategy is embedded in the smart meters and transmits data in the form of frames with a T_{max} interval which does not exceed 5 min. This interval is completely adaptive and related to the amount of consumed water [7]. Higher water consumption results in more data frames. To guarantee the reception of frames with no missing data, a sliding window is proposed which consists of $R_E = 6$ packages. These packages are numbered and can be considered to be independent broadcasts in the transmitted frame. This ensures the redundancy of the data through successive frames. This principle is illustrated in Figure 2. The maximum length of a frame, $L_f = R_E \times l_p$ with l_p the length of a package, is set depending on the radio technology and frequency that are used. In our AMI, we chose a maximum value of 120 bytes for L_f which is the limit of the frame size.

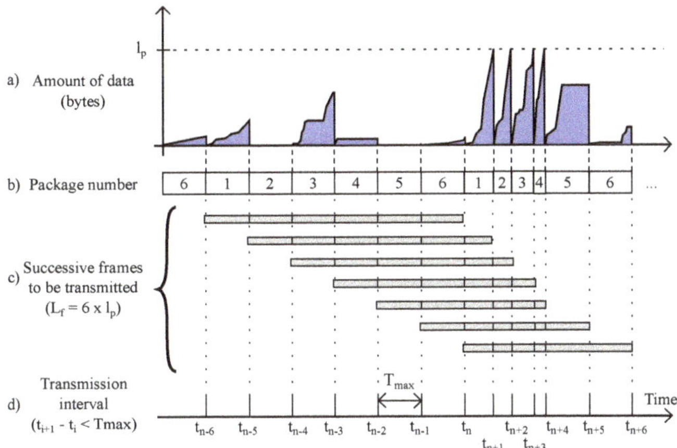

Figure 2. Operating principle of the sliding window for ensuring the redundancy of transmitted data from smart water meters through successive frames [7].

A web server receives all the transmitted frames from several smart meters. Here, a script receives, decompresses, and retrieves the data from the frames for storage in an SQL database [10]. This process runs continuously since 2014 and allows completion of the database in real time and under real operating conditions. The database contains raw data for each individual smart meter, i.e., the index which represents the volume of consumed water in liter and the instant when each liter has been consumed in millisecond. This instant is called a pulse or an event [7]. It is obvious that the data collected and stored according to this platform are of high resolution and therefore precisely represent consumption habits.

At any time, it is possible to extract from the database with another script, the data related to a well-defined smart meter by specifying the beginning and the end of a period. This is called a set of row data.

3.2. Data Description

The collected data are of a great value and must be analyzed. For this, the raw data must interpreted and therefore associated with some theoretical concepts and models. Among them are unevenly spaced time series or Load Curves (LC).

3.2.1. Water Consumption Time Series

A time series is a sequence of temporal data [25]. The time stamp of the series can be explicit such that a date is given for each data value or controlled by the appearance of the data represented by events perfectly dated. This is referred to as an unevenly spaced time series [9] defined by S in (7). In the context of water consumption, an event corresponds to each consumed liter and S is thus a sequence of scalar values of an incremented variable $Y_{i+1} = Y_i + 1$. S therefore corresponds to the raw data extracted from the previously described platform for one smart meter and is the result of a process observed during a period T. The platform and AMI proposed by [7] offer the possibility of recording the instants of consumption of each liter.

$$S = [Y_1(t_1), Y_2(t_2), ..., Y_i(t_i), ..., Y_T(t_T)] \tag{7}$$

3.2.2. Cumulated Water Consumption: The Index and the Load Curve

Each Y_i represents the index of a smart meter which is the cumulated volume of consumed water at each instant t_i. The time between two instants t_i and t_{i-1} is not constant. The evolution of Y_i during a period T is called a cumulative LC. An example is provided by Figure 3, it is an alternative representation of S. LC are very useful for analyzing and comparing consumption over days, weeks, months. We than speak of daily LC, weekly LC or monthly LC.

3.2.3. Sampled Water Consumption Data Series

The data collected from the platform is unevenly spaced in time. Each consumed liter represents an event, the process of water consumption can also be seen as a process generating dated event. To make the data compliant with most of the popular data analysis tools and concepts, a sampling is proposed to make the series evenly spaced in time. The sampling can be made in minutes or in hours and results in a sequence of 1440 data per day or 24 data per day.

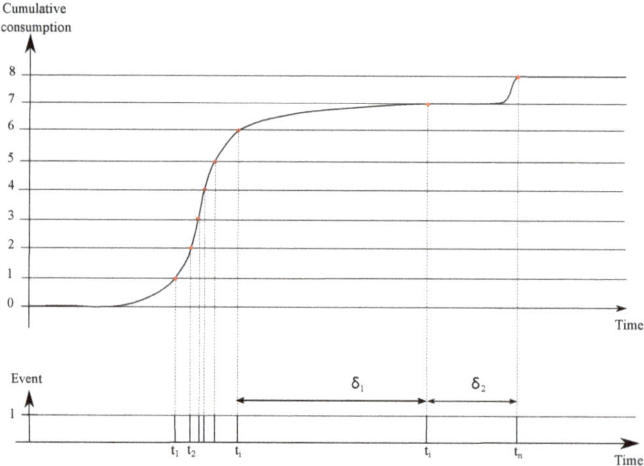

Figure 3. Example of a cumulative load curve (LC) which shows the raw data by red dots unevenly spaced in time as recorded and transmitted by a smart sensor (the black curve is an interpolation) and with results from the sequence of events corresponding to each consumed liter.

We also chose to derivate the cumulative LC in order work with sequences of n data that represents the number of liters consumed in each minute or hour. Consequently, a natural order of appearance constitutes an implicitly sampled chronological time series such as:

$$C = [y_1, y_2, y_2, ... y_i, ... y_n] \tag{8}$$

3.3. Data Integrity Checking and Interpolation

Under real operating conditions, the integrity of the data must be checked. Indeed, failures or malfunctions can lead to missing raw measurements in the database. We therefore propose a preprocessing step of the raw data to verify the data and to complete by interpolation eventually missing data. The whole proposed preprocessing strategy is represented by Figure 4. The raw time series is extracted from the database for each day. Since a forecast of water consumption is targeted with an accuracy of one hour, the data are sampled with a resolution of minute (i.e., 1440 mn per day). This preprocessing is achieved separately for each day. Then, periods without consumed liters, i.e., events, are identified and corrected by interpolation.

Data analysis and forecasting with ML algorithms needs to be achieved with no missing or inconsistent values. It is thus necessary to identify and separate abnormal consumption (such as water leakage, occasional consumption) which can influence water consumption) from normal and usual consumption. Abnormal water consumption is always due to an unusual and occasional behavior from the users [25]. The detection of abnormal water consumption is achieved as follows. A reference cumulative LC is calculated for each day of the week. This reference LC is completed with a minimum LC and a maximum LC for each day. Generally, a load profile for one day j is strongly correlated [26] with that for the previous day $j - 1$ and to the day for the previous week $(d - 7)$. The reference cumulative LC is calculated with:

$$C_j(t_i) = avg(C_{j-7}(t_i), C_{j-1}(t_i)) \tag{9}$$

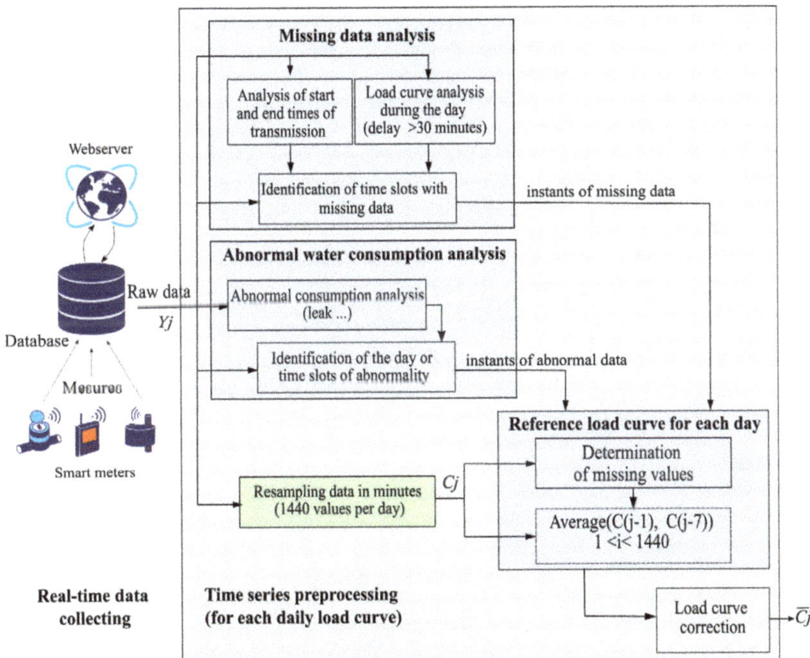

Figure 4. Global architecture of the water consumption LC preprocessing.

The minimum and a maximum LC for each day are calculated by the same way by changing the average $avg()$ function in (9) by $min()$ and $max()$ functions. The detection of normal consumption is based on the criteria given by:

$$abs\left[y_j(t) - avg(\sum_{i=1}^{n}(y_j(t)))\right] \geq \alpha \times std(\sum_{i=1}^{n}(y_j(t))) \tag{10}$$

where $std()$ is standard deviation for each value of the LC and α is a numerical variable chosen empirically, in our case $\alpha = 5$. Additional tests can be achieved to see if the instantaneous consumption is out of the range defined by the minimum and maximum LC for the same day of the week and allow the detection of any additional consumption that deviates significantly from the "normal consumption" [10]. It can be noticed that the detection of abnormal and unusual consumptions is only based on water consumption data and some statistical indicators [10]. Abnormal and unusual consumptions are corrected by an interpolation during their duration and will not be taken into account in the learning processes. At the end, we obtain a time series $\bar{C}_j$ sampled in minutes which corresponds to the LC C_j without loss of data and without abnormal and unusual consumptions.

4. Water Consumption Forecasting

To evaluate the efficiency of the platform and the ML techniques, we focused on the water consumption of a private building. The water consumption is collected from a smart meter which is a single measurement point for the whole building. These are the only data available from the building and the users. The objective consists of forecasting the number of liters of consumed water with a horizon of one hour and to predict the instant of the next consumed liter by different ML approaches.

All the algorithms have been developed with the Matlab R2018b environment on a desktop computer with 4 cores (Intel i7 processors at 3.6 GHz) and 16 GB of memory. Experiments and tests have been carried out under the same conditions to find the values of the learning parameters by trial and error (learning rate value, number of neurons, number of hidden layers, type of activation function) to provide the smallest error.

4.1. Hourly Water Consumption Forecasting

A three-month database (from October 2018 to December 2018) has been chosen to forecast the number of consumed water liters in the next coming hour. The data sequence is resampled with a resolution of one hour and is represented by Figure 5. This consumption has been recorded in a domestic house in France occupied by two people who consume on average 194 L per day (l/d). Household information will not be used by the ML approaches.

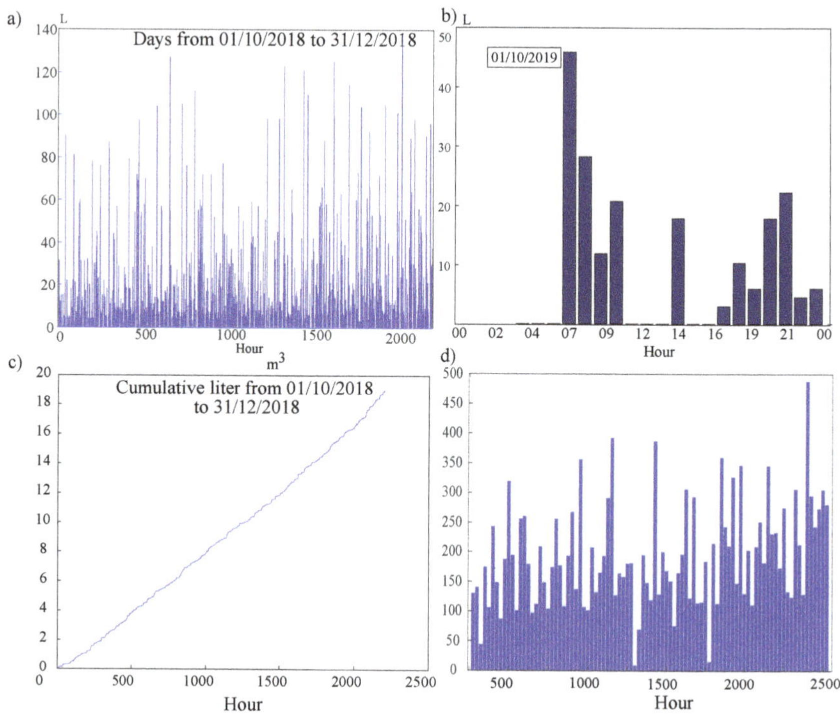

Figure 5. Water consumption time series: (**a**) LC from 1 October to 31 December 2018, (**b**) close-up view of the same time series for the first 24 h, (**c**) cumulative water LC over the whole period, (**d**) number of liters consumed per day.

Two ML approaches have been implemented for a one-hour water consumption forecasting, the LSTM and the BPNN. For this case, the series represented by Figure 5c is the input of the forecast approaches. With the LSTM, input $x(t)$ in Equations (1), (2),

(4) and (5) is the preprocessed cumulative LC C_j. We use the Adam algorithm, i.e., an optimization stochastic gradient descent for training deep learning approaches [27] to handle the noisy data. Indeed, the Adam algorithm is suitable for data with a lot of noise. We chose a learning rate value of 10^{-4} for the LSTM and 10^{-5} for the BPNN model and the training ends with a maximum number of epochs chosen at 100.

The forecast performances with the two learning approaches are evaluated with the RMSE and results are presented in Table 1. It can be seen that the LSTM can forecast the water consumption in the next hour with a precision of 6 L while the BPNN predicts the future consumption with a precision of 24 L (the consumption range is approximately between 1 to 50 L per days).

Table 1. Hourly prediction of water consumption in liters with the LSTM and BPNN.

	LSTM	BPNN
Hidden Layer number	2	3
Number of neurons	100/100	200/100/100
Activation function	relu/relu	relu/relu/relu
Train RMSE (l)	0.19	3.54
Test RMSE (l)	6.05	20.19
Total execution time (ms)	19.81	24.05

4.2. Forecasting Events of Water Consumption in Milliseconds

We also forecast the coming events, i.e., the instants when the next liters will be consumed. For this purpose, we chose a dataset composed of 4321 events dated in milliseconds, each representing the time difference between two consecutive liters. Obviously, this dataset provides more detailed information about the water consumption than in the previous experiment. The dataset has been recorded between December the 2nd to the 20st, 2018 and is represented by Figure 6. The dataset is divided into three subsets for the learning of the LSTM and the BPNN, the training, validation and test subsets which are respectively distributed in a percentage of the dataset: 60%, 0.3% and 40%. The parameters of two learning approaches are summarized in Table 2. Their input vector $x(t)$ is composed of the time difference between two successive consumed liters, i.e., the values of δ_i represented by Figure 3. The Adam algorithm is also used her to optimize the learning of the LSTM and BPNN which use the same parameters as in the previous experiment optimization because the data are noisy. The learning rate is lr = 10^{-4}. The training ends when the maximum number of epochs, 100 in our case, has been reached.

The results of two learning approaches are provided in Table 2. The instant of the next consumed liter of water is predicted respectively with an error (test RMSE) of 13 ms and 48 ms respectively with the LSTM and the BPNN. In addition, the forecast of the instant of the 5 next liters have also been calculated and are respectively estimated to occur at instants 450,925, 450,800, 451,200, 451,500 and 451,300 milliseconds. In other words, the next consumed liters have been correctly predicted on 21 December 21 (2018) at 00:08:07.487, 00:15:38.287, 00:23:09.487, 00:30:40,987 and at 00:38:12.287. The accuracy objective of the predicted instants is justified by industrial specifications.

Table 2. Event prediction of water consumption in ms with the LSTM and BPNN.

	LSTM	BPNN
Hidden Layer number	2	1
Number of neurons	200/120	150
Activation function	relu/relu	relu
Train RMSE (10^6 ms)	0.33	0.39
Test RMSE (10^6 ms)	0.13	0.48
Total execution time (s)	37.73	24.71

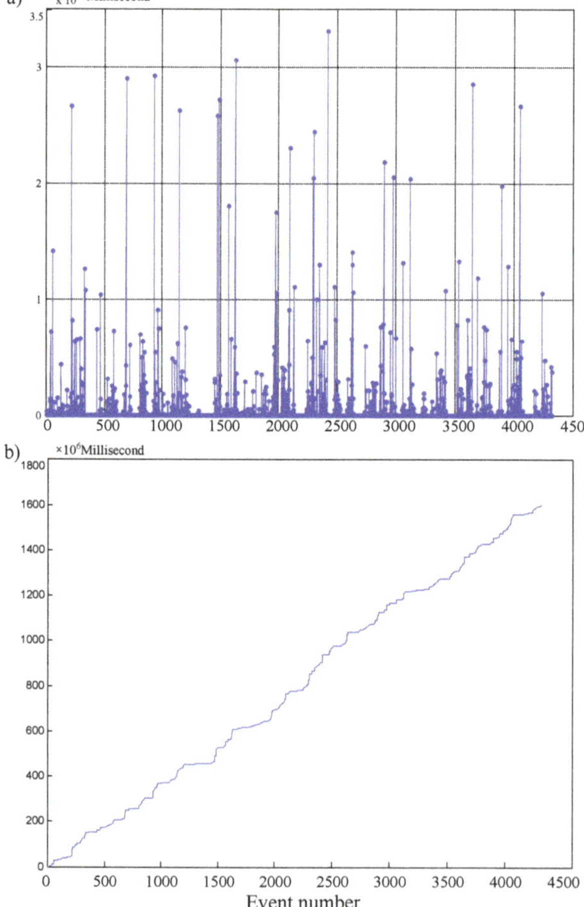

Figure 6. Time representation of the water consumption, (**a**) Time gap between 4321 events (i.e., consumed liters) from 02/12/2018 09:11:21.750 until 20/12/2018 23:23:40.625, (**b**) Cumulated duration δ as a function of consumed liters.

4.3. Discussion on the Hourly and Events Water Consumption Forecasting

Two forecasting tests have been experimented with the proposed water consumption collecting platform, i.e., hourly and event forecasting. The first case consists of predicting the amount of water consumed during the first hour that follows the period of the collected dataset. In the second case, the instant of the next consumed liters is predicted. In both cases, an LSTM and a BPNN architectures have been designed. Their performance has been evaluated under the same conditions and have been compared in terms of precision, computational resources and execution time. With very close resources and approximately the same execution time, the forecasting error obtained with LSTM is 3 times lower than with the BPNN. In both experiments, the LSTM is more appropriate than the BPNN to grab the temporality of the data sting tests have been experimented with the proposed water consumption collecting platform, i.e., hourly and event forecasting. This is because of its property of selectively remembering patterns in time series for long durations of time. Another reason is that the LSTM can better take into account the time-dependent structure of the data, i.e., the non-stationarity of the water data. The LSTM is therefore well suited to handle precise datasets over large periods of time such as water consumption.

5. Conclusions

In this study, we presented a web-oriented platform to collect in real-time water consumption data and to predict them with machine-learning approaches. The data are issued from smart meters and are transmitted to a server to be handled as unevenly spaced time series with high resolution, i.e., in milliseconds. Data sets are then extracted, preprocessed and eventually sampled to be used by machine-learning algorithms to predict the next consumptions. The preprocessing of the data consists of detecting missing values and in identifying abnormal consumption using a reference load curve for each day of the week. Then, machine-learning approaches such as the LSTM and BPNN have been implemented to forecast the next consumption. Two tests have been experimented for hourly and event water consumption forecasting in a private building. The first case consists of predicting the amount of water consumed during the hour that follows the period of the collected data. In the second case, the instants of the next consumed liters are predicted. By evaluating the performance of the LSTM and BPNN, it can be seen that the LSTM is more accurate than the BPNN. Indeed, the LSTM can predict the amount of consumed water in the next coming hour with an error of less than 6 L and is able to predict the instants of the 5 next consumed liters with an error of less than 15 ms. This can be considered to be very accurate prediction in the context of water consumption measurement and forecasting. This web-oriented platform endowed by its learning capabilities is generic and can be extended to other additional smart meters to measure and predict other variables such as power or gas consumptions.

Author Contributions: Conceptualization, A.B.; Funding acquisition, P.W.; Methodology, A.B., P.W.; Supervision, P.W.; Validation, P.W.; Visualization, P.W.; Writing— original draft preparation, A.B.; and Writing—review and editing, P.W. All authors have read and agreed to the published version of the manuscript.

Funding: This research received no external funding.

Institutional Review Board Statement: Not applicable.

Informed Consent Statement: Not applicable.

Data Availability Statement: Data were obtained from our IUT in Mulhouse, France.

Conflicts of Interest: The authors declare no conflict of interest.

Abbreviations

The following abbreviations are used in this manuscript:

AMI	Advanced Metering Infrastructure
BPNN	Back-Propagation Neural Network
BPTT	Back-Propagation Through Time
LC	Load Curve
LSTM	Long Short-Term Memory
ML	Machine Learning
RMSE	Root Mean Square Error
SQL	Structured Query Language

References

1. Roccetti, M.; Delnevo, G.; Casini, L.; Salomoni, P. A Cautionary Tale for Machine Learning Design: Why we Still Need Human-Assisted Big Data Analysis. *Mob. Netw. Appl.* **2020**, *25*, 1075–1083. [CrossRef]
2. Walker, D.; Creaco, E.; Vamvakeridou-Lyroudia, L.; Farmani, R.; Kapelan, Z.; Savić, D. Forecasting Domestic Water Consumption from Smart Meter Readings Using Statistical Methods and Artificial Neural Networks. *Procedia Eng.* **2015**, *119*, 1419–1428. [CrossRef]
3. Petropoulos, F.; Spiliotis, E. The Wisdom of the Data: Getting the Most Out of Univariate Time Series Forecasting. *Forecasting* **2021**, *3*, 478–497. [CrossRef]
4. de Souza Groppo, G.; Costa, M.A.; Libânio, M. Predicting water demand: A review of the methods employed and future possibilities. *Water Supply* **2019**, *19*, 2179–2198 [CrossRef]
5. Boudhaouia, A.; Wira, P. *Power and Water Consumption Monitoring with IoT Devices and Machine Learning Methods in a Smart Building*; Presses Universitaires de Strasbourg: Strasbourg, France, 2019; Volume 346.
6. Yang, L.; Yang, S.H. Domestic water consumption monitoring and behaviour intervention by employing the internet of things technologies. *Procedia Comput. Sci.* **2017**, *111*, 367–375. [CrossRef]
7. Spiegel, J. Nouvelle Stratégie de Collecte de Données Pour les Compteurs d'eau Communicants. Ph.D. Thesis, Université de Haute Alsace, Mulhouse, France, 2019.
8. Kong, W.; Dong, Z.Y.; Jia, Y.; Hill, D.J.; Xu, Y.; Zhang, Y. Short-Term Residential Load Forecasting Based on LSTM Recurrent Neural Network. *IEEE Trans. Smart Grid* **2017**, *10*, 841–851. [CrossRef]
9. Rehfeld, K.; Marwan, N.; Heitzig, J.; Kurths, J. Comparison of correlation analysis techniques for irregularly sampled time series. *Nonlinear Process. Geophys.* **2011**, *8*, 389–404. [CrossRef]
10. Boudhaouia, A.; Wira, P. Water Consumption Analysis for Real-Time Leakage Detection in the Context of a Smart Tertiary Building. In Proceedings of the 2018 International Conference on Applied Smart Systems (ICASS), Medea, Algeria, 24–25 November 2018; pp. 1–6. [CrossRef]
11. Boudhaouia, A.; Wira, P. Comparison of machine learning algorithms to predict daily water consumptions. In Proceedings of the 2021 International Conference on Design & Test of integrated micro & nano-Systems (DTS), Sfax, Tunisia, 7–10 June 2021; pp. 1–6.
12. Ali, Z.; Hussain, I.; Faisal, M.; Nazir, H.M.; Hussain, T.; Shad, M.Y.; Mohamd Shoukry, A.; Hussain Gani, S. Forecasting Drought Using Multilayer Perceptron Artificial Neural Network Model. *Adv. Meteorol.* **2017**, *2017*, 1–9. [CrossRef]
13. Candelieri, A.; Soldi, D.; Archetti, F. Short-term forecasting of hourly water consumption by using automatic metering readers data. *Procedia Eng.* **2015**, *119*, 844–853. [CrossRef]
14. Deb, C.; Zhang, F.; Yang, J.; Lee, S.E.; Shah, K.W. A review on time series forecasting techniques for building energy consumption. *Renew. Sustain. Energy Rev.* **2017**, *74*, 902–924. [CrossRef]
15. Liu, M.; Liu, D.; Sun, G.; Zhao, Y.; Wang, D.; Liu, F.; Fang, X.; He, Q.; Xu, D. Deep Learning Detection of Inaccurate Smart Electricity Meters: A Case Study. *IEEE Ind. Electron. Mag.* **2020**, *14*, 79–90. [CrossRef]
16. Szoplik, J. Forecasting of natural gas consumption with artificial neural networks. *Energy* **2015**, *85*, 208–220. [CrossRef]
17. Abadi, M.L.; Same, A.; Oukhellou, L.; Cheifetz, N.; Mandel, P.; Feliers, C.; Chesneau, O. Predictive Classification of Water Consumption Time Series Using Non-homogeneous Markov Models. In Proceedings of the IEEE International Conference on Data Science and Advanced Analytics (DSAA), Tokyo, Japan, 19–21 October 2017; pp. 323–331. [CrossRef]
18. Huntra, P.; Keener, T.C. Evaluating the Impact of Meteorological Factors on Water Demand in the Las Vegas Valley Using Time-Series Analysis: 1990–2014. *ISPRS Int. J. Geo-Inf.* **2017**, *6*, 249. [CrossRef]
19. Alvisi, S.; Franchini, M.; Marinelli, A. A short-term, pattern-based model for water-demand forecasting. *J. Hydroinformatics* **2007**, *9*, 39–50. [CrossRef]
20. de Souza Groppo, G.; Costa, M.A.; Libânio, M. Forecasting Water Demand in Residential, Commercial, and Industrial Zones in Bogotá, Colombia, Using Least-Squares Support Vector Machines. *Math. Probl. Eng.* **2016**, *2016*. [CrossRef]
21. Kenney, D.S.; Goemans, C.; Klein, R.; Lowrey, J.; Reidy, K. Residential Water Demand Management: Lessons from Aurora, Colorado. *JAWRA J. Am. Water Resour. Assoc.* **2008**, *44*, 192–207. [CrossRef]
22. Saigal, S.; Mehrotra, D. Performance comparison of time series data using predictive data mining techniques. *Adv. Inf. Min.* **2012**, *4*, 57–66.

23. Cominola, A.; Giuliani, M.; Piga, D.; Castelletti, A.; Rizzoli, A. Benefits and challenges of using smart meters for advancing residential water demand modeling and management: A review. *Environ. Model. Softw.* **2015**, *72*, 198–214. [CrossRef]
24. Spiegel, J.; Hermann, G.; Wira, P. A Comparative Experimental Study of Compression Algorithms for Enhancing Energy Efficiency in Smart Meters. In Proceedings of the IEEE 16TH International Conference of Industrial Informatics (INDIN 2018), Porto, Portugal, 18–20 July 2018.
25. Benkabou, S.E. Détection d'Anomalies dans les séries Temporelles: Application aux Masses de Données sur les Pneumatiques. Ph.D. Thesis, Université Claude Bernard, Lyon, France, 2018.
26. Lee, J.; Kim, J.; Ko, W. Day-Ahead Electric Load Forecasting for the Residential Building with a Small-Size Dataset Based on a Self-Organizing Map and a Stacking Ensemble Learning Method. *Appl. Sci.* **2019**, *9*, 1231. [CrossRef]
27. Hochreiter, S.; Schmidhuber, J. Long Short-Term Memory. *Neural Comput.* **1997**, *9*, 1735–1780. [CrossRef] [PubMed]

forecasting

Article

Battery Sizing for Different Loads and RES Production Scenarios through Unsupervised Clustering Methods

Alfredo Nespoli [1], Andrea Matteri [1,*], Silvia Pretto [2], Luca De Ciechi [3] and Emanuele Ogliari [1,*]

1   Politecnico di Milano, Dipartimento di Energia, Via La Masa 34, 20156 Milan, Italy; alfredo.nespoli@polimi.it
2   Equienergia S.r.l., c.so Sempione 62, 20153 Milan, Italy; silvia.pretto@equienergia.it
3   Helexia Energy Services S.r.l., Strada 8, Palazzo N, Rozzano, 20089 Milan, Italy; luca.de-ciechi@helexia.eu
*   Correspondence: andrea.matteri@polimi.it (A.M.); emanuelegiovanni.ogliari@polimi.it (E.O.)

Abstract: The increasing penetration of Renewable Energy Sources (RESs) in the energy mix is determining an energy scenario characterized by decentralized power production. Between RESs power generation technologies, solar PhotoVoltaic (PV) systems constitute a very promising option, but their production is not programmable due to the intermittent nature of solar energy. The coupling between a PV facility and a Battery Energy Storage System (BESS) allows to achieve a greater flexibility in power generation. However, the design phase of a PV+BESS hybrid plant is challenging due to the large number of possible configurations. The present paper proposes a preliminary procedure aimed at predicting a family of batteries which is suitable to be coupled with a given PV plant configuration. The proposed procedure is applied to new hypothetical plants built to fulfill the energy requirements of a commercial and an industrial load. The energy produced by the PV system is estimated on the basis of a performance analysis carried out on similar real plants. The battery operations are established through two decision-tree-like structures regulating charge and discharge respectively. Finally, an unsupervised clustering is applied to all the possible PV+BESS configurations in order to identify the family of feasible solutions.

Keywords: battery energy storage system; battery sizing; photovoltaic power production; performance ratio; electrical load; decision tree; k-means clustering

Citation: Nespoli, A.; Matteri, A.; Pretto, S.; De Ciechi, L.; Ogliari, E. Battery Sizing for Different Loads and RES Production Scenarios through Unsupervised Clustering Methods. *Forecasting* **2021**, *3*, 663–681. https://doi.org/10.3390/forecast3040041

Academic Editor: Cong Feng

Received: 6 August 2021
Accepted: 21 September 2021
Published: 24 September 2021

Publisher's Note: MDPI stays neutral with regard to jurisdictional claims in published maps and institutional affiliations.

1. Introduction

The rising penetration of Renewable Energy Sources (RESs), together with the progressive digitization of grids, is leading to an energy scenario where power production is increasingly decentralized [1] and those who were once only energy consumers become producers themselves and are called "prosumers" [2,3].

Nowadays, RESs are widely connected to distribution grids thanks to the advantages they offer: clean energy and additional generation to address the ever increasing electricity demand [4]. Between RESs power generation technologies, solar PhotoVoltaic (PV) systems are a promising option offering a significant potential for providing energy in a sustainable way [5], directly generating it onsite [6]. However, solar energy is, by nature, intermittent and not programmable [7]. For this reason, energy storage systems, endowed of a proper management software, are needed [8].

Among all possible storage systems, the electrochemical ones represent an attractive option [9]. Electrochemical technologies store energy through specific chemical components. Being available in modules, the desired voltages and currents can be achieved by connecting single modules in series and/or in parallel [10]. Currently, a growing fraction of installed utility-scale PV systems incorporates Battery Energy Storage Systems (BESS) [11,12]. This allows to achieve a flexibility improvement in power generation by shifting production from the peak of non-programmable solar energy towards hours of large consumption [13,14].

When coupling a BESS with a PV power production system, a key design consideration is constituted by the selection between DC- and AC-coupling. AC-coupled systems have largely independent PV and batteries, each using its own inverter, and the coupling is located on the AC side of the inverters. On the contrary, DC-coupled systems, where the PV field the and battery share a common inverter, have the advantage of potentially reducing costs from shared components [15,16].

In general, the design phase of PV+BESS hybrid systems requires a large number of decisions due to the large number of possible configurations in terms of overall system architecture as well as the sizing of various components [17]. Before constructing a new PV power production facility, feasibility studies are needed to assess its viability from both financial and technical perspectives [18]. In detail, simulations are carried out to assess the energy production permitted by a given plant configuration in a given geographical position [19] and to evaluate the expected investment costs [20].

The main objective of the present work is to provide a preliminary forecast that identifies a family of batteries which is suitable, from both a technical and a financial point of view, for a given scenario. Techno-economical simulations are carried out for new grid-connected PV+BESS hybrid power production plants. Several scenarios are considered in terms of PV plant configuration, load curves and battery technologies available on the market.

2. Case Study and Procedure

In this paper, a procedure is proposed to forecast a family of batteries which are suitable to be coupled with a given PV plant configuration.

The proposed procedure is applied to new hypothetical PV facilities installed on the rooftop of two different buildings: a single-brand point of sale and a ceramics factory. According the analyzed buildings, two different load types will be considered, namely a commercial and an industrial load curves. The energy production is simulated on the basis of an analysis carried out on real PV plants and thanks to irradiance databases available online. The battery operation is managed by means of a specific control logic defined in decision-tree-like diagrams considering all possible operating conditions for both charge and discharge. Several PV+BESS configurations are simulated and, for each one, a set of performance and economic indicators are computed. In the end, an unsupervised clustering algorithm is applied to all the analyzed PV+BESS configurations, aimed at detecting the family of battery solutions which are the most suitable according to the considered scenario.

In the following Sections, all aspects of the proposed procedure are thoroughly discussed: Section 2.1 analyzes the performance of several real PV plants in order to compute a proper value of Performance Ratio to be used during the following power production simulations; Section 2.2 describes the load curves corresponding to the industrial and the commercial buildings involved in the analysis; Section 2.3 explains how to simulate the PV power production; Section 2.4 provides a list of all the battery technologies considered in couple with the PV plant; Section 2.5 displays and discusses two decision-tree-like structures providing indications about the control logic of batteries during both charge and discharge; Section 2.6 describes a set of useful parameters used to evaluate the technical and economical viability of the considered PV+BESS configurations; Section 2.7 discusses how to apply a clustering method to all the possible PV+BESS configurations in order to find a group of batteries that are suitable for coupling with a given PV plant.

2.1. Plant Monitoring and Performance Ratio Calculation

The first part of the present study takes into account 22 monitored PV plants distributed all over the Italian territory with a total peak power installed of about 7 MW. These facilities can be divided based on four different types of installation:

- Fixed tilt: the solar field presents a fixed tilt angle. In general, the modules are installed either on concrete ballasts or metal structures placed on flat roofs and convection is allowed on their back surface.

- Flush mount: the modules are integrated on building roofs presenting a tilt larger than 4–5° and convection is not allowed on their back surface.
- East-West: the solar field is halved in two sections: one exposed toward East and the other towards West. The modules are generally installed on concrete ballasts.
- Carport: the modules are installed on parking structures and convection on their back is allowed.

A single plant can be composed of multiple sections with different tilt, azimuth or type of installation, that are considered independently.

Figure 1 reports the location of the considered PV facilities on the Italian territory considering five different regions: North-West, North-East, Center, South and Islands. Moreover, the chart highlights the fraction of plants corresponding to each installation type.

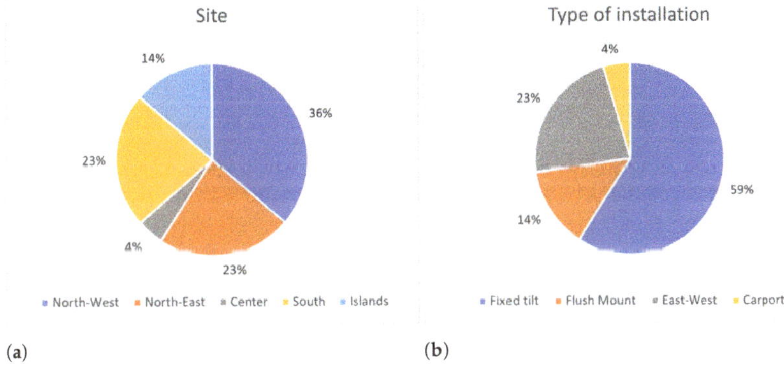

Figure 1. Fraction of plants located in a certain region (**a**) and with a given configuration (**b**).

For each plant, the following characteristics are known: the nominal power, the peak power of all plant sections, the tilt and azimuth angles of the modules, the temperature coefficient of the module (accounting for temperature-related power losses) and the degradation factor. Monitoring campaigns carried out for each of the considered facilities allowed to collect hourly-basis data about the active energy produced at Alternating Current (AC) side, the solar irradiation on module's plane, the cell's temperature on the back side of the module and the ambient temperature. In case of plant sections with different exposure, the monitored parameters are recorded independently for each section.

Different plants started their operation in different years. However, the start of operation period does not always correspond with the starting date of monitoring: for instance, the oldest facility started to produce in August 2012, while its monitoring started in 2018.

Data from each PV facility are properly cleared out of inconsistent and unreliable samples determined by erroneous measurement, like negative values of produced energy, values of produced energy exceeding the corresponding value of irradiation, values of produced energy larger than the maximum feasible ones (computed on the basis of the plant nominal power increased by 5% to account for inverter overpower) and values of solar irradiation lower than lunar irradiation (4 W/m^2) or larger than 1200 W/m^2.

The data available allow to compute a performance index which is crucial for further analyses: the Performance Ratio (PR), which allows to compare the performance of PV facilities with different configurations and geographical location [21]. PR represents the overall effect of losses on the array's rated output, due to array temperature, incomplete utilization of the irradiation (soiling and shading losses) and system component inefficiencies and failures [22]:

$$\mathrm{PR} = \frac{Y_{f,t}}{Y_{r,t}} \tag{1}$$

In the equation: $Y_{f,t}$ represents the final PV system yield in the time interval t, hence the portion of net energy output of the entire PV plant which was supplied by the array per kW installed; $Y_{r,t}$ corresponds to the reference yield in the time interval t, hence the ratio between total in-plane irradiation and module's reference in-plane irradiance [23].

Starting from the historical data available, PRs are computed for each of the analyzed plants, first on a daily basis and then on a yearly basis (starting from the daily values). Then, the average value of both daily and yearly PR is computed for all the plants sharing the same type of installation. In the present work, the yearly PR values will be useful to provide some considerations about the performances of different types of plant, while the averaged daily PR values are crucial in estimating the power production of new plants.

2.2. Load Curves

In the present work, two different types of building are chosen to hypothetically install a new PV+BESS facility on their rooftop: one dedicated to commercial activities and the other devoted to industrial production. The power requirements of the two structures, given their different purposes, are described by distinct load curves.

The commercial load curve considered corresponds to a single-brand point of sale, whose building covers an area of about 6100 m^2. It is located in Italy, in the region of Piemonte, in climatic zone E, where the heating system start-up is allowed from 15 October to 15 April. The annual consumption of electric energy in 2019 (chosen as reference year) is equal to 828 MWh. The hourly consumption is visualized in Figure 2, in form of heat map covering all the hours and all the days of the reference year. Moreover, the seasonal loads during a typical week are plotted in Figure 3.

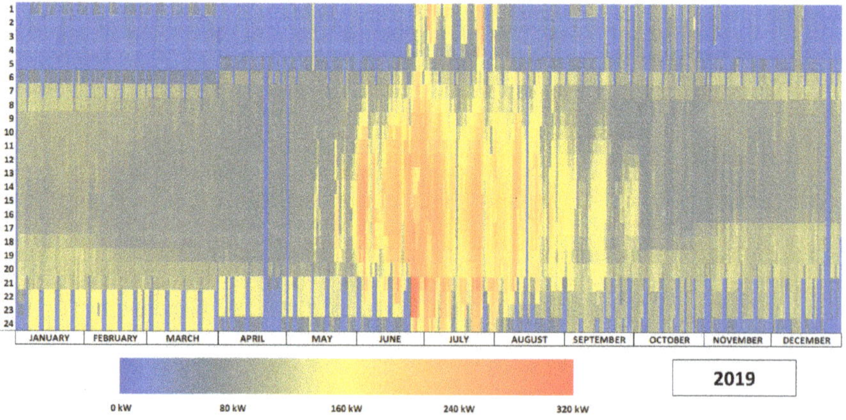

Figure 2. Heat map corresponding to commercial loads.

As shown in the chart, the building is closed on the first day of the Year, on Easter, on the 1st of May, in mid-August and on Christmas. During these periods, the photovoltaic energy self-consumed onsite is expected to be very low because only related to security equipment and perimeter lights. The maximum power absorbed is about 320 kW in summer due to chillers operation. In general, among the seasons, the PV production fits well the load: both the peaks in energy production and consumption are expected during summer, while the lowest values are registered in winter. The daily load curves present a peak in the late afternoon. During autumn and winter, another peak is observed also in early morning, due to HVAC machines start-up.

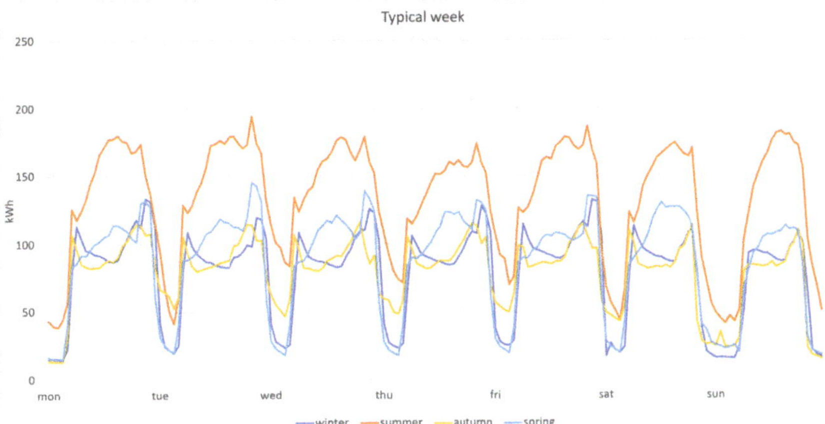

Figure 3. Seasonal commercial loads during a typical week.

The industrial load curve considered corresponds to a ceramics factory, whose structure covers an area of about 17,300 m^2. It is located in Italy, in the region of Emilia-Romagna, in climatic zone F. The industrial process covers the entire day and the corresponding consumption is much larger the one related to conditioning and lighting systems. The annual consumption of electric energy in 2019 (chosen once again as reference year) is 7.5 GWh. The hourly consumption is visualized in Figure 4, in form of heat map covering all the hours and all the days of the reference year. Moreover, the seasonal loads during a typical week are plotted in Figure 5.

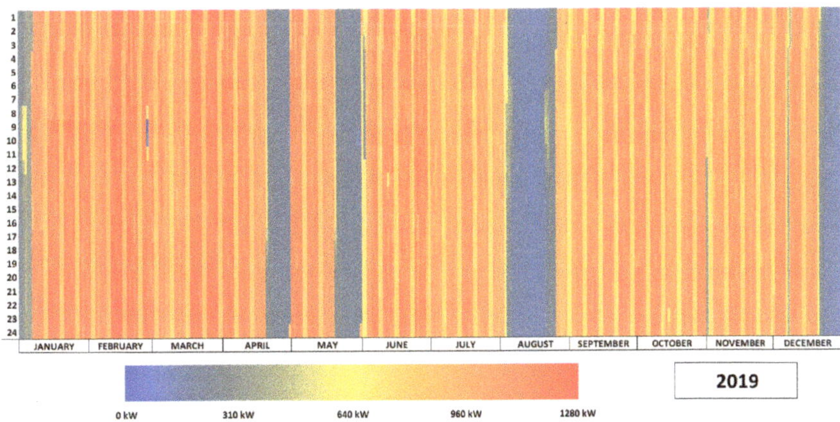

Figure 4. Heat map corresponding to industrial loads.

The power consumption ranges from 0 to 1280 kW. Saturdays and Sundays correspond to the yellow lines, representing a power absorption of about 650 kWp. The production is stopped during some periods in April, May, August and December. The load curve is constant among weeks and the electric consumption is generally constant among all the working days. The PV power production does not fit this type of load curve as good as the commercial one.

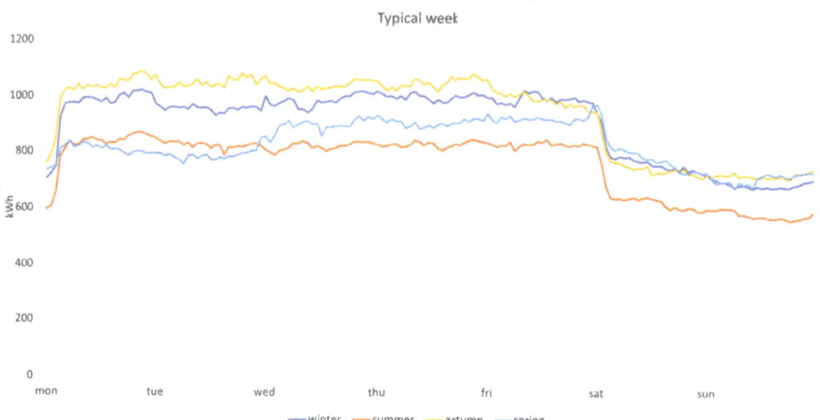

Figure 5. Seasonal industrial loads during a typical week.

2.3. PV Energy Production Simulation

A preliminary study on new PV plants is needed in order to estimate their potential energy production. In this analysis, the input variables are: the plant geographical coordinates, the peak power installable on a roof or on a specific area, the type of installation, the tilt and the azimuth of the roof. Notice that, in case of a PV facility where different sections present different exposures, the last four variables are considered independently for each exposure. Different sections may differ also in the type of installation and, consequently, in the mean daily PR. Finally, hourly irradiation data from the first to the last day of the considered reference year are acquired from SoDa Helioclim database for each section of the new plant, exploiting the information about the geographic coordinates, the tilt angle and the azimuth angle.

The energy production is calculated hour by hour using the solar irradiation data and the performance ratio:

$$E_{PV,i} = \frac{H_i}{1000} \cdot PR_{daily,i} \cdot P_i \tag{2}$$

In the equation: i stands for a generic plant section; H_i is the hourly solar irradiation on the surface of the modules in a given section; $PR_{daily,i}$ is the daily Performance Ratio derived from the monitoring of real PV plants; P_i is the total peak power installed for a given section. The total plant production in each hour is given by the sum of the energy produced by each section.

The simulations are performed under the assumption of ideal rooftop, where either fixed tilt, flush mount or East-West installations are possible. A total of six cases are considered, one for each combination between the three different PV plant configurations and the two possible load curves.

The tilt angle, the azimuth angle and the exposure are set for each configuration and thus they are independent from the load curve. In detail:

- Fixed tilt plant exposure is set toward South.
- Flush mount configuration is divided in two sections with different exposure: the first one is set toward South-West and the second one toward South-East.
- East-West type of plant is divided in two sections with opposite exposure.

The peak power of the plant is fixed: in case of commercial load, the peak power is 500 kWp, while in the case of industrial load the peak power is 2 MWp. The characteristics of each configuration are summarized in Table 1.

Table 1. New PV plants characteristics: (a) fixed tilt; (b) flush-mount; (c) East-West.

(a)				
			Fixed Tilt	
	Exposure	P_{DC} **[kWp]**	**Tilt [deg]**	**Azimuth [deg]**
Commercial	1	500	30	0
	2	-	-	-
Industrial	1	2000	30	0
	2	-	-	-

(b)				
			Flush Mount	
	Exposure	P_{DC} **[kWp]**	**Tilt [deg]**	**Azimuth [deg]**
Commercial	1	250	10	45
	2	250	10	−45
Industrial	1	1000	10	45
	2	1000	10	−45

(c)				
			East-West	
	Exposure	P_{DC} **[kWp]**	**Tilt [deg]**	**Azimuth [deg]**
Commercial	1	250	10	90
	2	250	10	−90
Industrial	1	1000	10	90
	2	1000	10	−90

2.4. Battery Energy Storage System Models

A list of the battery models to be analyzed is obtained choosing between the products available on the market: different brands, sizes and technologies are adopted and compared in the simulations. All the batteries considered present the possibility to be recharged from the grid. All the batteries useful parameters are retrieved from catalogs. Two main technologies are considered: LiFePo and Li-ion NMC batteries.

The maximum volume of the technical room where batteries are installed is arbitrarily set at 50 m^3: this constitutes an upper limit to the maximum number of battery modules installable. The volume occupied from each battery pack accounts for the dimension of the battery and the minimum space necessary for heat dissipation, reported in the data sheets. The weight of the system is kept into account.

The batteries that are simulated in combination with the PV system are listed in Table 2.

A total of 14 different battery models are chosen, and their corresponding 242 feasible configurations are simulated in couple with each considered PV facility. Remembering that 3 type of PV installation and 2 type of load are considered, a total of 1452 PV+BESS systems are evaluated.

Table 2. List of battery models considered and corresponding characteristics (extracted from data sheets).

Battery	Capacity [kWh]	P_{nom} [kW]	Efficiency [%]	Technology	Max Series	Price [€/kWh]
SonnenBatterie 10/11	10	4.6	0.98	LiFePo	9	650
SonnenBatterie10/27.5	25	4.6	0.98	LiFePo	9	650
Tesla PowerPack	232	130	0.89	N.A.	20	600
LG Chem R1000 M48189P3B	166.4	102	0.96	Li-ion NMC	30	500
LG Chem R1000 M48126P3B	110.9	135	0.96	Li-ion NMC	30	500
LG Chem R800 M48189P3B	137	84	0.96	Li-ion NMC	30	500
LG Chem R800 M48126P3B	91.3	112	0.96	Li-ion NMC	30	500
Pylontech Force H1	24.9	5	0.96	LiFePo	1	500
Pylontech Force H2	14.2	2.8	0.96	LiFePo	1	500
Kokam high energy rack	139	75	0.95	Li-ion	30	600
Kokam high energy 2P20S	13.9	7.5	0.95	Li-ion	12	600
Kokam high power 2P20S	11.5	12.5	0.95	Li-ion	12	600
Kokam ultra-high power 2P20S	10.2	11.1	0.95	Li-ion	12	600
BYD B-Box LVS	15.4	12	0.95	LiFePo	16	450

2.5. Battery Energy Storage System Control Logic

In the simulations, batteries are evaluated in terms of model and number, assuming that more packs of the same model can be considered in series or parallel connection. The battery simulation starts from a single pack of the first model of battery and ends at the maximum number of packs of the last type. A BESS configuration is simulated only if its volume is lower than the maximum volume of the technical room.

The batteries are connected to the grid, and therefore it is evaluated the convenience of recharging the battery when price of energy is lower. In order not to have the battery fully charged at the morning of a sunny day, the maximum state of charge achievable in F3 band is limited to the monthly difference between load and PV production divided by the number of days in that month. Moreover, the batteries are assumed to be AC-coupled with the PV system.

Real charge/discharge operations are always constrained by technical limits. However, in a preliminary battery assessment like the one proposed here, there is no need to account for these constraints. In real applications there is the necessity to identify as soon as possible a group of batteries suitable for a given application. Then and only then a specific battery model is chosen between the possible one (the choice is most of the times constrained by the availability of the different models) and further detailed analyses are carried out by means of specific software.

The battery operation is based on a precise control logic, capable of optimally managing the system. Decision-tree-like structure are constructed to visually represent the BESS control logic adopted. In Table 3, the terms adopted in the decision-tree-like structures are listed and explained.

The control logic of BESS charge is defined in the decision-tree-like structure reported in Figure 6.

In particular, the charge is permitted in three different modes:

- I charge: if the battery is not fully charged and the PV energy surplus is larger than the power of the battery, the state of charge in that hour increase of a quantity equal to the max power of charge; in case the capacity is exceeded, the state of charge is set at unity.
- II charge: if the battery is not fully charged and the PV energy surplus is positive and smaller than the power of charge, the battery is charged with the available PV energy in surplus; in case the capacity is exceeded, the state of charge is set at unity.
- III charge: if the battery is connected to the grid, the considered time window belongs the F3 band, it is not Sunday or holiday, the PV surplus is equal to zero and the SoC (State of Charge) is lower than the maximum SoC reachable in that month in F3 band, the battery is charged from the grid at maximum power.

Table 3. Notation adopted in decision-tree-like structures defining BESS charge and discharge control logic.

Symbol	Unit	Description
h	h	Analyzed hour
h-1	h	Previous hour
$C_{battery}$	kWh	Nominal capacity of the battery
$P_{battery}$	kW	Nominal power of charge and discharge
P_{loss}	%	Power losses during charge and discharge
C^*_h	kWh	Battery capacity at the analyzed hour
C^*_{h-1}	kWh	Battery capacity at the previous hour
$C^*_{max,F3}$	kWh	Maximum capacity allowed with grid charging
$E_{PV,Surplus,h}$	kWh	PV energy that remains available for storage at the analyzed hour
$E_{PV+BESS,Surplus,h}$	kWh	PV energy injected to the grid at the analyzed hour
$E_{grid,PV}$	kWh	Load demand after instantaneous self-consumption that is requested from the battery at the analyzed hour
$E_{grid,PV+BESS}$	kWh	Load demand after self-consumption that is requested from the grid at the analyzed hour

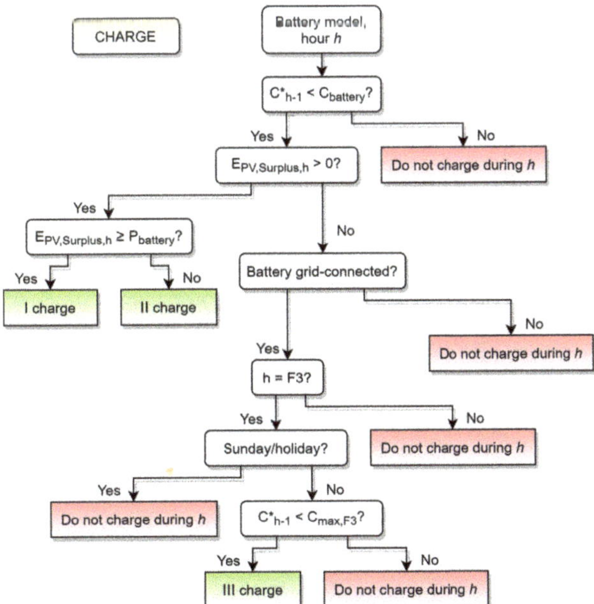

Figure 6. BESS charge control logic.

The control logic of BESS discharge is defined in the decision-tree-like structure reported in Figure 7.

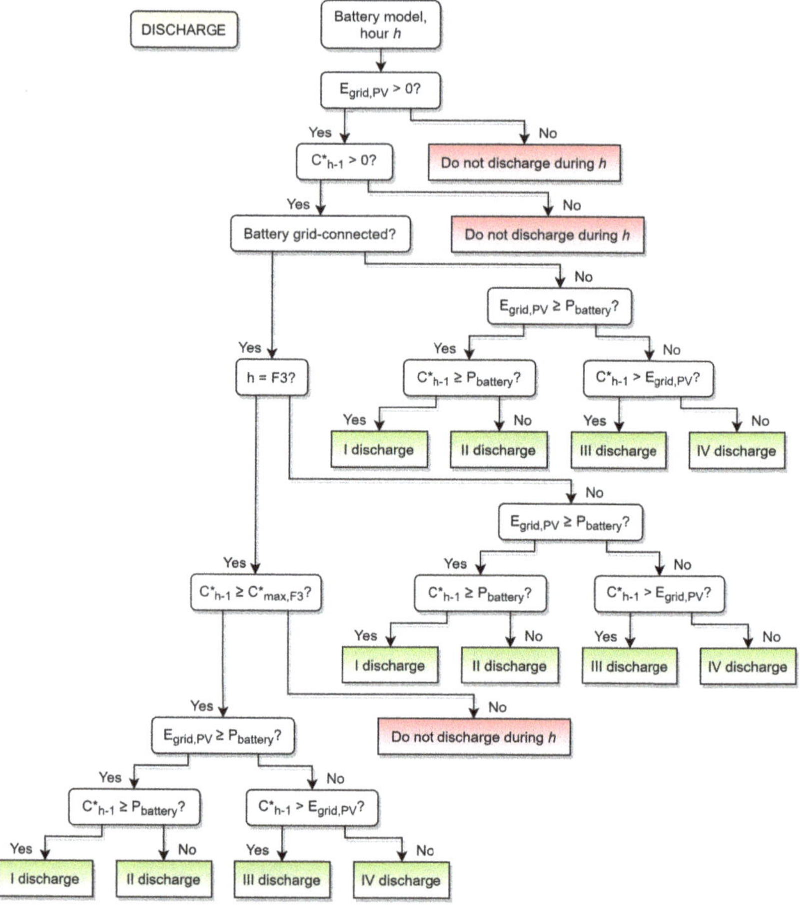

Figure 7. BESS discharge control logic.

The battery discharge takes place in four different modes:

- I discharge: if the battery is grid-connected, the discharge is allowed only if the hour is in F1 band, in F2 band or if the SoC exceeds the maximum SoC reachable in the specific month in F3 band. Then, it is evaluated if the load, after the self-consumption, needs power greater than the maximum power removable from the battery and if the available SoC of the battery is enough to fulfill the demand. In the case the battery is not grid-connected, only these last two conditions are evaluated. If these requirements are verified, the battery is discharged at maximum power.
- II discharge: if the load requires more energy than the ones produced by the PV plant and it is greater than the power of the battery, while the available SoC is not enough, the battery is fully discharged and the load withdraws also energy from the grid. This discharge mode is always allowed if the battery is not grid-connected, while, if it is connected, it is also checked that the hour does not belong to F3 band.
- III discharge: this type of discharge is allowed at the same conditions of the first mode of discharge, but accounts for load lower than the battery maximum power. When the energy stored is enough, the load is balanced discharging the battery.
- IV discharge: if the load, after the self-consumption, is lower than the maximum discharge power of the battery and the energy stored is lower than the requirement, the battery is fully discharged and the remaining energy required for balancing the

load is taken from the grid. This discharge mode is allowed evaluating if the battery is connected to the grid or not, as discussed in the first type of discharge.

2.6. Characteristic Features for PV+Bess Configurations

The prediction of the feasible BESS configurations accounts for some key indicators: the PayBack Time (PBT) of the battery capital expenditure, to be minimized; the number of residual cycles at end of life, to be minimized; the self-consumption, the coverage and the on-site self-production, to be maximized.

The Self-Consumption (SC) is defined as [24]:

$$SC = \frac{E_{PV \to load}}{E_{PV,y}} \tag{3}$$

In the equation: $E_{PV \to load}$ is the PV energy consumed by the load; $E_{PV,y}$ is the total annual PV production.

The coverage, sometimes also called self-sufficiency, is defined as [25]:

$$cov = \frac{E_{PV \to load}}{E_{load,y}} \tag{4}$$

In the equation: $E_{PV \to load}$ is the PV energy consumed by the load; $E_{load,y}$ is the total annual energy consumption.

The Self-Production (SP) is defined as [24]:

$$SP = \frac{E_{PV,y}}{E_{load,y}} \tag{5}$$

In the equation: $E_{PV,y}$ is the annual PV production; $E_{load,y}$ is the total annual energy consumption.

The PayBack Time (PBT) is computed as [26]:

$$PBT = \frac{BESS\ investment\ cost}{Annual\ economic\ saving} \tag{6}$$

The annual economic saving is the amount of money saved thanks to the presence of the BESS with respect to the same facility without any energy storage. In order to calculate it, a database with hundreds of electricity bills is exploited. The bills are divided according to zone, voltage (medium or low) and type of contract (peak-off peak, monorary, fixed multi-hourly and variable multi-hourly). Then, economic savings are calculated on the basis of the mean value of bills expenditures varying in function of energy.

Notice that the computed values of PBT refer only to the storage system and not to the entire power generation facility, including the PV plant. The OPEX (OPerating EXpense) related to the storage system consist of batteries O&M (Operation & Maintenance) costs (for instance related to maintenance interventions, remote monitoring etc.) and insurance costs. However, considering the purpose of the current preliminary analysis, all those factors can be neglected: they would be estimated equally for all the considered battery models and therefore they would not have any influence on the identification of the optimal capacity.

2.7. Battery Sizing Optimization by Means of Unsupervised Clustering

Finally, an unsupervised clustering based on k-means algorithm [27] is applied to all the analyzed BESS configurations. This final step aims at identifying a family containing all the feasible BESS solutions. K-means divides the dataset into a fixed number (k) of clusters according to some feature variables describing each sample. In this analysis, each sample corresponds to a possible BESS configuration. The feature variables chosen to fulfill the above-mentioned task are: the total photovoltaic energy stored in the battery within one year, the self-consumption, the number of residual cycles and the payback time.

In order to properly choose the number of clusters k, the Silhouette index [28] is exploited. This index provides a measure of how similar each sample is to samples in its own cluster, when compared to samples in other clusters and thus constitutes a tool to evaluate the quality several possible partitions of the available dataset. In practical terms, the Silhouette index is computed in function of the number of cluster k, and then the k corresponding to the highest Silhouette value is selected as number of clusters to be identified with k-means clustering.

The Silhouette plot for the BESS configurations coupled with the commercial load, representing the Silhouette value in function of the number of clusters k, is displayed in Figure 8. The number of clusters to be identified by k-means algorithm is equal to 2, coinciding with the largest Silhouette value.

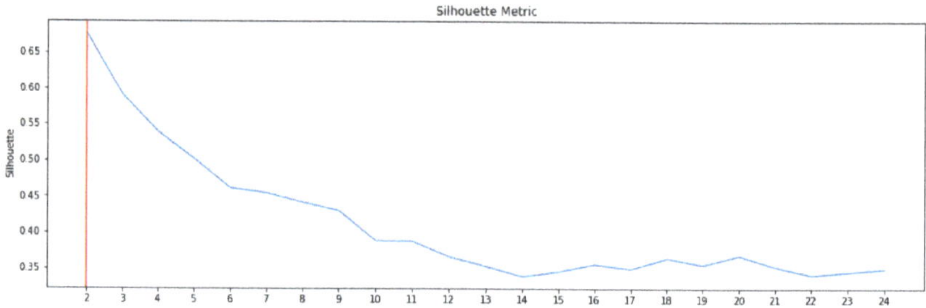

Figure 8. Silhouette plot for the commercial load scenario.

The silhouette plot for the BESS configurations coupled with the industrial load, representing the Silhouette value in function of the number of clusters k, is displayed in Figure 9. The number of clusters to be identified by k-means algorithm is equal to 3, coinciding with the largest Silhouette value.

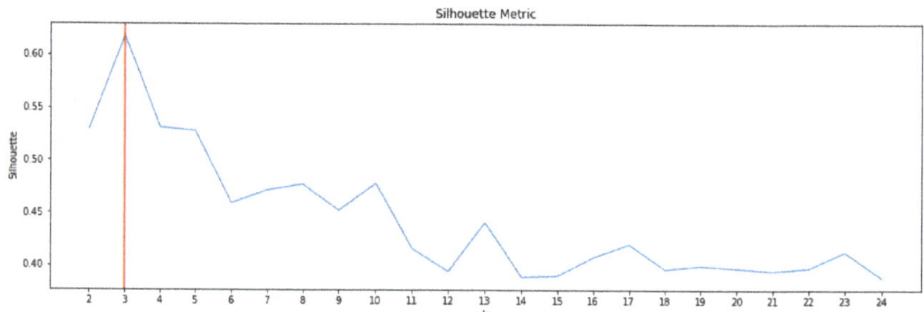

Figure 9. Silhouette plot for the industrial load scenario.

3. Results and Discussion

All the considered plants, their geographical position, their type of installation and their annual PR value are listed in Table 4. The observed annual PR ranges from 0.69 to 0.91. The two types of installation generally showing better performances are the fixed tilt and the East-West configurations, except for some outliers. The case of carport installation has little relevance in the current analysis: data are available only for one plant and the PR is calculated over a time period of only eight months.

Table 4. Location, type of installation and annual Performance Ratio for each of the considered PV facilities.

Plant Name	Location	Installation Type	Annual PR [%]
01_TARANTO	South	Fixed tilt	74
02_PALERMO	Islands	Fixed tilt	84
03_CUNEO	North-West	Flush Mount	80
04_CAGLIARI	Islands	Carport	77
05_CASAMASSIMA	South	Fixed tilt	81
06_SAN ROCCO AL PORTO	North-West	Fixed tilt	74
07_SERIATE	North-West	Fixed tilt	79
08_LISSONE	North-West	Fixed tilt	76
09_VICENZA_01	North-East	Flush Mount	70
10_VICENZA_02	North-East	Fixed tilt	86
11_CASALECCHIO DI RENO	North-East	Fixed tilt	75
12_PALERMO FORUM	Islands	Flush Mount	74
13_MESAGNE	South	East-West	91
14_CURNO	North-West	East-West	85
15_ROZZANO	North-West	East-West	83
16_VERONA	North-East	East-West	84
17_SAVIGNANO SUL RUBICONE	North-East	Fixed tilt	87
18_ROMA	Center	East-West	76
19_SANTA CATERINA	South	Fixed tilt	84
20_S. GIOVANNI TEATINO	South	Fixed tilt	81
21_CARUGATE	North-West	Fixed tilt	69
22_SOLBIATE ARNO	North-West	Fixed tilt	84

As already described, the PR value for single plants is averaged over all the plants characterized by a specific type of installation. The result of this operation is reported in Figure 10. The box plot shows, for both fixed tilt and East-West configurations, an average annual PR higher than 0.80. However, the variability of the performances observed with fixed tilt PV facilities is much larger than that of East-West PV plants.

Finally, Figure 11 displays an heat map representing the daily average values of PR in the reference year in function of the plant configuration, computed averaging the daily PR values of single plants.

The new PV+BESS hybrid plants simulations return the forecast of the total amount of energy self-consumed, sold to the grid, stored in the battery or acquired from the grid in order to balance the demand. It is then possible to discuss the results in terms of PBT of the battery. As expected, increasing the number of battery packs in series, thus the capacity of the storage system, the energy self-consumed by the load grows but also the PBT increases significantly.

The results reported in Table 5 identify the BESS configuration that minimizes the PBT for each PV system configuration.

Most of the configurations identified result in a PBT approximately equivalent to the lifetime of the battery, equal to 15 years. In the last two cases, the PBT that is even larger than the battery lifetime. The cost of energy storage technologies is still too high to conclude that nowadays it is convenient to install a BESS system for large buildings. However, if the investment cost per kWh of capacity decreases, it will be possible to install a large capacities and to achieve a significant advantage also in terms of additional self-consumption. Changing the PV installation type for the commercial load, the choice of battery models remains unchanged, as well as the PBT. In case of industrial load, the same battery model with the same number of modules shows a decrease in PBT for the fixed tilt configuration thanks to higher annual economic savings.

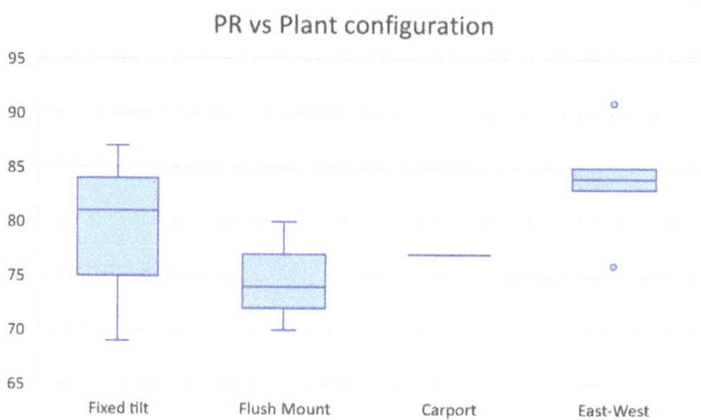

Figure 10. Annual PR averaged in function of the PV plant configuration.

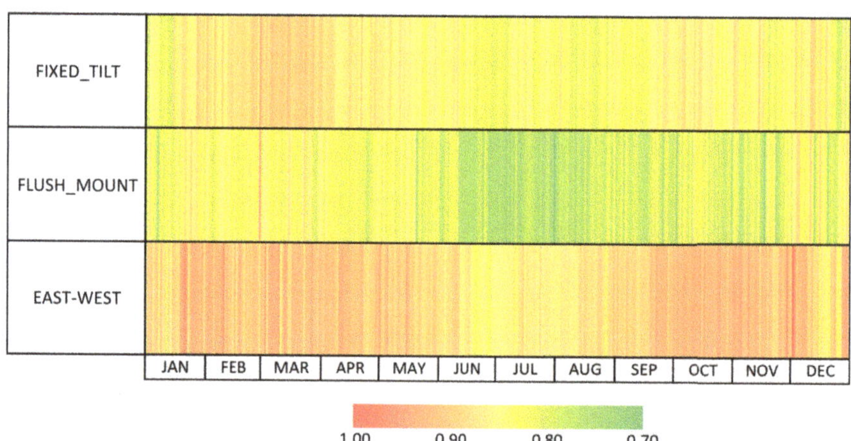

Figure 11. Daily PR (computed on the chosen reference year) averaged over all the plants characterized by a given configuration.

Table 5. BESS configurations with minimum PBT.

| | Commercial | | | Industrial | | |
	Fixed Tilt	Flush Mount	East-West	Fixed Tilt	Flush Mount	East-West
Model	Kokam ultra-high power 2P20S	Sonnen Batterie10/11	Kokam ultra-high power 2P20S	LG Chem R800 M48126P3B	Kokam ultra-high power 2P20S	Kokam high energy rack
N° modules	2	2	2	5	5	5
Residual cycles	960	945	991	980	408	443
Self-consumption [%]	55	58	59	74	80	79
Coverage [%]	44	43	44	28	26	26
PBT [y]	14.5	14.5	14.6	14.3	18.7	18.3

The results reported in Table 6 identify the BESS configuration that minimizes the number of residual cycles for each PV system configuration.

Most of battery models optimized in terms of number of residual cycles are different from the ones optimized in terms of PBT. Focusing on the industrial load case, the fixed tilt configuration with the battery storage could be an interesting solution in case of decreasing in investment cost for batteries, because it has the minimum PBT between batteries with the optimal value of residual cycles. The last configuration has a PBT which is way too high for the feasibility of the investment.

Table 6. BESS configurations with minimum number of residual cycles.

| | Commercial | | | Industrial | | |
	Fixed Tilt	Flush Mount	East-West	Fixed Tilt	Flush Mount	East-West
Model	Kokam high power 2P20S	Kokam high power 2P20S	Kokam high power 2P20S	BYD B-Box LVS	BYD B-Box LVS	BYD B-Box LVS
N° modules	10	7	3	2	9	12
Residual cycles	22	33	45	2	16	1
Self-consumption [%]	58	60	59	74	80	83
Coverage [%]	45	44	44	28	26	27
PBT [y]	20.5	23.7	21.4	16.8	25.9	51.6

The results obtained from k-means clustering application are reported in the following. As already discussed, the clustering procedure exploits the total photovoltaic energy stored in the battery within one year, the self-consumption, the number of residual cycles and the payback time as relevant features to characterize each possible BESS configuration.

Figure 12 represents all the 242 possible BESS configurations for the case of commercial load with a fixed tilt PV installation divided in two clusters. The clustering results does not show significant differences for other types of installation. As discussed before, the number of clusters is chosen on the basis of the Silhouette index and is equal to 2. The feature space is represented by three different points of view: on the PBT/residual cycles plan, on the self-consumption/residual cycles plan and on the self-consumption/PBT plan. The last diagram represents the overlap between clusters in terms of PBT. All values on the axes are standardized in the range between −1 and 1.

The purple cluster represents the family of BESS that are best suited to be coupled with the analyzed PV facility configuration. The trend of self-consumption over the PBT confirms what stated before: increasing the capacity of the battery, the self-consumption increases but, as a drawback, the PBT increases as well.

Figure 13 represents all the 242 possible BESS configurations for the case of industrial load with a fixed tilt PV installation divided in two clusters. Even when k-means is applied to the industrial scenario, the results are similar among different PV installation types, as observed for the commercial case. The number of clusters is chosen on the basis of the Silhouette index and is equal to 3. The feature space is represented in the same way as the commercial case and all values on the axes are standardized in the range between −1 and 1.

The large number of batteries with high capacity (and consequently high PBT) and low number of residual cycles, in the top left region of the upper diagrams, is related to the high electric consumption typical of an industrial load. Once again, the purple cluster represents the family of BESS that are best suited to be coupled with the analyzed PV facility configuration. The green cluster correspond to BESS configurations with low number of residual cycles and high PBT, while the orange cluster represents high-capacity batteries that are strongly oversized and thus not suited for the considered PV facility.

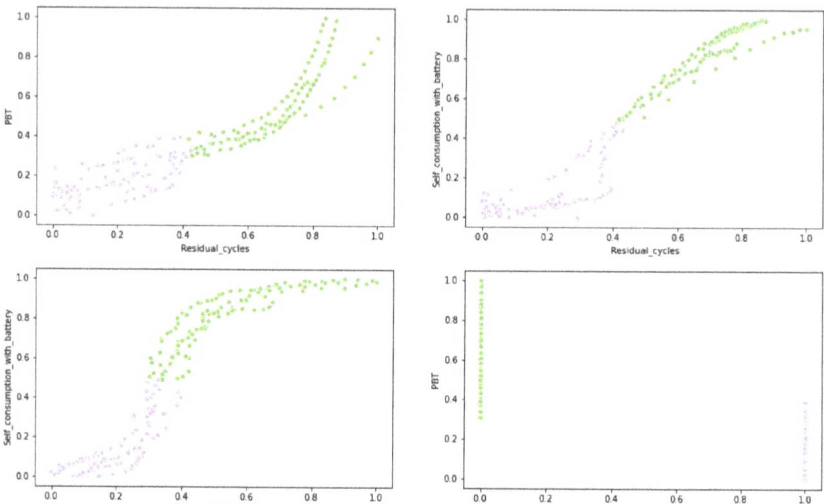

Figure 12. Possible BESS configurations divided in clusters (commercial load).

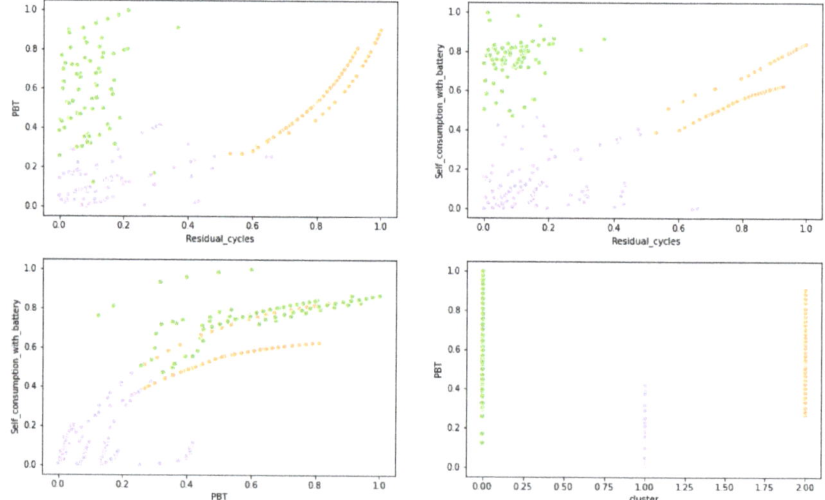

Figure 13. Possible BESS configurations divided in clusters (industrial load).

4. Conclusions

In recent years, the technological development and the increasing market competitiveness of RESs-based power production systems determined favorable conditions to switch from electricity generation in large centralized facilities to small decentralized energy systems.

In this scenario, PV facilities find profitable conditions for the grid connected users when the produced energy is self-consumed. Due to the intermittent and stochastic nature of the solar source, PV plants require the addition of an energy storage system to compensate fluctuations and to meet the energy demand even during night hours.

In this paper, a procedure is developed to forecast a family of batteries which is suitable to be coupled with a given PV plant configuration and is applied to some new PV facilities.

The PV+BESS hybrid plant energy production simulation is possible by:

- Knowing the geographical coordinates of the installation site and tilt and azimuth of the roof.
- Assuming the peak power installable on a roof and an installation type.
- Estimating a proper PR value, computed through data from real operating plants with similar installation characteristics and size.

Two different types of load curve are considered in the current work, namely:

- A commercial load curve, corresponding to a single-brand point of sale.
- An industrial load curve, corresponding to a ceramics factory.

The battery operations are managed by means of a control logic defined in decision-tree-like diagrams. The two diagrams, provided in the current work, consider all possible operating conditions during both charge and discharge. The main strategies behind the defined control logic are:

- Optimizing PV self-consumption, beneficial in markets whose value of electricity (€/kWh) is high.
- Charging the battery in the time bands with lower price of electric energy.

For each possible PV+BESS configuration, performance features, like the number of residual cycles at the end of lifetime and the self-consumption, and economic features, as the payback time, are computed. The self-consumption is defined as the ratio between PV energy consumed by the load and total annual PV production. On the other hand, PBT is based on the annual economic savings allowed by the presence of an energy storage system compared to the case of PV plant without battery.

The following observations are derived from the analysis performed:

- The knowledge of the annual distribution of electrical loads is crucial to determine which season or time window with high power demand justifies the existence of the storage, reducing the energy purchased from the grid. Energy-intensive applications, characterized by high loads even during night, enhance the profitability of the PV+BESS configuration.
- At present, the billing savings in themselves might not be enough to encourage the use of PV+BESS hybrid systems. Besides, their profitability strongly depends on the electricity tariff structure and energy policy of a country, in addition to PV and storage systems costs.

Finally, a clustering algorithm based on k-means algorithm is applied to all the considered PV+BESS configurations, aimed at detecting the family of battery solutions which is the most suitable according to the scenario considered. The number of clusters to be identified is established by means of the Silhouette index. As expected, the cluster of the best solutions contains all those configurations characterized by low PBT and number of residual cycles.

Possible future developments of the present work consist in adopting different clustering criteria and different features to possibly improve the identification of the family of batteries that are suitable for a given application.

Author Contributions: Data curation, S.P. and L.D.C.; Formal analysis, A.N., A.M., S.P., L.D.C. and E.O.; Investigation, A.N., A.M., S.P., L.D.C. and E.O.; Methodology, A.N., A.M., S.P., L.D.C. and E.O.; Software, L.D.C.; Writing—original draft, A.M. and L.D.C.; Writing—review & editing, A.N., S.P. and E.O. All authors have read and agreed to the published version of the manuscript.

Funding: This research received no external funding.

Conflicts of Interest: The authors declare no conflict of interest.

Abbreviations

The following abbreviations are used in this manuscript:

AC	Alternating Current
BESS	Battery Energy Storage System
DC	Direct Current
DoD	Depth of Discharge
HVAC	Heating, Ventilation and Air Conditioning
NMC	Nickel Manganese Cobalt
O&M	Operation & Maintenance
OPEX	OPerating EXpense
PBT	PayBack Time
PR	Performance Ratio
PV	PhotoVoltaic
RES	Renewable Energy Source
SC	Self-Consumption
SoC	State of Charge
SP	Self-Production

References

1. Roffel, B.; de Boer, W. Analysis of power and frequency control requirements in view of increased decentralized production and market liberalization. *Control. Eng. Pract.* **2003**, *11*, 367–375. [CrossRef]
2. Wu, C.; Zhou, D.; Lin, X.; Wei, F.; Chen, C.; Ma, Y.; Huang, Y.; Li, Z.; Dawoud, S.M. A novel energy cooperation framework for community energy storage systems and prosumers. International *J. Electr. Power Energy Syst.* **2022**, *134*, 107428. [CrossRef]
3. Botelho, D.F.; Dias, B.H.; de Oliveira, L.W.; Soares, T.A.; Rezende, I.; Sousa, T. Innovative business models as drivers for prosumers integration-Enablers and barriers. *Renew. Sustain. Energy Rev.* **2021**, *144*, 111057. [CrossRef]
4. Boonluk, P.; Siriratiwat, A.; Fuangfoo, P.; Khunkitti, S. Optimal Siting and Sizing of Battery Energy Storage Systems for Distribution Network of Distribution System Operators. *Batteries* **2020**, *6*, 56. [CrossRef]
5. Al-Aboosi, F.Y.; Al-Aboosi, A.F. Preliminary Evaluation of a Rooftop Grid-Connected Photovoltaic System Installation under the Climatic Conditions of Texas (USA). *Energies* **2021**, *14*, 586. [CrossRef]
6. Alghamdi, A.S. Performance Enhancement of Roof-Mounted Photovoltaic System: Artificial Neural Network Optimization of Ground Coverage Ratio. *Energies* **2021**, *14*, 1537. [CrossRef]
7. Mellit, A.; Massi Pavan, A.; Ogliari, E.; Leva, S.; Lughi, V. Advanced Methods for Photovoltaic Output Power Forecasting: A Review. *Appl. Sci.* **2020**, *10*, 487. [CrossRef]
8. Kosmadakis, I.E.; Elmasides, C. A Sizing Method for PV–Battery–Generator Systems for Off-Grid Applications Based on the LCOE. *Energies* **2021**, *14*, 1988. [CrossRef]
9. Rashid, K. Design, Economics, and Real-Time Optimization of a Solar/Natural Gas Hybrid Power Plant. Ph.D. Thesis, The University of Utah, Salt Lake City, UT, USA, 2019.
10. Tabares, A.; Martinez, N.; Ginez, L.; Resende, J.F.; Brito, N.; Franco, J.F. Optimal Capacity Sizing for the Integration of a Battery and Photovoltaic Microgrid to Supply Auxiliary Services in Substations under a Contingency. *Energies* **2020**, *13*, 6037. [CrossRef]
11. Attya, A.B.; Vickers, A. Operation and Control of a Hybrid Power Plant with the Capability of Grid Services Provision. *Energies* **2021**, *14*, 3928. [CrossRef]
12. Javeed, I.; Khezri, R.; Mahmoudi, A.; Yazdani, A.; Shafiullah, G.M. Optimal Sizing of Rooftop PV and Battery Storage for Grid-Connected Houses Considering Flat and Time-of-Use Electricity Rates. *Energies* **2021**, *14*, 3520. [CrossRef]
13. Mair, J.; Suomalainen, K.; Eyers, D.M.; Jack, M.W. Sizing domestic batteries for load smoothing and peak shaving based on real-world demand data. *Energy Build.* **2021**, *247*, 111109. [CrossRef]
14. Pena-Bello, A.; Burer, M.; Patel, M.K.; Parra, D. Optimizing PV and grid charging in combined applications to improve the profitability of residential batteries. *J. Energy Storage* **2017**, *13*, 58–72. [CrossRef]
15. He, J.; Yang, Y.; Vinnikov, D. Energy Storage for 1500 V Photovoltaic Systems: A Comparative Reliability Analysis of DC- and AC-Coupling. *Energies* **2020**, *13*, 3355. [CrossRef]
16. Sandelic, M.; Sangwongwanich, A.; Blaabjerg, F. Reliability Evaluation of PV Systems with Integrated Battery Energy Storage Systems: DC-Coupled and AC-Coupled Configurations. *Electronics* **2019**, *8*, 1059. [CrossRef]
17. Jufri, F.H.; Aryani, D.R.; Garniwa, I.; Sudiarto, B. Optimal Battery Energy Storage Dispatch Strategy for Small-Scale Isolated Hybrid Renewable Energy System with Different Load Profile Patterns. *Energies* **2021**, *14*, 3139. [CrossRef]
18. Rashid, K.; Safdarnejad, S.M.; Ellingwood, K.; Powell, K.M. Techno-Economic Evaluation of Different Hybridization Schemes for a Solar Thermal/Gas Power Plant. *Energy* **2019**, *181*, 91–106. [CrossRef]
19. Ma, D.; Pan, G.; Xu, F.; Sun, H. Quantitative Analysis of the Impact of Meteorological Environment on Photovoltaic System Feasibility. *Energies* **2021**, *14*, 2893. [CrossRef]

20. Aguilar-Jiménez, J.A.; Hernández-Callejo, L.; Alonso-Gómez, V.; Velázquez, N.; López-Zavala, R.; Acuña, A.; Mariano-Hernández, D. Techno-economic analysis of hybrid PV/T systems under different climate scenarios and energy tariffs. *Sol. Energy* **2020**, *212*, 191–202. [CrossRef]
21. Khalid, A.M.; Mitra, I.; Warmuth, W.; Schacht, V. Performance ratio – Crucial parameter for grid connected PV plants. *Renew. Sustain. Energy Rev.* **2016**, *65*, 1139–1158. [CrossRef]
22. Lee, C.-S.; Lee, H.-M.; Choi, M.-J.; Yoon, J.-H. Performance Evaluation and Prediction of BIPV Systems under Partial Shading Conditions Using Normalized Efficiency. *Energies* **2019**, *12*, 3777. [CrossRef]
23. Reich, N.; Müller, B.; Armbruster, A.; van Sark, W.; Kiefer, K.; Reise, C. Performance ratio revisited: Is PR > 90% realistic? *Prog. Photovoltaics Res. Appl.* **2012**, *20*, 717–726. [CrossRef]
24. Pena-Bello, A.; Barbour, E.; Gonzalez, M.C.; Yilmaz, S.; Patel, M.K.; Parra, D. How Does the Electricity Demand Profile Impact the Attractiveness of PV-Coupled Battery Systems Combining Applications? *Energies* **2020**, *13*, 4038. [CrossRef]
25. Fachrizal, R.; Munkhammar, J. Improved Photovoltaic Self-Consumption in Residential Buildings with Distributed and Centralized Smart Charging of Electric Vehicles. *Energies* **2020**, *13*, 1153. [CrossRef]
26. Kharseh, M.; Wallbaum, H. How Adding a Battery to a Grid-Connected Photovoltaic System Can Increase its Economic Performance: A Comparison of Different Scenarios. *Energies* **2019**, *12*, 30. [CrossRef]
27. Xu, R.; Wunsch, D. Survey of clustering algorithms. *IEEE Trans. Neural Networks* **2005**, *16*, 645–678. [CrossRef]
28. Shutaywi, M.; Kachouie, N.N. Silhouette Analysis for Performance Evaluation in Machine Learning with Applications to Clustering. *Entropy* **2021**, *23*, 759. [CrossRef]

forecasting

Article

Influence of the Characteristics of Weather Information in a Thunderstorm-Related Power Outage Prediction System

Peter L. Watson *, Marika Koukoula and Emmanouil Anagnostou

Department of Civil & Environmental Engineering, University of Connecticut, Storrs, CT 06269, USA;
MARIKA.KOUKOULA@uconn.edu (M.K.); emmanouil.anagnostou@uconn.edu (E.A.)
* Correspondence: peter.watson@uconn.edu

Abstract: Thunderstorms are one of the most damaging weather phenomena in the United States, but they are also one of the least predictable. This unpredictable nature can make it especially challenging for emergency responders, infrastructure managers, and power utilities to be able to prepare and react to these types of events when they occur. Predictive analytical methods could be used to help power utilities adapt to these types of storms, but there are uncertainties inherent in the predictability of convective storms that pose a challenge to the accurate prediction of storm-related outages. Describing the strength and localized effects of thunderstorms remains a major technical challenge for meteorologists and weather modelers, and any predictive system for storm impacts will be limited by the quality of the data used to create it. We investigate how the quality of thunderstorm simulations affects power outage models by conducting a comparative analysis, using two different numerical weather prediction systems with different levels of data assimilation. We find that limitations in the weather simulations propagate into the outage model in specific and quantifiable ways, which has implications on how convective storms should be represented to these types of data-driven impact models in the future.

Keywords: power outages; machine learning; thunderstorms; numerical weather prediction

Citation: Watson, P.L.; Koukoula, M.; Anagnostou, E. Influence of the Characteristics of Weather Information in a Thunderstorm-Related Power Outage Prediction System. *Forecasting* **2021**, *3*, 541–560. https://doi.org/10.3390/forecast3030034

Academic Editor: Sonia Leva

Received: 27 June 2021
Accepted: 3 August 2021
Published: 5 August 2021

Publisher's Note: MDPI stays neutral with regard to jurisdictional claims in published maps and institutional affiliations.

1. Introduction

Weather-related power outages, and the severe weather events that cause them, pose a persistent threat to the functioning of the infrastructure and economy of the United States. These types of power outages affect millions of people and cost the U.S. economy tens of billions of dollars every year; moreover, the rate at which they occur appears to be increasing [1]. Anticipating the damages that storms can cause is a critical step in electrical utility managers' storm outage management process. They need reliable information before a storm to be able to stage repair crews and effectively prepare for the damages that the storm will cause [2]. As such, there has been a recent surge in research and development activity into methods to predict storm damages and weather-related power outages.

Arguably, the most destructive types of storms in the United States are thunderstorms, including the associated convective phenomena (tornadoes, microbursts, hail, etc). While hurricanes often receive special attention because they are larger and more dramatic, thunderstorms are more common and cause more damage to the electrical infrastructure every year than any other type of weather. Indeed, investigations of major outage events reported to the Department of Energy have found that convective storms are responsible for the majority of weather-related outage events, the greatest number of customer outages, and the most outage hours [3,4]. Additionally there is every indication that the severity of thunderstorms is going to increase in the future. Changes in the climatic patterns of thunderstorms can already be seen in a time series analysis [5], and long-term climate projections suggest that, because of climate change, thunderstorms are likely to become stronger, more frequent, and more damaging [6,7].

Despite the demonstrated risk that thunderstorms present to the electrical infrastructure, they have not received much attention in the recent research for modeling weather-related power outages. While there are some outage modeling approaches that are generalized to a range of types of weather [8–11], much of research in this field has been focused on other types of storms. The vast majority of the work has focused on tropical storms and hurricanes, which can have particularly dramatic impacts [12–16], but several mature modeling approaches, specifically for extratropical storms [17–19], have also been developed.

In the existing general outage models, thunderstorms are sometimes included in the analysis [9,10,20,21], but the weather characteristics of these storms are treated in a similar fashion to other, more structured types of weather. There are also some studies that infer a focus on thunderstorms by including information about lightning strikes [11,22,23], but do not have an explicit focus on thunderstorms because they also include other types of weather events in their analysis.

This lack of focus on thunderstorms may be a result of the technical difficulty associated with describing and simulating them. Convective storms are particularly challenging for established numerical weather prediction (NWP) models and meteorological forecasts. While the increased horizontal resolution of convective-allowing configurations can lead to improved simulations, even with state-of-the-art high-resolution NWP models, reliable deterministic forecasts of thunderstorms longer than several hours are elusive [24–26]. As Yano et al. describe, there may be limitations to modern NWP models' ability to simulate convective storms because of the wide-spread use of assumptions and parameterizations that are reasonable for synoptic-scale weather patterns but are much less applicable to more complex convective phenomena [26]. These potential limitations of NWP simulations are long standing, and multiple strategies for mitigating them have emerged. Assimilating radar or even lightning observations into initial conditions of simulations can be used to improve short-term predictions [27,28]; forecasting systems that leverage this type of data assimilation for rapidly-updating nowcasts are currently operational [29]. In addition, for forecasts longer than several hours, stochastic predictions from convective-allowing ensembles have shown an improved forecasting skill by being able to capture the range of potential outcomes, instead of one deterministic scenario [30–32].

Similar approaches and findings can be seen in the few studies in the literature that specifically focus on predicting thunderstorm-related power outages. In Alpay et al., the authors take a rapid-refresh nowcasting approach to modeling thunderstorm-related outages, using an LSTM neural network trained on data from a rapidly updating radar-ingesting weather model from NOAA [33]. The works of Shield and Kabir et al. both describe a thunderstorm outage prediction system trained on weather data from the National Digital Forecast Database for an area in Alabama [34,35]. Shields investigates the limitations of the model he develops and identifies that it has better skill at a synoptic scale, which illustrates the difficulty of forecasting with thunderstorms [34]. Kabir et al. take a more stochastic approach and develop a quantile regression model, which allows the communication of the significant uncertainties associated with predicting the impacts of thunderstorms [35].

While this previous work attempts to manage the known limitations of weather simulations of thunderstorms, how these limitations propagate from weather simulations into machine-learning based impact models remains poorly described. The problem of poor inputs for a computational algorithm has been recognized since the dawn of computation [36], but the effects in this context are not fully understood. In this paper, we attempt to shed light on this matter by analyzing the quality of the weather data from two different weather simulation systems with differing amounts of data assimilation, determine how outage models trained on these different sets of weather data differ in skill and accuracy, and what information the outage models learn from. This knowledge is critical to build an understanding of the limitations of the data used to build impact models for thunderstorms and to suggest how improved representations of weather will improve the quality of the insights that can be derived from them.

2. Materials and Methods

This study involved the creation and comparison of two separate machine-learning models designed to predict thunderstorm-related power outages, using data from NWP-based weather simulations and a wide range of other data sources in a region covering three states: Connecticut, Massachusetts, and New Hampshire, and five distinct electrical utility service territories: Eversource Connecticut (CT), Eversource Western Massachusetts (WMA), Eversource Eastern Massachusetts (EMA), Eversource New Hampshire (NH), and AVANGRID United Illuminating (UI). For geographical details of the modeling domain, refer to Figure 1.

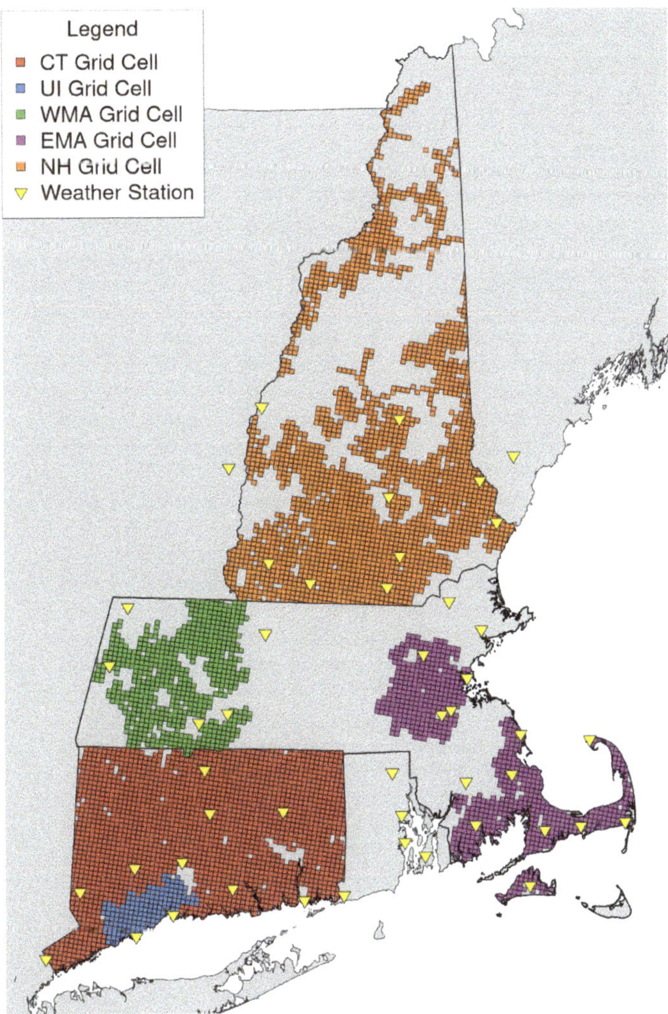

Figure 1. The location of the outage model grid cells by service territory as well as the location of the airport weather stations used in the meteorological analysis.

2.1. Data

The outage models developed in this analysis use data describing 372 thunderstorm events that occurred in the utility service territories from 2016 to 2020, as well as a range of environmental characteristics, such as vegetation and drought status, as well as proprietary outage and infrastructure data provided by the power utilities, aggregated to the grid cells of the weather simulations. We included as many thunderstorm events that could be observed in weather station reports from each utility service territory, and aggregated the data to the RTMA grid cells of each service territory for each thunderstorm event. For details about the amount of data used from each territory, see Table 1.

Table 1. The amount of data available for training the thunderstorm-related outage models.

	CT	WMA	EMA	NH	UI	Total
Number of Storms	74	82	69	91	56	372
Territory Grid Cells	2019	638	820	2128	169	5774
Total Entries	149,406	52,316	56,580	193,648	9464	461,414

2.1.1. Weather

The core of the analysis centers around datasets produced by two separate NWP gridded weather simulation systems: a hybrid NOAA analysis system, and a WRF 2 km simulation system. The NOAA analysis dataset is a combination of data from the Real-Time Mesoscale Analysis (RTMA) [37] and Stage IV Quantitative Precipitation Estimates (Stage IV) [38]. RTMA is a weather analysis product that produces a gridded estimate of weather conditions by statistically downscaling a 1 h short-term forecast and adjusting it with weather station observations. It produces a high-resolution, near real-time estimate of temperature, humidity, dew point, wind speed and direction, wind gusts, and surface pressure for the entire United States. The RTMA data were sourced from the archive hosted on the Google Earth Engine [39]. Stage IV is a Quantitative Precipitation Estimate (QPE) dataset created by the National Weather Service and the National Centers for Environmental Prediction (NWS, NCEP), using a blend of NEXRAD radar and the NWS River Forecast Center precipitation processing system [40]. It takes gridded precipitation estimates derived from radar scans, adjusts the values based on rain gauge data, and aggregates the data to produce gridded hourly estimates of precipitation for the continental United States. It is popular for analytical purposes and is often used as a reference to analyze the accuracy of satellite and other precipitation estimates [38]. By using a blend of RTMA and Stage IV, we are able to have a reasonable estimate of the average hourly weather conditions in each grid cell during each storm used in this analysis. For the sake of brevity, this dataset will sometimes be referred to as the "RTMA" system.

We compare this hybrid NOAA analysis dataset with another weather dataset developed from a configuration of the Weather Research and Forecasting Model (WRF), similar to one that was used in several outage predictions models [17,18], but with an increased horizon resolution to potentially help resolve convection. This model is initialized with the North American Mesoscale Forecast System analysis [41], which has 2 km horizontal grid spacing with one 6 km external domain. For configuration details, please see Table 2. These WRF simulations use a different projection than the RTMA system, so the results were resampled with bilinear interpolation to match the spatial characteristics of the RTMA analysis product.

Table 2. Details of the WRF simulation configuration.

Horizontal Resolution	2 km
Vertical Levels	51
Horizontal Grid Scheme	Arakawa C Grid
Nesting	One 6km Nested Domain
Microphysics Option	Thompson Graupel Scheme [42]
Longwave Radiation Option	RRTM Scheme [43]
Shortwave Radiation Option	Goddard Shortwave Scheme [44]
Surface-Layer Option	Revised MM5 Scheme [45]
Land-Surface Option	Noah Land-Surface Model [46]
Planetary Boundary Layer	Yonsei Scheme [47]

For outage modeling purposes, 24 h time series of a common set of weather variables generated from both weather simulation systems were processed to generate descriptive data features for each thunderstorm in this analysis. The weather variables considered are dew point temperature, specific humidity, air temperature, surface pressure, wind speed, wind gust speed, wind direction, and hourly precipitation rate. Established weather parameters that directly describe convective potential, such as CAPE and CIN, were unfortunately not available for this study because they are not published in RTMA, which is primarily a surface analysis product. For each of the included variables, the mean, max, minimum, standard deviation, 4 h mean during peak winds, and total were calculated for each storm, except for wind direction for which we took the median value. The median was taken to limit its sensitivity to outliers. Several additional features were calculated: the number of hours of winds above various wind speeds, calculated using various thresholds applied to wind speeds and gusts; typical wind direction by taking the mean of the median wind direction of included storms; and the difference between the typical wind direction and the median wind direction for that storm. To preserve its characteristics, all computation and analysis of wind direction was performed via the `circular` library in R [48]. Additionally, we included an additional set of features describing the time series of wind stress exerted on the trees by taking the product of the leaf area index (see below) and the square of the wind speed. For details, please see Appendix A, which contains a detailed table of all data features used for modeling.

2.1.2. Infrastructure and Outage Data

Proprietary data of the infrastructure and historical outages are made available for this study for the five utility service territories. Using `rgdal` and `rgeos` [49,50] for the area within each outage model grid cell, we calculated the length of overhead power lines, the number of utility poles, the number of fuses and cutouts, and the number of circuit reclosers.

The historic outage data describes the time and location of where damage occurred to the power distribution grid for a period of five years (2016 to 2020). Based on this information, we were able to calculate the number of damage locations within each outage model grid cell associated with each storm. A damage location is a physical location where repair crews are dispatched to repair damage after a storm. In the vast majority of cases, this meant counting the damage locations that were identified in the 24 h storm period, but in several cases, additional "nested" storm-related outages were recognized by utility operators after the storm period, so a longer window was sometimes used. These damage data were extracted from the utility outage management system, which is a software tool used by most large utilities to identify outages and dispatch repair crews.

2.1.3. Environmental Data

Because weather-related power outages are the result of interactions between the weather, the infrastructure and the environment, a range of environmental information was considered for this analysis. We processed the environmental data in several different ways depending on spatial resolution. When working with datasets with a resolution higher than the 2.5 km RTMA grid, for each grid cell, the raster data were sampled from a 60 m buffer around the overhead lines in that cell, and we calculated the representative percentage for the categorical data, or the average and standard deviation for the numerical data. We applied this process to a range of datasets, including the following: categorical land cover from the 2016 National Land Cover Database (NLCD) [51], 2016 NLCD Tree Canopy Coverage [52], vegetation height estimates from the Global Ecosystem Dynamics Investigation (GEDI) lidar instrument on the International Space Station [53], USGS 3DEP DEM elevation [54], and several other datasets, which required special processing. For example, we sampled the soils dataset developed by Watson et al. [18] from the USDA SSURGO database [55] to describe the soil characteristics (density, porosity, hydraulic conductivity, composition, and saturation). Additionally, because that previous work suggests that systemic biases caused by differences in the elevations of the weather predictions and the infrastructure may be present, we used the difference between those two elevations as an additional feature, `elvDiff`.

As seen in other outage modeling work [15,16], high-resolution data from the Individual Tree Species Parameter Maps (developed to support the USDA National Insect and Disease Risk Map) were used to calculate information about the density of the forest and the presence of various tree species [56]. However, because these data contain information about 264 individual tree species, we aggregated the basal area and stand density index of the species data by wood type (hardwood or softwood). Additionally, we were able to calculate the mean and standard deviation of the basal area (BA), stand density index (SDI), quadratic mean diameter (DQ), total frequency (TF), and trees per acre (TPA) for all trees, and generate statistics for the area around the infrastructure as described in the previous paragraph.

Data at the courser resolutions were handled more simply by sampling the data using the centroid of the grid cell. This included data describing the climatological leaf area index generated by Cerrai et al. [9], and a collection of drought indices published by the West Wide Drought Tracker [57]. While drought data was used in outage modeling in the past [12,15], we included more information, including the 1, 3 and 12 month Standardized Precipitation Index (SPI) of the month of the storm, as well as 12 month SPI from 1 to 5 years before the storm occurred. This information was included to capture not only the immediate drought conditions, but also any lingering effects of long-term drought stress on the vegetation.

2.2. Outage Modeling

To generate a robust outage prediction system based on the 131 data features, generated via the processes described in the previous section, additional steps were taken to confirm each variable's importance for the modeling outage, tune the model's hyperparameters, and test the system's performance via cross-validation. All modeling processes were coded in R [58], using a range of support libraries.

Variable importance for modeling was initially confirmed via a Boruta variable selection process. This process involves calculating the variable importance in a random forest model, and comparing each variable's importance against the importance of a randomized variable with the same distribution of values. Over many iterations, this process can confirm the importance of each variable in a dataset in comparison to random noise [59]. This was implemented via the `Boruta` R library [60].

Based on experience and the previous literature [9,10,18], we chose the Bayesian Additive Regression Tree (BART) model for this analysis [61], implemented via the BART R library [62]. While this is a quantile regression algorithm, we simplified outputs to deterministic predictions for each storm by taking the mean of the outputs of the model. The hyperparameters used by the BART algorithm (sparse parameters a and b, shrinkage parameter k, the number of trees, the number of posterior draws, and the number of iterations used to initialize the Monte-Carlo Markov Chains) were tuned for this dataset via differential evolution [63] implemented via the DEoptim library [64]. It was used to find the optimal configuration of the BART algorithm based on the mean root mean square logarithmic error (RMSLE) of a fixed 5-fold cross-validation of the RTMA system dataset. To maintain comparability, these optimized hyperparameter values were consistently applied to all models and experiments in this analysis. RMSLE was chosen because it is less sensitive to extreme errors.

2.3. Analysis

To understand the differences between the hybrid NOAA analysis dataset, the WRF simulation dataset, and the outage prediction models built on them, we evaluated each weather simulation's ability to represent the local weather conditions by comparing its predictions against weather station observations. Then, to understand the different qualities of the two outage models, as well as evaluate the importance of individual and groups of variables in the outage models, we compared the cross-validation results, using traditional and spatial error metrics.

More specifically, to evaluate the two-gridded weather simulations, data were collected from METAR and SPECI reports via the Integrated Surface Data archive maintained by the National Centers for Environmental Information [65]. Any data flagged with quality issues were removed, and all observations reported were averaged for every hour to produce a 24 h time series. Any station or variable with more than two hours of missing data were removed from the analysis. Then, the same summary statistics used to generate the outage model features (mean, minimum, maximum, standard deviation, total, 4 h mean during peak winds) were calculated based on the weather station observations. Any mean or maximum gust values reported as zero by the weather stations were also removed from consideration.

For this analysis, all weather stations in the proximity of the outage prediction service territories were considered, with the exception of Northern New Hampshire. We removed that area from consideration because it is dominated by the White Mountains, and the complex topography would cause biased results. See Figure 1 for the detailed weather station location information used in this analysis. While additional data cleaning steps are common when this process is used for weather model evaluation, we determined that this would not be appropriate because the localized differences between the weather station observations and gridded NWP data are of interest.

The outage model performance was evaluated using leave-one-date-out cross-validation. This validation process simulates the operational predictability of the outages caused by each weather event by iteratively isolating the information of each storm event, and testing the model's ability to predict it. More specifically, for each storm date and time present in the database of storms, we reserved the data from that date and time, trained the outage model on the remaining data, and tested that trained model on the reserved data. This way, we had a comprehensive evaluation of all storms in our database, but prevented any spatial or temporal correlations in the weather data from influencing the model performance. While 372 thunderstorm events were considered in this analysis, because of overlapping times, each outage model was only trained and tested 226 times for this cross-validation. To evaluate the overall cross-validation results, we calculated the median absolute percent error (MdAPE), mean absolute percent error (MAPE), centered root mean squared error (CRMSE), correlation coefficient (R^2), and the Nash–Suttcliffe efficiency (NSE) [66]. For definitions of these error metrics, please see Appendix C.

Because the spatial predictability of thunderstorm outages is also of interest, we also applied the fraction skill score (FSS) to evaluate the spatial skill of the outage models. FSS uses a threshold, or a series of thresholds, to generate binary rasters of predictions and actual values, and compares the two within a series of neighborhoods [67]. A skillful model is able to predict a similar fraction of values above the threshold as the actual in a small area. This metric is becoming a widely accepted method to evaluate the spatial skill of precipitation forecasts, especially in the U.S. [68]. Under ideal conditions, an FSS value greater than 0.5 indicates a "useful" skill, but depending on the conditions of the baseline performance ($FSS_{uniform}$), it is subject to change as defined by the following equation:

$$FSS_{uniform} = 0.5 + FSS_{random}/2 \tag{1}$$

where FSS_{random} is the total of the derived binary raster, divided by the number of cells in the domain [67]. For precipitation, the threshold tends to increase with smaller domains and as the prevalence of precipitation increases [69]. For this analysis, we calculated the FSS for each storm by service territory for a range of scales (3×3 to 21×21 cells), and outage thresholds between upscaled outage predictions and actual outages via the `validation` library [70]. Upscaling the predicted and actual values for the FSS calculation was important because the resolution of our model and the frequency of actual damages is such that the actual values are extremely zero-inflated and very sparse (96.3% zeros, and mean of 0.048 damages per grid cell). The outage model predictions however, tend to be small (median of 0.0292 and 0.0314 for RTMA and WRF systems respectively) and are more evenly distributed. This difference in spatial distribution was minimized by applying boxcar smoothing to a small 3×3 neighborhood on both the actual and predicted outages for each event and territory via the `SpatialVx` library [71]. While this process effectively degrades the precision of the analysis, it generates more continuously distributed values that are more comparable, while not affecting the total number of damage locations for each event.

To measure the variable importance of each outage model, we applied the variable permutation technique described by Fisher et al. [72] via the `DALEX` library in R [73]. This technique is model agnostic and uses a loss function to measure model performance as the input variables are perturbed. This allows for a quantitative understanding of each variable's influence on the model performance. Doing this evaluation via cross-validation would be prohibitively complex and computationally expensive, so to evaluate the variable importance within the outage models, all available data were used to train the models before variable importance was measured. In addition, because there is a significant random component in this analysis, we calculated this variable importance over ten iterations for both outage models, and calculated the confidence intervals. The loss metric used to evaluate variable importance, root mean squared logarithmic error (RMSLE), was chosen because it is robust to the inclusion of zeros and is less sensitive to rare cases of extreme errors, which can be present because of the statistical distribution of actual outages as described above. However, because it is a logarithmic error metric, differences in RMSLE can often appear small, despite being significant.

3. Results

3.1. Weather Analysis

As demonstrated in Figure 2, the NOAA analysis dataset represents almost all weather parameters used in the outage models more accurately than the WRF simulation dataset. Very significant differences are seen between the quality of the precipitation parameters, as well as several wind and gust features. Both systems are able to represent parameters associated with synoptic scale processes, such as temperature, humidity, and surface pressure dynamics, much more accurately than mesoscale and microscale processes, such as wind and precipitation. Some surface pressure parameters appear to be poorly captured, but this is likely due to differences in elevation between the NWP data and weather station data, which are not accounted for in this evaluation. In general, these results are quite consistent with what we would expect from the state of the art of NWP of a deterministic 24 h simulation of thunderstorms. For detailed metrics, see Appendix B.

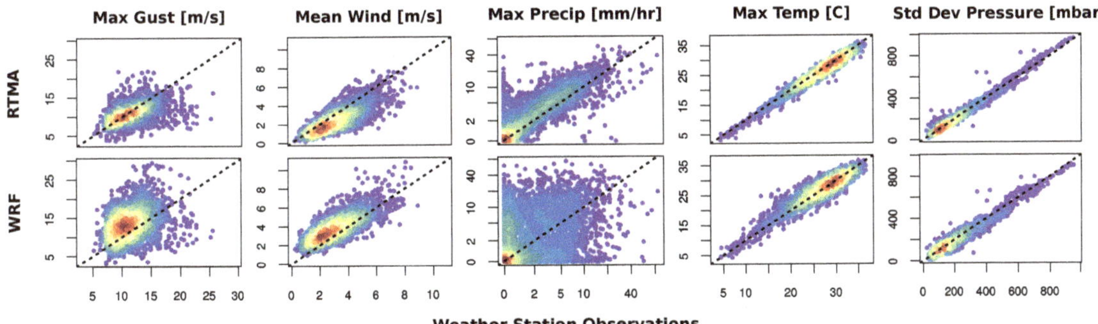

Figure 2. Point-to-point comparison of the NOAA analysis parameters (RTMA, (**Top**)) and the WRF simulation parameters (WRF, (**Bottom**)) versus weather station observations for select variables, describing 24 h thunderstorm events.

3.2. The Outage Models

The RTMA-based outage model performs slightly better than the WRF-based model based on all metrics used in our analysis as seen in Table 3 and Figure 3.

Table 3. Error metrics of the event-level performance of the cross-validation of the outage prediction systems.

	MdAPE	MAPE	CRMSE	R^2	NSE
RTMA	31%	46%	50	0.39	0.37
WRF	35%	50%	51	0.36	0.35

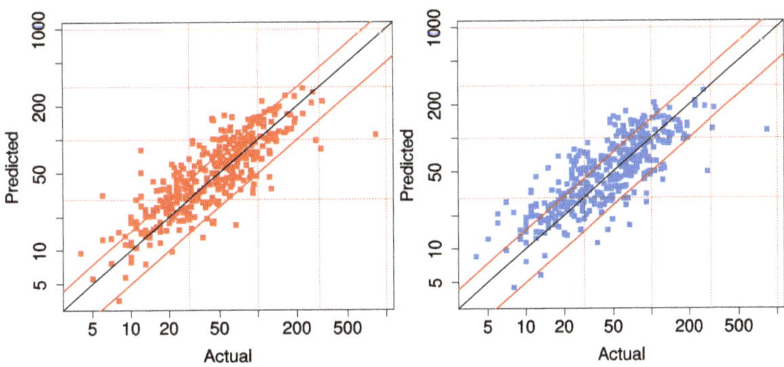

Figure 3. Scatterplots of cross-validation predictions versus actual outages for all thunderstorm events for RTMA- (red, (**left**)) and WRF (blue, (**right**))-based outage prediction systems.

While a direct comparison is not particularly fair because of the differences in the events used in the analysis and the domains of the models, both outage models presented here perform reasonably well in comparison to other outage prediction models of a similar architecture. Wanik et al. [10] describe a warm weather outage model that has a slightly higher MdAPE (35.1 to 38.7%). In Cerrai et al. [9], the best overall outage model has an overall MdAPE of 43%, a MAPE of 59% and an NSE of 0.53. In Yang et al. [17], their conditional outage prediction system designed for severe events has a MdAPE of 38%, MAPE of 46%, and NSE of 0.79. In Watson et al. [18], their best performing rain/wind storm model has a MdAPE of 38%, MAPE of 57%, and an NSE of 43%. The thunderstorm outage models described here have competitive APE metrics, but have a comparatively low NSE, in part because of one under-predicted extreme event.

Overall, the cross-validation results indicate that the models presented here are sensitive to the overall severity of the different thunderstorms. The model has a good dynamic range, especially if one considers that the median daily outages for CT, WMA, EMA, NH, and UI are 35, 6, 20, 22, and 25, respectively. The models shown here demonstrate a dynamic range of around 10 times the typical daily outage level for each service territory, depending on storm severity.

3.2.1. Spatial Skill

As seen in Figure 4, the RTMA-based outage model has slightly better spatial performance than the WRF-based model, but the differences between the outage models are small in comparison to the differences between the events and territories. While many thresholds were evaluated, we show the results for a threshold of 0.111 damage locations, which correspond to having one damage location smoothed out in a 3×3 pixel area (approximately 7.5 km^2).

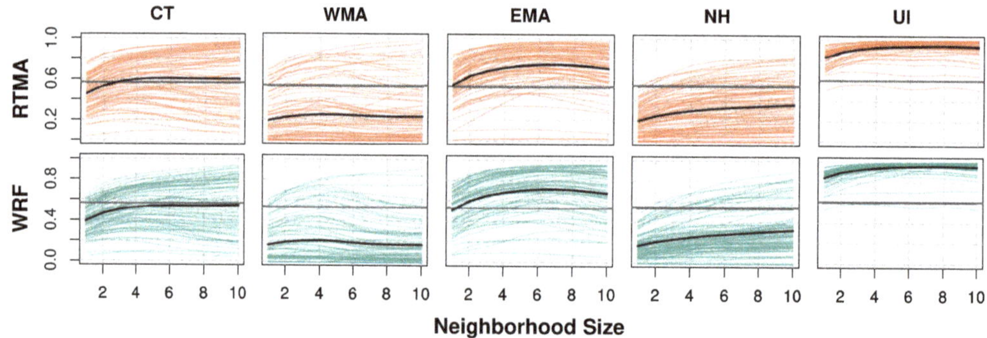

Figure 4. FSS for all events by territory for the RTMA- and WRF-based outage models with a moderate outage risk threshold (0.111 damage locations), plotted for neighborhood sizes 3×3 to 21×21 grid cells. The colored lines are FSS values for each event; the black line indicates the average FSS over all events; and the horizontal dark grey line indicates the average FSS$_{uniform}$.

3.2.2. Outage Model Variable Importance

The grouped variable importance analysis of the outage models in Figure 5 shows that, while infrastructure-related variables are the most important by far, there are differences between the two models as to which weather parameters contribute the most to the models. While the RTMA-based system finds precipitation information to be very useful, the WRF-based system has stronger preference for winds, temperature, and humidity than the RTMA model. The WRF model also appears to fit more on such environmental variables as land cover, vegetation, and elevation, which do not vary storm-by-storm in a given service territory. The results of an individual variable importance analysis is displayed in Appendix A. Although the importance of any one variable to the model is relatively small, given the large number of variables used, and the logarithmic error metric used to measure the dropout loss only makes the differences appear smaller, there are some interesting differences between the two models. Most notably, the maximum precipitation rate is one of the least important variables in the WRF model but is the second most important variable in the RTMA model.

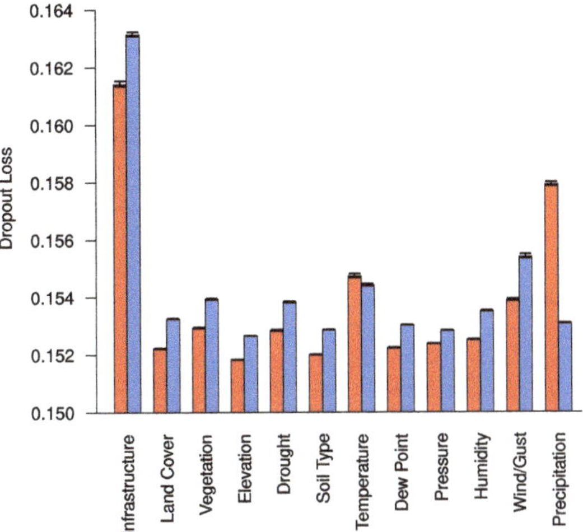

Figure 5. Grouped variable importance as measured by dropout loss (RMSLE) over 10 iterations of permuted groups of variables. The 95% confidence intervals are also shown for both the RTMA-based outage model (red, **(left)**), and the WRF-based outage model (blue, **(right)**).

4. Discussion

Based on these results, several conclusions can be made about the predictability of thunderstorm-related power outages. Firstly, while the NOAA analysis data represent local weather conditions more accurately than the WRF simulation, many weather features used in the outage prediction models have significant errors in both systems. Rather than these errors being simulation or forecasting errors, because of the amount of observational data assimilated into the NOAA analysis system, they are likely due to the representativeness error caused by depicting complex and locally variable phenomena as deterministic and uniform in the 2.5 km × 2.5 km area. This type of error has been documented in the literature for precipitation and winds [74–77], and the errors in the RTMA data for winds and the Stage IV are comparable to the magnitude of representativeness error found in these works.

Secondly, because the NOAA analysis data have higher quality weather data than the WRF simulations, it is unsurprising that the RTMA outage model is more accurate than the WRF-based one. However, what is surprising is how modest the performance differences between these outage models are. Even with the large amount of observational data incorporated into the RTMA and Stage IV analysis products, which have much fewer simulation errors present than the WRF simulations, the outage model is unable to predict thunderstorm-related outages with greater accuracy.

This suggests that the randomness of storm damages is quite significant, and more precise outage predictions may require significantly more precise information. One possibility is that additional factors that are not considered in this study, such as the age of the infrastructure, limit the outage model. However, there are also differences between the two models that suggest other possibilities. As described above, the spatial resolution of the representation of the weather data is a readily apparent source of imprecision in our data. Although all data used in these models, including the environmental and infrastructure information, may suffer from similar representativeness errors, we can see that some weather variables are better represented at 2.5 km × 2.5 km than others. How the precision of the weather data affects the outage models can be understood with a more detailed analysis of the variable importance.

By comparing the R^2 values of the weather feature evaluation and the importance of the weather variables in the outage models, we find that there is a weak but real correlation between the two (0.23 ± 0.07 for RTMA, 0.29 ± 0.07 for WRF). This indicates that the outage models have a preference for precise and accurate weather information. This may be obvious, but this preference also appears regardless of whether or not the weather phenomena directly causes power outages. Both RTMA and WRF outage systems find temperature and humidity to be somewhat important to its predictions, although these variables are not direct causes of outages in thunderstorms. They are more indicators of convective potential and are, thus, indirectly related to power outages, but because of their accurate representation, the machine learning algorithms of the outage models find them useful for understanding the risk of weather-related damage.

At the same time, there is also a distinct preference for variables that have a more direct causative relationship with weather-related outages. This can best be seen in how the RTMA system has a strong preference for precipitation variables. maxPREC is the 2nd most important variable of all for that model, despite it having only a moderate correlation with local conditions (R^2 of 0.5298). It can also be seen in how both models find useful information in wind and gust variables, despite the most precisely predicted variable in that group, avgWIND, only having a moderate correlation with local conditions (R^2 of 0.6346 and 0.5879 for RTMA and WRF, respectively). This is because both wind and precipitation are good indicators of the location and intensity of a convective storm, and more direct indicators of the risk of weather-related damages. Indeed, in the case of the RTMA system, the strong preference for precipitation information comes with a comparatively weaker preference for most other variable groups.

This suggests that if the precision of the precipitation and wind information could be increased further, we can expect corresponding increases in the accuracy of outage prediction models for thunderstorms. Additionally, if we consider how the apparent lack of precision in these data is likely from representativeness error, as described above, future directions for research become apparent.

Lastly, the spatial skill of the outage prediction system appears to vary significantly from storm-to-storm as well as territory-by-territory. It is beyond the scope of this paper to speculate about the storm-to-storm variability in the FSS scores, which may also be a function of the accuracy and precision of the weather simulations, but the distinct differences in spatial predictability of outages in different service territories is suggestive of distinct differences between them. It has been documented for precipitation that the FSS calculations change significantly depending on the size of the domain. However, in the case of outages, this effect is likely only moderate because the average value of $FSS_{uniform}$ does not vary much between territories. The most apparent and potentially impactful difference for outage models between the territories is the densities of the infrastructure. As seen in Figure 5 and Appendix A, infrastructure is a very influential variable for outage modeling, and while all the service territories included in this study contain some urban areas, some are much more consistently urbanized than others. As such, the mean density of overhead lines for each territory varies widely with a minimum of 8.5km per grid cell in WMA, and a maximum 27.5 km per grid cell in UI. If the mean density of the overhead lines and mean FSS as shown in Figure 4 for each territory are compared, we see that the Pearson correlations between the two are 0.927 for RTMA, and 0.946 for WRF: a very strong correlation between the overall spatial predictability of outages and the density of the infrastructure in the region. This is a clear indication of the influence that the infrastructure density has over the spatial predictability of power outages. However, this also may be an indication of over-fitting on the infrastructure features. Infrastructure is by far the most important variable group in this analysis, but in the case of the RTMA outage model, better spatial skill comes with a corresponding lower importance of infrastructure.

5. Conclusions

While the two thunderstorm-related outage models shown here are acceptably skilled at predicting the total number of damages for each storm event, they have difficulty predicting the location of storm impacts. Both the models based on the NOAA analysis dataset and the WRF simulation dataset appear to fit strongly on the amount of infrastructure present in an area and a combination of weather variables that are either directly related to storm damages but imprecisely represented (precipitation, winds), or are more general indicators of convective potential but more precisely represented (temperature, humidity).

Because predictions of the weather conditions and power outages appear to have similar limitations for thunderstorms, there are established analytical methods that could be readily applied to improve the modeling of power outages and other impacts associated with thunderstorms. Just as weather ensembles allow meteorologists to predict the potential intensity of thunderstorms beyond the capabilities of deterministic forecasts, an outage model coupled to a weather ensemble may allow us to predict the potential impacts in a similar way. Because of the high uncertainties, rapidly-refreshing outage models, such as that described in Alpay et al. [33], may be more useful in an operational decision-making context for thunderstorm preparedness.

If one considers how strong convective storms are an increasing threat, globally, there is an implicit call to accelerate investment in global weather prediction and the observation infrastructure. The impact models presented here, even with their limitations, are only possible because of the availability of high-resolution nowcasting products in the United States. While recent developments in global convective-allowing NWP systems are encouraging [78], for this type of impact modeling to be applied in other countries, more work in this space is needed.

Based on our findings, we can expect that as better representations of local weather conditions during thunderstorms are developed both in the United States and globally, outage model accuracy, overall as well as spatially, will improve; the outage models will learn more and more of the phenomena directly linked to weather-related power outages, such as strong winds and extreme precipitation, instead of the synoptic patterns that are correlated to them. To progress along that path, a more granular understanding of the weather conditions that cause damage in convective storms and how they can be represented is needed. Further research involving an analysis or modeling of storm impacts based on microscale numerical weather prediction, large eddy simulations, or even observations from radar or lidar instruments could be very informative about how weather information can be generated in a way that improves our ability to understand and anticipate the impacts of convective storms.

Author Contributions: Conceptualization, P.L.W.; methodology, P.L.W.; software, P.L.W. and M.K.; validation, P.L.W.; formal analysis, P.L.W.; investigation, P.L.W. and E.A.; resources, E.A.; data curation, P.L.W. and M.K.; writing—original draft preparation, P.L.W.; writing—review and editing, E.A. and M.K.; visualization, P.L.W.; supervision, E.A.; project administration, E.A.; funding acquisition, E.A. All authors have read and agreed to the published version of the manuscript.

Funding: This research was funded by Eversource Energy.

Institutional Review Board Statement: Not Applicable.

Informed Consent Statement: Not Applicable.

Data Availability Statement: Restrictions apply to the availability of these data. Data were obtained from Eversource Energy and United Illuminating. They are available from the authors with the permission of Eversource Energy and United Illuminating.

Conflicts of Interest: The authors declare no conflict of interest. The funders had no role in the design of the study; analyses, or interpretation of data; in the writing of the manuscript; or in the decision to publish the results.

Abbreviations

The following abbreviations are used in this manuscript:

NWP	Numerical Weather Prediction
WRF	Weather Research and Forecasting model
RTMA	Real-Time Mesoscale Analysis

RMSLE	Root Mean Squared Logarithmic Error
NSE	Nash–Sutcliffe Efficiency
MAPE	Mean Absolute Percent Error
CRMSE	Centered Root Mean Squared Error
FSS	Fraction Skill Score
CT	Eversource Connecticut
WMA	Eversource Western Massachusetts
EMA	Eversource Eastern Massachusetts
NH	Eversource New Hampshire
UI	AVANGRID United Illuminating
SPI	Standardized Precipitation Index
LAI	Leaf Area Index
DEM	Digital Elevation Model
NLCD	National Land Cover Database
3DEP	3D Elevation Program
GEDI	Global Ecosystem Dynamics Investigation
SSURGO	Soil Survey Geographic Database
ITSP	Individual Tree Species Parameter
WWDT	West Wide Drought Tracker
MODIS	Moderate Resolution Imaging Spectroradiometer
METAR	Meteorological Aerodrome Reports
SPECI	Aviation Selected Special Weather Report
NOAA	National Oceanic and Atmospheric Administration
NCEP	National Centers for Environmental Prediction
USDA	United States Department of Agriculture
USGS	United States Geological Survey
MRLC	Multi-Resolution Land Characteristics

Appendix A. Data Features

Table A1. Description of variables used in outage prediction models. The dropout loss of the top ten variables for each model are in bold. Higher dropout loss indicates greater importance.

Name	Description	Source	Variable Group	RTMA Drp. Loss	WRF Drp. Loss
ohLength	Length of Overhead Line	Utility Company	Infrastructure	**0.153695**	**0.155182**
poleCount	Number of Utility Poles	Utility Company	Infrastructure	**0.152473**	**0.153222**
fuseCount	Number of Fuses	Utility Company	Infrastructure	0.152181	**0.153253**
reclrCount	Number of Reclosers	Utility Company	Infrastructure	**0.152233**	**0.153057**
prec11	Percent NLCD 11—Open Water	NLCD 2016 [51]	Land Cover	0.151933	0.152713
prec21	Percent NLCD 21—Developed, Open	NLCD 2016 [51]	Land Cover	0.152056	0.152876
prec22	Percent NLCD 22—Developed, Low	NLCD 2016 [51]	Land Cover	0.151910	0.152749
prec23	Percent NLCD 23—Developed, Medium	NLCD 2016 [51]	Land Cover	0.152079	**0.152963**
prec24	Percent NLCD 24—Developed, High	NLCD 2016 [51]	Land Cover	0.151989	0.152700
prec31	Percent NLCD 31—Barren	NLCD 2016 [51]	Land Cover	0.151927	0.152714
prec41	Percent NLCD 41—Deciduous Forest	NLCD 2016 [51]	Land Cover	0.151974	0.152783
prec42	Percent NLCD 42—Evergreen Forest	NLCD 2016 [51]	Land Cover	0.151936	0.152731
prec43	Percent NLCD 43—Mixed Forest	NLCD 2016 [51]	Land Cover	0.151861	0.152732
prec52	Percent NLCD 52—Shrub	NLCD 2016 [51]	Land Cover	0.151933	0.152704
prec71	Percent NLCD 71—Grassland	NLCD 2016 [51]	Land Cover	0.151928	0.152699
prec82	Percent NLCD 82—Cultivated Crops	NLCD 2016 [51]	Land Cover	0.151933	0.152715
prec95	Percent NLCD 95—Herbaceous Wetlands	NLCD 2016 [51]	Land Cover	0.151934	0.152713
avgCanopy	Mean Percent Tree Canopy Cover	NLCD Tree Canopy 2016 [52]	Vegetation	**0.152329**	**0.152956**
stdCanopy	Standard Deviation of Canopy Cover	NLCD Tree Canopy 2016 [52]	Vegetation	0.151968	0.152736
avgVegHgt	Mean Vegetation Height	GEDI 2019 [53]	Vegetation	0.152037	0.152906
stdVegHgt	Standard Deviation of Vegetation Height	GEDI 2019 [53]	Vegetation	0.151945	0.152771

Table A1. *Cont.*

Name	Description	Source	Variable Group	RTMA Drp. Loss	WRF Drp. Loss
avgHardBA	Mean Hardwood Basal Area	ITSP [56]	Vegetation	0.151903	0.152714
stdHardBA	Standard Deviation of Hardwood BA	ITSP [56]	Vegetation	0.151939	0.152722
avgHardSDI	Mean Hardwood Stand Density Index	ITSP [56]	Vegetation	0.151963	0.152686
stdHardSDI	Standard Deviation of Hardwood SDI	ITSP [56]	Vegetation	0.151919	0.152698
avgSoftBA	Mean Softwood Basal Area	ITSP [56]	Vegetation	0.151914	0.152702
stdSoftBA	Standard Deviation of Softwood BA	ITSP [56]	Vegetation	0.151881	0.152691
avgSoftSDI	Mean Softwood Stand Density Index	ITSP [56]	Vegetation	0.151927	0.152677
stdSoftSDI	Standard Deviation of Softwood SDI	ITSP [56]	Vegetation	0.151891	0.152684
avgBA	Mean Total Basal Area	ITSP [56]	Vegetation	0.151950	0.152737
stdBA	Standard Deviation of Total Basal Area	ITSP [56]	Vegetation	0.151910	0.152776
avgSDI	Mean Total Stand Density Index	ITSP [56]	Vegetation	0.151951	0.152698
stdSDI	Standard Deviation of Total SDI	ITSP [56]	Vegetation	0.151981	0.152767
avgDQ	Mean Total Quadratic Mean Diameter	ITSP [56]	Vegetation	0.151929	0.152718
stdDQ	Standard Deviation of Total DQ	ITSP [56]	Vegetation	0.151936	0.152760
avgTF	Mean of Total Frequency	ITSP [56]	Vegetation	**0.152247**	**0.153137**
stdTF	Standard Deviation of TF	ITSP [56]	Vegetation	0.151984	0.152783
avgTPA	Mean of Trees per Acre	ITSP [56]	Vegetation	0.151881	0.152653
stdTPA	Standard Deviation of TPA	ITSP [56]	Vegetation	0.151932	0.152690
LAI	Leaf Area Index	MODIS [9,79]	Vegetation	0.152083	0.152848
avgDEM	Mean Elevation	3DEP [54]	Elevation	0.151837	0.152660
stdDEM	Standard Deviation of Elevation	3DEP [54]	Elevation	0.151924	0.152720
elvDiff	Difference of avgDEM and weather elevation	3DEP [54], RTMA [37], WRF [80]	Elevation	0.151931	0.152706
spi1	One Month Standardized Precipitation Index	WWDT [57]	Drought	0.151987	**0.153005**
spi3	Three Month Standardized Precipitation Index	WWDT [57]	Drought	0.152001	0.152830
spi12_0	12 Month SPI, current	WWDT [57]	Drought	0.151998	0.152773
spi12_1	12 Month SPI, 1 year prior	WWDT [57]	Drought	0.152137	0.152853
spi12_2	12 Month SPI, 2 years prior	WWDT [57]	Drought	0.152075	0.152831
spi12_3	12 Month SPI, 3 years prior	WWDT [57]	Drought	0.152155	0.152853
spi12_4	12 Month SPI, 4 years prior	WWDT [57]	Drought	0.152027	0.152809
spi12_5	12 Month SPI, 5 years prior	WWDT [57]	Drought	0.151939	0.152801
hydNo	Percent not hydric soils	SSURGO [55]	Soil Type	0.151954	0.152771
siltTotal	Percent Silt Content	SSURGO [55]	Soil Type	0.151943	0.152747
clayTotal	Percent Clay Content	SSURGO [55]	Soil Type	0.151929	0.152717
rockTotal	Percent of Rock Content	SSURGO [55]	Soil Type	0.151966	0.152738
soilDepth	Depth of Soil	SSURGO [55]	Soil Type	0.151847	0.152661
orgMat	Percent of Organic Material	SSURGO [55]	Soil Type	0.151949	0.152743
soilDens	Soil Density	SSURGO [55]	Soil Type	0.151950	0.152730
kSat	Saturated Hydraulic Conductivity	SSURGO [55]	Soil Type	0.151945	0.152732
satP	Soil Porosity	SSURGO [55]	Soil Type	0.151961	0.152716
avgTMP	Mean Air Temperature	RTMA [37], WRF [80]	Temperature	0.152191	0.152650
stdTMP	Standard Deviation of Air Temp	RTMA [37], WRF [80]	Temperature	0.152156	0.152819
maxTMP	Maximum Air Temperature	RTMA [37], WRF [80]	Temperature	**0.152685**	**0.153235**
minTMP	Minimum Air Temperature	RTMA [37], WRF [80]	Temperature	0.151925	0.152873
sumTMP	Sum of Air Temperatures	RTMA [37], WRF [80]	Temperature	0.152029	0.152741
peakTMP	Mean Temp during peak winds	RTMA [37], WRF [80]	Temperature	0.152020	0.152780
avgDPT	Mean Dew Point Temperature	RTMA [37], WRF [80]	Dew Point	0.151976	0.152767
stdDPT	Standard Deviation of Dew Point	RTMA [37], WRF [80]	Dew Point	0.152013	0.152804
maxDPT	Maximum Dew Point Temperature	RTMA [37], WRF [80]	Dew Point	0.151926	0.152687
minDPT	Minimum Dew Point Temperature	RTMA [37], WRF [80]	Dew Point	0.151941	0.152832
sumDPT	Sum of Dew Point Temperatures	RTMA [37], WRF [80]	Dew Point	0.152012	0.152792
peakDPT	Mean Dew Point during peak winds	RTMA [37], WRF [80]	Dew Point	0.152007	0.152723
avgPRES	Mean Surface Pressure	RTMA [37], WRF [80]	Pressure	0.151914	0.152716
stdPRES	Standard Deviation of Pressure	RTMA [37], WRF [80]	Pressure	**0.152297**	0.152797
maxPRES	Maximum Surface Pressure	RTMA [37], WRF [80]	Pressure	0.151950	0.152735
minPRES	Minimum Surface Pressure	RTMA [37], WRF [80]	Pressure	0.151946	0.152737
sumPRES	Sum of Surface Pressures	RTMA [37], WRF [80]	Pressure	0.151943	0.152706
peakPRES	Mean Pressure during peak winds	RTMA [37], WRF [80]	Pressure	0.151960	0.152694
avgSPFH	Mean Specific Humidity	RTMA [37], WRF [80]	Humidity	0.152062	0.152817
stdSPFH	Standard Deviation of Spec. Humidity	RTMA [37], WRF [80]	Humidity	0.152018	0.152836
maxSPFH	Maximum Specific Humidity	RTMA [37], WRF [80]	Humidity	0.151949	0.152751
minSPFH	Minimum Specific Humidity	RTMA [37], WRF [80]	Humidity	0.152082	0.152905
sumSPFH	Sum of Specific Humidities	RTMA [37], WRF [80]	Humidity	0.152163	0.152752

Table A1. *Cont.*

Name	Description	Source	Variable Group	RTMA Drp. Loss	WRF Drp. Loss
peakSPFH	Mean of Spec. Humidity during peak winds	RTMA [37], WRF [80]	Humidity	0.151984	0.152767
avgWIND	Mean 10m Wind Speed	RTMA [37], WRF [80]	Wind/Gust	0.151961	0.152710
stdWIND	Standard Deviation of 10m Wind Speed	RTMA [37], WRF [80]	Wind/Gust	0.151954	0.152750
maxWIND	Maximum 10m Wind Speed	RTMA [37], WRF [80]	Wind/Gust	0.151977	0.152748
minWIND	Minimum 10m Wind Speed	RTMA [37], WRF [80]	Wind/Gust	0.151997	0.152745
sumWIND	Sum of Wind Speeds	RTMA [37], WRF [80]	Wind/Gust	0.151972	0.152716
peakWIND	Mean wind speed during peak winds	RTMA [37], WRF [80]	Wind/Gust	0.151948	0.152742
avgGUST	Mean Wind Gust Speed	RTMA [37], WRF [80]	Wind/Gust	0.152045	0.152836
stdGUST	Standard Deviation of Wind Gust Speed	RTMA [37], WRF [80]	Wind/Gust	0.151985	0.152769
maxGUST	Maximum Wind Gust Speed	RTMA [37], WRF [80]	Wind/Gust	0.152040	0.152752
minGUST	Minimum Wind Gust Speed	RTMA [37], WRF [80]	Wind/Gust	0.152089	0.152746
sumGUST	Sum of Wind Gusts	RTMA [37], WRF [80]	Wind/Gust	0.151988	0.152746
peakGUST	Mean Wind Gust Speed during peak winds	RTMA [37], WRF [80]	Wind/Gust	0.152039	0.152748
avgLFSH	Mean Leaf Stress	MODIS [9,79], RTMA [37], WRF [80]	Wind/Gust	0.151991	0.152744
stdLFSH	Standard Deviation of Leaf Stress	MODIS [9,79], RTMA [37], WRF [80]	Wind/Gust	0.151961	0.152738
maxLFSH	Maximum Leaf Stress	MODIS [9,79], RTMA [37], WRF [80]	Wind/Gust	0.151980	0.152743
minLFSH	Minimum Leaf Stress	MODIS [9,79], RTMA [37], WRF [80]	Wind/Gust	0.151963	0.152755
sumLFSH	Sum of Leaf Stresses	MODIS [9,79], RTMA [37], WRF [80]	Wind/Gust	0.152024	0.152760
peakLFSH	Mean Leaf Stress during peak winds	MODIS [9,79], RTMA [37], WRF [80]	Wind/Gust	0.151961	0.152826
wgt5	Hours of Winds >5 m/s	RTMA [37], WRF [80]	Wind/Gust	0.151974	0.152793
cowgt5	Continuous Hours of Winds >5 m/s	RTMA [37], WRF [80]	Wind/Gust	0.151952	0.152770
ggt13	Hours of Gusts >13 m/s	RTMA [37], WRF [80]	Wind/Gust	0.151967	**0.152997**
ggt17	Hours of Gusts >17 m/s	RTMA [37], WRF [80]	Wind/Gust	0.151932	0.152729
ggt22	Hours of Gusts >22 m/s	RTMA [37], WRF [80]	Wind/Gust	0.151934	0.152717
coggt13	Continuous Hours of Gusts >13 m/s	RTMA [37], WRF [80]	Wind/Gust	0.151935	0.152804
coggt17	Continuous Hours of Gusts >17 m/s	RTMA [37], WRF [80]	Wind/Gust	0.151940	0.152736
coggt22	Continuous Hours of Gusts >22 m/s	RTMA [37], WRF [80]	Wind/Gust	0.151945	0.152719
typWDIR	Typical (mean) wind direction of all storms	RTMA [37], WRF [80]	Wind/Gust	0.152005	0.152712
medWDIR	Median Wind direction of storm	RTMA [37], WRF [80]	Wind/Gust	0.152002	0.152806
difWDIR	Difference between typWDIR and medWDIR	RTMA [37], WRF [80]	Wind/Gust	0.151966	0.152745
avgPREC	Mean Hourly Precipitation Rate	Stage IV [38], WRF [80]	Precipitation	0.152209	0.152784
stdPREC	Standard Deviation of Precip. Rate	Stage IV [38], WRF [80]	Precipitation	**0.152403**	0.152731
maxPREC	Maximum Hourly Precipitation Rate	Stage IV [38], WRF [80]	Precipitation	**0.152844**	0.152773
sumPREC	Total Precipitation	Stage IV [38], WRF [80]	Precipitation	0.152187	0.152746
peakPREC	Mean Precip. Rate during peak winds	Stage IV [38], WRF [80]	Precipitation	**0.152311**	0.152726

Appendix B. Weather Correlations

Table A2. Correlation between RTMA and WRF weather datasets, and METAR and SPECI observations.

Name	Variable Group	RTMA—METAR R^2	WRF—METAR R^2
avgTMP	Temperature	0.9836	0.9129
stdTMP	Temperature	0.9119	0.6448
maxTMP	Temperature	0.9707	0.8686
minTMP	Temperature	0.9443	0.8592
sumTMP	Temperature	0.9119	0.8459
peakTMP	Temperature	0.7814	0.6480
avgDPT	Dew Point	0.9798	0.9461
stdDPT	Dew Point	0.9092	0.7349
maxDPT	Dew Point	0.9608	0.8966
minDPT	Dew Point	0.9511	0.8897
sumDPT	Dew Point	0.9234	0.8921
peakDPT	Dew Point	0.8348	0.7189
avgPRES	Pressure	0.1700	0.1588
stdPRES	Pressure	0.9766	0.9392
maxPRES	Pressure	0.1498	0.1363
minPRES	Pressure	0.2200	0.2038
sumPRES	Pressure	0.0015	0.0013
peakPRES	Pressure	0.1708	0.1469
avgSPFH	Humidity	0.9735	0.9274
stdSPFH	Humidity	0.8878	0.6932

Table A2. *Cont.*

Name	Variable Group	RTMA—METAR R^2	WRF—METAR R^2
maxSPFH	Humidity	0.9470	0.8648
minSPFH	Humidity	0.9500	0.8799
sumSPFH	Humidity	0.9204	0.8735
peakSPFH	Humidity	0.8219	0.7002
avgWIND	Wind/Gust	0.6346	0.5879
stdWIND	Wind/Gust	0.3217	0.1736
maxWIND	Wind/Gust	0.3327	0.2667
minWIND	Wind/Gust	0.5053	0.3046
sumWIND	Wind/Gust	0.6057	0.5643
peakWIND	Wind/Gust	0.3632	0.3246
avgGUST	Wind/Gust	0.5915	0.5056
stdGUST	Wind/Gust	0.1411	0.0627
maxGUST	Wind/Gust	0.2484	0.1067
minGUST	Wind/Gust	0.0060	0.0091
sumGUST	Wind/Gust	0.5789	0.4957
peakGUST	Wind/Gust	0.1487	0.0625
avgLFSH	Wind/Gust	0.5512	0.5444
stdLFSH	Wind/Gust	0.3583	0.2756
maxLFSH	Wind/Gust	0.2845	0.2249
minLFSH	Wind/Gust	0.4397	0.2735
sumLFSH	Wind/Gust	0.5382	0.5385
peakLFSH	Wind/Gust	0.2939	0.2786
wgt5	Wind/Gust	0.4230	0.4820
cowgt5	Wind/Gust	0.3837	0.3517
ggt13	Wind/Gust	0.4432	0.2149
ggt17	Wind/Gust	0.0352	0.0137
ggt22	Wind/Gust	NA[1]	0.0000
coggt13	Wind/Gust	0.4110	0.1665
coggt17	Wind/Gust	0.0396	0.0105
coggt22	Wind/Gust	NA[1]	0.0000
typWDIR	Wind/Gust	0.0054	0.1378
medWDIR	Wind/Gust	0.3357	0.0304
difWDIR	Wind/Gust	0.2362	0.0219
avgPREC	Precipitation	0.6056	0.0886
stdPREC	Precipitation	0.5589	0.0622
maxPREC	Precipitation	0.5298	0.0538
sumPREC	Precipitation	0.5585	0.0862
peakPREC	Precipitation	0.1989	0.0279

[1] Not enough variance to compute.

Appendix C. Error Metrics

$$\text{MAPE} = \frac{1}{N} \sum_{i=1}^{N} \frac{|P - A|}{A} \times 100 \tag{A1}$$

$$\text{CRMSE} = \sqrt{\frac{1}{N} \sum_{i=1}^{N} [(P_i - \bar{P}) - (A_i - \bar{A})]^2} \tag{A2}$$

$$\text{NSE} = 1 - \frac{\sum_{i=1}^{N} (P_i - A_i)^2}{\sum_{i=1}^{N} (P_i - \bar{A})^2} \tag{A3}$$

$$R^2 = \left(\frac{\sum_{i=1}^{N} (P_i - \bar{P})(A_i - \bar{A})}{\sqrt{\sum_{i=1}^{N} (P_i - \bar{P})^2 \sum_{i=1}^{N} (A_i - \bar{A})^2}} \right)^2 \tag{A4}$$

$$\text{RMSLE} = \sqrt{\frac{1}{N} \sum_{i=1}^{N} (log(P_i + 1) - log(A_i + 1))^2} \tag{A5}$$

References

1. *Economic Benefits of Increasing Electric Grid Resilience to Weather Outages*; Technical Report; Executive Office of the President: Washington, DC, USA, 2013.
2. Lubkeman, D.; Julian, D. Large scale storm outage management. In Proceedings of the IEEE Power Engineering Society General Meeting, Denver, CO, USA, 6–10 June 2004; Volume 2, pp. 16–22. [CrossRef]
3. Hall, K.L. *Out of Sight, Out of Mind*; Technical Report; Edison Electric Institute: Washington, DC, USA, 2012.
4. Mukherjee, S.; Nateghi, R.; Hastak, M. A multi-hazard approach to assess severe weather-induced major power outage risks in the U.S. *Reliab. Eng. Syst. Saf.* **2018**, *175*, 283–305. [CrossRef]
5. Sander, J.; Eichner, J.F.; Faust, E.; Steuer, M. Rising Variability in Thunderstorm-Related U.S. Losses as a Reflection of Changes in Large-Scale Thunderstorm Forcing. *Weather. Clim. Soc.* **2013**, *5*, 317–331. [CrossRef]
6. Diffenbaugh, N.S.; Scherer, M.; Trapp, R.J. Robust increases in severe thunderstorm environments in response to greenhouse forcing. *Proc. Natl. Acad. Sci. USA* **2013**, *110*, 16361–16366. [CrossRef]
7. Scaff, L.; Prein, A.F.; Li, Y.; Liu, C.; Rasmussen, R.; Ikeda, K. Simulating the convective precipitation diurnal cycle in North America's current and future climate. *Clim. Dyn.* **2020**, *55*, 369–382. [CrossRef]
8. Li, Z.; Singhee, A.; Wang, H.; Raman, A.; Siegel, S.; Heng, F.L.; Mueller, R.; Labut, G. Spatio-temporal forecasting of weather-driven damage in a distribution system. In Proceedings of the 2015 IEEE Power & Energy Society General Meeting, Denver, CO, USA, 26–30 July 2015; pp. 1–5. [CrossRef]
9. Cerrai, D.; Wanik, D.W.; Bhuiyan, M.A.E.; Zhang, X.; Yang, J.; Frediani, M.E.B.; Anagnostou, E.N. Predicting Storm Outages Through New Representations of Weather and Vegetation. *IEEE Access* **2019**, *7*, 29639–29654. [CrossRef]
10. Wanik, D.W.; Anagnostou, E.N.; Hartman, B.M.; Frediani, M.E.B.; Astitha, M. Storm outage modeling for an electric distribution network in Northeastern USA. *Nat. Hazards* **2015**, *79*, 1359–1384. [CrossRef]
11. Kankanala, P.; Das, S.; Pahwa, A. AdaBoost^{+}: An Ensemble Learning Approach for Estimating Weather-Related Outages in Distribution Systems. *IEEE Trans. Power Syst.* **2014**, *29*, 359–367. [CrossRef]
12. Han, S.R.; Guikema, S.D.; Quiring, S.M.; Lee, K.H.; Rosowsky, D.; Davidson, R.A. Estimating the spatial distribution of power outages during hurricanes in the Gulf coast region. *Reliab. Eng. Syst. Saf.* **2009**, *94*, 199–210. [CrossRef]
13. Quiring, S.M.; Zhu, L.; Guikema, S.D. Importance of soil and elevation characteristics for modeling hurricane-induced power outages. *Nat. Hazards* **2011**, *58*, 365–390. [CrossRef]
14. Guikema, S.D.; Nateghi, R.; Quiring, S.M.; Staid, A.; Reilly, A.C.; Gao, M. Predicting Hurricane Power Outages to Support Storm Response Planning. *IEEE Access* **2014**, *2*, 1364–1373. [CrossRef]
15. McRoberts, D.B.; Quiring, S.M.; Guikema, S.D. Improving Hurricane Power Outage Prediction Models Through the Inclusion of Local Environmental Factors. *Risk Anal.* **2018**, *38*, 2722–2737. [CrossRef]
16. D'Amico, D.F.; Quiring, S.M.; Maderia, C.M.; McRoberts, D.B. Improving the Hurricane Outage Prediction Model by including tree species. *Clim. Risk Manag.* **2019**, *25*, 100193. [CrossRef]
17. Yang, F.; Watson, P.; Koukoula, M.; Anagnostou, E.N. Enhancing Weather-Related Power Outage Prediction by Event Severity Classification. *IEEE Access* **2020**, *8*, 60029–60042. [CrossRef]
18. Watson, P.L.; Cerrai, D.; Koukoula, M.; Wanik, D.W.; Anagnostou, E. Weather-related power outage model with a growing domain: Structure, performance, and generalisability. *J. Eng.* **2020**, *2020*, 817–826. [CrossRef]
19. Tervo, R.; Láng, I.; Jung, A.; Mäkelä, A. Predicting power outages caused by extratropical storms. *Nat. Hazards Earth Syst. Sci.* **2021**, *21*, 607–627. [CrossRef]
20. Singhee, A.; Wang, H. Probabilistic forecasts of service outage counts from severe weather in a distribution grid. In Proceedings of the 2017 IEEE Power & Energy Society General Meeting, Chicago, IL, USA, 16–20 July 2017; pp. 1–5. [CrossRef]
21. Yue, M.; Toto, T.; Jensen, M.P.; Giangrande, S.E.; Lofaro, R. A Bayesian Approach-Based Outage Prediction in Electric Utility Systems Using Radar Measurement Data. *IEEE Trans. Smart Grid* **2018**, *9*, 6149–6159. [CrossRef]
22. Zhou, Y.; Pahwa, A.; Yang, S.S. Modeling Weather-Related Failures of Overhead Distribution Lines. *IEEE Trans. Power Syst.* **2006**, *21*, 1683–1690. [CrossRef]
23. Kankanala, P.; Pahwa, A.; Das, S. Regression models for outages due to wind and lightning on overhead distribution feeders. In Proceedings of the 2011 IEEE Power and Energy Society General Meeting, San Detroit, MI, USA, 24–28 July 2011; pp. 1–4. [CrossRef]
24. Hohenegger, C.; Schar, C. Atmospheric Predictability at Synoptic Versus Cloud-Resolving Scales. *Bull. Am. Meteorol. Soc.* **2007**, *88*, 1783–1794. [CrossRef]
25. Sun, J.; Xue, M.; Wilson, J.W.; Zawadzki, I.; Ballard, S.P.; Onvlee-Hooimeyer, J.; Joe, P.; Barker, D.M.; Li, P.W.; Golding, B.; et al. Use of NWP for Nowcasting Convective Precipitation: Recent Progress and Challenges. *Bull. Am. Meteorol. Soc.* **2014**, *95*, 409–426. [CrossRef]
26. Yano, J.I.; Ziemiański, M.Z.; Cullen, M.; Termonia, P.; Onvlee, J.; Bengtsson, L.; Carrassi, A.; Davy, R.; Deluca, A.; Gray, S.L.; et al. Scientific Challenges of Convective-Scale Numerical Weather Prediction. *Bull. Am. Meteorol. Soc.* **2018**, *99*, 699–710. [CrossRef]
27. Papadopoulos, A.; Chronis, T.G.; Anagnostou, E.N. Improving Convective Precipitation Forecasting through Assimilation of Regional Lightning Measurements in a Mesoscale Model. *Mon. Weather Rev.* **2005**, *133*, 1961–1977. [CrossRef]
28. Hu, M.; Xue, M. Impact of Configurations of Rapid Intermittent Assimilation of WSR-88D Radar Data for the 8 May 2003 Oklahoma City Tornadic Thunderstorm Case. *Mon. Weather Rev.* **2007**, *135*, 507–525. [CrossRef]

29. Benjamin, S.G.; Weygandt, S.S.; Brown, J.M.; Hu, M.; Alexander, C.R.; Smirnova, T.G.; Olson, J.B.; James, E.P.; Dowell, D.C.; Grell, G.A.; et al. A North American Hourly Assimilation and Model Forecast Cycle: The Rapid Refresh. *Mon. Weather Rev.* **2016**, *144*, 1669–1694. [CrossRef]

30. Clark, A.J.; Gallus, W.A.; Xue, M.; Kong, F. A Comparison of Precipitation Forecast Skill between Small Convection-Allowing and Large Convection-Parameterizing Ensembles. *Weather Forecast.* **2009**, *24*, 1121–1140. [CrossRef]

31. Roberts, B.; Gallo, B.T.; Jirak, I.L.; Clark, A.J. The High Resolution Ensemble Forecast (HREF) system: Applications and Performance for Forecasting Convective Storms. *Meteorology* **2019**. [CrossRef]

32. Bouttier, F.; Marchal, H. Probabilistic thunderstorm forecasting by blending multiple ensembles. *Tellus A Dyn. Meteorol. Oceanogr.* **2020**, *72*, 1–19. [CrossRef]

33. Alpay, B.A.; Wanik, D.; Watson, P.; Cerrai, D.; Liang, G.; Anagnostou, E. Dynamic Modeling of Power Outages Caused by Thunderstorms. *Forecasting* **2020**, *2*, 151–162. [CrossRef]

34. Sheild, S.A.; Quiring, S.M.; McRoberts, D.B. Development of a Thunderstorm Outage Prediction Model. Ph.D. Thesis, The Ohio State University, Columbus, OH, USA, 2018.

35. Kabir, E.; Guikema, S.D.; Quiring, S.M. Predicting Thunderstorm-Induced Power Outages to Support Utility Restoration. *IEEE Trans. Power Syst.* **2019**, *34*, 4370–4381. [CrossRef]

36. Babbage, C. *Passages from the Life of a Philosopher*; Longman, Green, Longman, Roberts, & Green: London, UK, 1864.

37. De Pondeca, M.S.F.V.; Manikin, G.S.; DiMego, G.; Benjamin, S.G.; Parrish, D.F.; Purser, R.J.; Wu, W.S.; Horel, J.D.; Myrick, D.T.; Lin, Y.; et al. The Real-Time Mesoscale Analysis at NOAA's National Centers for Environmental Prediction: Current Status and Development. *Weather Forecast.* **2011**, *26*, 593–612. [CrossRef]

38. Nelson, B.R.; Prat, O.P.; Seo, D.J.; Habib, E. Assessment and Implications of NCEP Stage IV Quantitative Precipitation Estimates for Product Intercomparisons. *Weather Forecast.* **2016**, *31*, 371–394. [CrossRef]

39. NOAA/NWS. *RTMA: Real-Time Mesoscale Analysis Data*; NOAA/NWS: Washington, DC, USA, 2015

40. Hudlow, M.D. Technological Developments in Real-Time Operational Hydrologic Forecasting in the United States. *J. Hydrol.* **1988**, *102*, 69–92. [CrossRef]

41. Environmental Modeling Center; National Centers for Environmental Prediction; National Weather Service; NOAA, U.S. Department of Commerce. *NCEP North American Mesoscale (NAM) 12 km Analysis*; U.S. Department of Commerce: Washington, DC, USA, 2015.

42. Morrison, H.; Thompson, G.; Tatarskii, V. Impact of Cloud Microphysics on the Development of Trailing Stratiform Precipitation in a Simulated Squall Line: Comparison of One- and Two-Moment Schemes. *Mon. Weather Rev.* **2009**, *137*, 991–1007. [CrossRef]

43. Mlawer, E.J.; Taubman, S.J.; Brown, P.D.; Iacono, M.J.; Clough, S.A. Radiative transfer for inhomogeneous atmospheres: RRTM, a validated correlated-k model for the longwave. *J. Geophys. Res. Atmos.* **1997**, *102*, 16663–16682. [CrossRef]

44. Chou, M.; Suarez, M. An Efficient Thermal Infrared Radiation Parameterization for Use in General Circulations Models. *NASA Tech. Memo.* **1994**, *3*, 1–85.

45. Jiménez, P.A.; Dudhia, J.; González-Rouco, J.F.; Navarro, J.; Montávez, J.P.; García-Bustamante, E. A Revised Scheme for the WRF Surface Layer Formulation. *Mon. Weather Rev.* **2012**, *140*, 898–918. [CrossRef]

46. Tewari, M.; Chen, F.; Wang, W.; Dudhia, J.; LeMone, M.; Mitchell, K.; Ek, M.; Gayno, G.; Wegiel, J.; Cuenca, R. Implementation and verification of the unified NOAH land surface model in the WRF model. In Proceedings of the 20th Conference on Weather Analysis and Forecasting/16th Conference on Numerical Weather Prediction, Seattle, WA, USA, 11–15 January 2004; Volume 1115, pp. 2165–2170.

47. Hong, S.; Noh, Y.; Dudhia, J. A New Vertical Diffusion Package with an Explicit Treatment of Entrainment Processes. *Mon. Weather Rev.* **2006**, *134*, 2318–2341. [CrossRef]

48. Agostinelli, C.; Lund, U. *R Package Circular: Circular Statistics (Version 0.4-93)*; Department of Environmental Sciences, Informatics and Statistics, Ca' Foscari University: Venice, Italy; UL: Department of Statistics, California Polytechnic State University: San Luis Obispo, CA, USA, 2017. Available online: https://cran.r-project.org/web/packages/circular/circular.pdf (accessed on 2 February 2021).

49. Bivand, R.; Keitt, T.; Rowlingson, B. rgdal: Bindings for the 'Geospatial' Data Abstraction Library. R Package Version 1.5-23. 2021. Available online: https://cran.r-project.org/web/packages/rgdal/index.html (accessed on 2 February 2021).

50. Bivand, R.; Rundel, C. rgeos: Interface to Geometry Engine—Open Source ('GEOS'). R Package Version 0.5-5. 2020. Available online: https://cran.r-project.org/web/packages/rgeos/index.html (accessed on 1 June 2020).

51. Jin, S.; Homer, C.; Yang, L.; Danielson, P.; Dewitz, J.; Li, C.; Zhu, Z.; Xian, G.; Howard, D. Overall Methodology Design for the United States National Land Cover Database 2016 Products. *Remote Sens.* **2019**, *11*, 2971. [CrossRef]

52. Coulston, J.W.; Moisen, G.G.; Wilson, B.T.; Finco, M.V.; Cohen, W.B.; Brewer, C.K. Modeling Percent Tree Canopy Cover: A Pilot Study. *Photogramm. Eng. Remote Sens.* **2012**, *78*, 715–727. [CrossRef]

53. Potapov, P.; Li, X.; Hernandez-Serna, A.; Tyukavina, A.; Hansen, M.C.; Kommareddy, A.; Pickens, A.; Turubanova, S.; Tang, H.; Silva, C.E.; et al. Mapping global forest canopy height through integration of GEDI and Landsat data. *Remote Sens. Environ.* **2021**, *253*, 112165. [CrossRef]

54. Gesch, D.; Evans, G.; Oimoen, M.; Arundel, S. *The National Elevation Dataset*; American Society for Photogrammetry and Remote Sensing: Bethseda, MD, USA, 2018; pp. 83–110.

55. Soil Survey Staff, Natural Resources Conservation Service. Soil Survey Geographic (SSURGO) Database. 2010. Available online: https://websoilsurvey.nrcs.usda.gov/ (accessed on 21 August 2020).
56. Individual Tree Species Parameter Maps. 2015. Available online: https://www.fs.fed.us/foresthealth/applied-sciences/mapping-reporting/indiv-tree-parameter-maps.shtml (accessed on 19 March 2021).
57. Abatzoglou, J.T.; McEvoy, D.J.; Redmond, K.T. The West Wide Drought Tracker: Drought Monitoring at Fine Spatial Scales. *Bull. Am. Meteorol. Soc.* **2017**, *98*, 1815–1820. [CrossRef]
58. R Core Team. *R: A Language and Environment for Statistical Computing*; R Foundation for Statistical Computing: Vienna, Austria, 2021.
59. Kursa, M.B.; Jankowski, A.; Rudnicki, W.R. Boruta—A System for Feature Selection. *Fundam. Inform.* **2010**, *101*, 271–285. [CrossRef]
60. Kursa, M.B.; Rudnicki, W.R. Feature Selection with the Boruta Package. *J. Stat. Softw.* **2010**, *36*, 1–13. [CrossRef]
61. Chipman, H.A.; George, E.I.; McCulloch, R.E. BART: Bayesian additive regression trees. *Ann. Appl. Stat.* **2010**, *4*, 266–298. [CrossRef]
62. Sparapani, R.; Spanbauer, C.; McCulloch, R. Nonparametric Machine Learning and Efficient Computation with Bayesian Additive Regression Trees: The BART R Package. *J. Stat. Softw.* **2021**, *97*, 1–66. [CrossRef]
63. Ardia, D.; Boudt, K.; Carl, P.; Mullen, K.M.; Peterson, B.G. Differential Evolution with DEoptim: An Application to Non-Convex Portfolio Optimization. *R J.* **2011**, *3*, 27–34. [CrossRef]
64. Mullen, K.; Ardia, D.; Gil, D.; Windover, D.; Cline, J. DEoptim: An R Package for Global Optimization by Differential Evolution. *J. Stat. Softw.* **2011**, *40*, 1–26. [CrossRef]
65. National Centers for Environmental Information. Integrated Surface Data (ISD) Archive. Available online: https://www.ncei.noaa.gov/data/global-hourly/access/ (accessed on 3 March 2021).
66. Nash, J.; Sutcliffe, J. River flow forecasting through conceptual models part I—A discussion of principles. *J. Hydrol.* **1970**, *10*, 282–290. [CrossRef]
67. Roberts, N.M.; Lean, H.W. Scale-Selective Verification of Rainfall Accumulations from High-Resolution Forecasts of Convective Events. *Mon. Weather Rev.* **2008**, *136*, 78–97. [CrossRef]
68. Gilleland, E.; Ahijevych, D.A.; Brown, B.G.; Ebert, E.E. Verifying Forecasts Spatially. *Bull. Am. Meteorol. Soc.* **2010**, *91*, 1365–1376. [CrossRef]
69. Mittermaier, M.; Roberts, N. Intercomparison of Spatial Forecast Verification Methods: Identifying Skillful Spatial Scales Using the Fractions Skill Score. *Weather Forecast.* **2010**, *25*, 343–354. [CrossRef]
70. Laboratory, N.R.A. Verification: Weather Forecast Verification Utilities. R Package Version 1.42. 2015. Available online: https://CRAN.R-project.org/package=verification (accessed on 3 March 2021).
71. Gilleland, E. SpatialVx: Spatial Forecast Verification. R Package Version 0.8. 2021. Available online: https://CRAN.R-project.org/package=SpatialVx (accessed on 3 March 2021).
72. Fisher, A.; Rudin, C.; Dominici, F. All Models are Wrong, but Many are Useful: Learning a Variable's Importance by Studying an Entire Class of Prediction Models Simultaneously. *J. Mach. Learn. Res.* **2019**, *20*, 1–81.
73. Biecek, P. DALEX: Explainers for Complex Predictive Models in R. *J. Mach. Learn. Res.* **2018**, *19*, 1–5.
74. Lee, T.R.; Buban, M.; Turner, D.D.; Meyers, T.P.; Baker, C.B. Evaluation of the High-Resolution Rapid Refresh (HRRR) Model Using Near-Surface Meteorological and Flux Observations from Northern Alabama. *Weather Forecast.* **2019**, *34*, 635–663. [CrossRef]
75. Pichugina, Y.L.; Banta, R.M.; Bonin, T.; Brewer, W.A.; Choukulkar, A.; McCarty, B.J.; Baidar, S.; Draxl, C.; Fernando, H.J.S.; Kenyon, J.; et al. Spatial Variability of Winds and HRRR–NCEP Model Error Statistics at Three Doppler-Lidar Sites in the Wind-Energy Generation Region of the Columbia River Basin. *J. Appl. Meteorol. Climatol.* **2019**, *58*, 1633–1656. [CrossRef]
76. Shucksmith, P.E.; Sutherland-Stacey, L.; Austin, G.L. The spatial and temporal sampling errors inherent in low resolution radar estimates of rainfall: Spatial and temporal sampling errors in low resolution radar estimates of rainfall. *Meteorol. Appl.* **2011**, *18*, 354–360. [CrossRef]
77. Moreau, E.; Testud, J.; Le Bouar, E. Rainfall spatial variability observed by X-band weather radar and its implication for the accuracy of rainfall estimates. *Adv. Water Resour.* **2009**, *32*, 1011–1019. [CrossRef]
78. Zhou, L.; Lin, S.J.; Chen, J.H.; Harris, L.M.; Chen, X.; Rees, S.L. Toward Convective-Scale Prediction within the Next Generation Global Prediction System. *Bull. Am. Meteorol. Soc.* **2019**, *100*, 1225–1243. [CrossRef]
79. Leaf Area Index (1 Month—Terra/MODIS). 2017. Available online: https://modis.gsfc.nasa.gov/data/dataprod/mod15.php (accessed on 17 December 2020).
80. Community, WRF. *Weather Research and Forecasting (WRF) Model*; UCAR/NCAR: Boulder, CO, USA, 2000. [CrossRef]

Article

Tobacco Endgame Simulation Modelling: Assessing the Impact of Policy Changes on Smoking Prevalence in 2035

Michael Chaiton [1,2,*], **Jolene Dubray** [1], **G. Emmanuel Guindon** [3] **and Robert Schwartz** [1,2]

1   Dalla Lana School of Public Health, University of Toronto, Toronto, ON M5T 3M7, Canada; jolene.dubray@utoronto.ca (J.D.); robert.schwartz@utoronto.ca (R.S.)
2   Centre for Addiction and Mental Health, Toronto, ON M5S 2S1, Canada
3   Department of Health Research Methods, Evidence and Impact, McMaster University, Hamilton, ON L8S 4K1, Canada; emmanuel.guindon@mcmaster.ca
*   Correspondence: michael.chaiton@camh.ca; Tel.: +1-416-978-7096

Abstract: Smoking causes substantial amount of mortality and morbidity. This article presents the findings from simulation models that projected the impact of five potential Tobacco Endgame strategies on smoking prevalence in Ontario by 2035 and expected impact of smoking prevalence "less than 5 by 35" on tax revenue. We used Ontario SimSmoke simulation for modelling the expected impact of four strategies: plain packaging, free cessation services, decreasing the number of tobacco outlets, and increasing tobacco taxes. Separate models were used to project the impact of increasing the minimum age to legally purchase tobacco to 21 years on smoking prevalence and impact of price and tax increase to achieve "less than 5 by 35" on taxation revenue. The combined effect of four strategies in Ontario SimSmoke Model are expected to reduce smoking prevalence by 8.5% in 2035. Increasing tobacco taxes had the greatest independent predicted decrease in smoking prevalence (2.8%) followed by raised minimum age for legal purchase to 21 years (2.4%), decreasing tobacco outlets (1.5%), free cessation services (0.7%), and plain packaging (0.6%). Increasing tobacco excise tax and prices are projected to have minimal impact on taxation revenue, with a decrease from 1.5 billion to 1.2 billion annual tax receipts.

Keywords: tobacco endgame; policy; simulation model; tobacco tax revenue

Citation: Chaiton, M.; Dubray, J.; Guindon, G.E.; Schwartz, R. Tobacco Endgame Simulation Modelling: Assessing the Impact of Policy Changes on Smoking Prevalence in 2035. *Forecasting* **2021**, *3*, 267–275. https://doi.org/10.3390/forecast3020017

Academic Editor: Konstantinos Nikolopoulos

Received: 17 March 2021
Accepted: 8 April 2021
Published: 13 April 2021

Publisher's Note: MDPI stays neutral with regard to jurisdictional claims in published maps and institutional affiliations.

1. Introduction

Great strides have been made in tobacco control in Canada and globally over the past few decades through implementation of various measures, including those endorsed by the international Framework Convention for Tobacco Control [FCTC] [1]. Nevertheless, smoking prevalence remains substantial: 18.1% of Canadians over 12 years of age, representing 5.4 million Canadians, were current smokers in the year 2014 [2]. The overall burden of smoking related illness and death from cancer and from respiratory and cardiovascular diseases continues to be devastating. In 2002, 37,000 Canadians died from tobacco associated illnesses–the size of a small town being wiped off the map each year [3]. Canadians lose an estimated 515,607 person years of life every year as a result of premature mortality from tobacco smoking [3]. The idea of a "Tobacco Endgame" is based on the perspective that "control" of tobacco will never be enough to deal with the epidemic of tobacco related diseases and that the focus must be shifted to develop strategies to reach a future that is free of commercial tobacco. This notion of "endgame" is qualitatively different from tobacco control strategies currently in place. This recognition is becoming more widespread and is increasingly leading to the view that a strategy for an "endgame" for commercial tobacco is required.

In October 2016, a Tobacco Endgame for Canada Summit was convened with over 80 experts, researchers, government officials, advocates, and health professionals in attendance to discuss possible strategies to the target goal "less than 5 by 35"; that is, to achieve less than 5% smoking prevalence by 2035. In this report, we describe the findings from simulation models that assessed the impact in Ontario of five potential Tobacco Endgame strategies [4]. They include:

1. Plain packaging for all tobacco products.
2. Free cessation services for all (both pharmaceutical and behavioural therapy).
3. Decreasing the number of outlets selling tobacco products.
4. Increasing tobacco taxes.
5. Increasing the minimum age to legally purchase tobacco to 21 years.

In addition, we also modeled the impact of tax and price increase to achieve "less than 5 by 35" on government taxation revenue. Cigarette taxes bring in significant revenue to governments at the national and provincial level. Apart from sales taxes, in 2014–2015 Canadian Federal and Provincial governments received $8.2 billion from the sale of tobacco [5]. There is concern expressed by those opposed to tobacco elimination that reducing the number of smokers would decrease government revenue and that this would be of such a magnitude that it could not happen. However, there is overwhelming Canadian and international evidence that increases in tobacco taxes can reduce tobacco use and increase government tax revenue [6–13]. At current taxation and tobacco use rates, taxes on tobacco products have the dual effect of decreasing the demand for tobacco and increasing government revenue. In fiscal year 2014–2015, the federal government collected more than $3 billion in cigarette taxes [14]. In Ontario and Québec, Canada's largest provinces, the provincial governments collected more than $1 billion each.

If Canada achieves "less than 5 by 35" through non-tax interventions, total taxes collected on the sale of tobacco products would dwindle substantially. Given that in 2014, 18.1% of Canadians aged 12 and older smoked either daily or occasionally [2], it could be expected that annual tobacco tax receipts decrease by as much as 75% from 2035. Moreover, during the period of transitioning from 18% to 5% smoking prevalence, the cumulative amount of tax losses year over year would be far from negligible. Achieving "less than 5 by 35", however, need not be achieved solely on the back of non-tax interventions. In the case, albeit extreme, that "less than 5 by 35" is achieved solely through tax and price increases, the cumulative tax revenue gains during the transition period could be considerable. Irrespective of the substantial cost savings gained from reductions in health care spending and reductions in indirect costs to society detailed above, there might be minimal changes in government revenue during the period of transition to "less than 5", if increased tax rates are a component of an endgame strategy.

The purpose of this paper is to evaluate the expected impact of endgame policies and understand the expected tax revenue impact of reducing smoking prevalence to less than 5%.

2. Materials and Methods

2.1. Ontario SimSmoke Model

Four of the Tobacco Endgame strategies were modelled using the Ontario SimSmoke simulation model. The Ontario SimSmoke model is adapted from the SimSmoke simulation model of tobacco control policies, previously developed for the U.S. and other countries [15–17]. The model uses population, smoking rates, and tobacco control policy data for Ontario. It assesses, individually and in combination, the effect of seven types of tobacco control policies (taxes, clean air, mass media, advertising bans, warning labels, cessation treatment, and youth access policies) on smoking prevalence and associated future premature mortality [18]. Each policy parameter in the model is accorded an effect size developed for the SIMSMOKE model based on literature reviews and expert panel. These existing parameters were then either maximized to represent full implementation of the intervention or the parameter effect sizes themselves were adapted according the new assumptions. Modifications were made to the Ontario SimSmoke policy levels or policy effect sizes to assess the impact of each Tobacco Endgame strategy on smoking prevalence in Ontario between 2019 and 2035. The following represent the changes in the SIMSMOKE model to represent the effect of the endgame scenarios.

To simulate the impact of plain packaging, the comprehensive marketing ban (both direct and indirect) policy level in Ontario SimSmoke was increased to 90% (up from 25%) as a proxy measure for plain packaging in which the package itself was assumed to be the primary method of direct consumer marketing in Ontario..

Free cessation services were modeled adapting two parameters in Ontario SimSmoke. The first parameter incorporated free cessation services (pharmacotherapy and behavioural therapy) in all primary care and hospital settings,. The second parameter expanded the number of settings offering free cessation to also include offices of health professionals, community, and 'other.' Free cessation services are currently limited in Ontario.

Analyses conducted by Chaiton, Mecredy, and Cohen [19] identified an increased risk of relapse among smokers who resided within 500 m from a tobacco outlet (Hazard ratio: 1.41) compared to those who lived further away. As a proxy measure for decreasing the number of outlets selling tobacco products, the policy effect sizes in Ontario SimSmoke for the five cessation treatment policies (treatment availability, treatment access, quitlines, quitlines with treatment access, and brief interventions) were increased by a value of 1.41.

Price elasticities were doubled in the Ontario SimSmoke model to assess the impact of increased tobacco taxes on smoking prevalence. Specifically, the policy effects were increased to −0.6 for youth less than 18 years (60% reduction in smoking), −0.4 for young adults aged 18 to 24 (40% reduction in smoking), −0.3 for adults aged 25 to 34 years (30% reduction in smoking) and −0.2 for adults aged 35 years or more (20% reduction in smoking).

2.2. Ontario Population Model

Our final endgame model, increasing the minimum age of legal purchase to 21 years and tax revenue, was modelled separately from the SIMSMOKE model. In this model, we simulated the impact of minimum age laws by using a population program in which the baseline status quo rate of change in smoking prevalence was estimated to be 1.1% per year. We adjusted our model for effects in age group less than 19 and eliminated the effect of cessation in our model. This model was also used to evaluate the effect of taxation using a separate model that simulates the impact of tax and price increases required to achieve "less than 5 by 35".

Based on the analyses conducted by Callaghan et al. [20], it was assumed that the rate of onset for new smokers aged 20–22 would be 2.7 percentage points lower than it would have been under the standard projection for each year if the minimum age ban took effect immediately. No changes in prevalence were modelled for older ages at the time on the implied onset of the law; however, the effect was carried through as the cohort aged. Additionally, it was assumed that the increased age of onset would be associated with increased cessation in this cohort (natural rate of decrease adjusted from 0.011 to 0.022). No adjustment was made for any effects in youth younger than 19 who might be affected by reduced access to tobacco. No adjustment was made for any additional social normative effects.

This model obtained smoking prevalence from 2014 Canadian Community Health Survey (CCHS) [2]. We used Statistics Canada medium growth population projection scenario (M1: medium-growth, 1991/1992 to 2010/2011 trend, CANSIM Table 052-0005) [21]. The number of people aged 20–22 was obtained from the Ontario Ministry of Finance for years 2018–2035 [22]. Smoking prevalence and daily number of cigarettes consumed per smoker, by age: We used the most recent cycle (2014) of a large national survey, the CCHS, and obtained point estimates for smoking prevalence and intensity. Excise tax rate and revenue: We obtained current tobacco excise tax rates and more recent estimates of tobacco excise tobacco tax revenue from provincial Ministries of Finance. Total cigarette tax paid sales: As a measure of tax-paid sales, we used cigarette wholesale data as reported by tobacco manufacturers to Health Canada. Underlying trend: Smoking prevalence in Canada has steadily decreased since the mid-1960s. In 1965 about half of all Canadians aged 15 and above smoked. By the early 2010s, only about 20% did [23]. This steady decline was due to many factors such as information on the harmful effects of active smoking and secondhand smoke, tobacco control policies such as smoke free policies, advertising bans and taxation, and changes in anti-smoking sentiment. Although it is difficult to disentangle the effects of each of these factors, it seems reasonable to assume that the downward trend in smoking prevalence observed between the early 2000s and the present would not abruptly end in the near future. In the last decade for which data are available, smoking prevalence, on average, declined annually by about 2% to 3% depending on the province. We assumed an underlying trend of 2.5% in annual decrease in both smoking prevalence and daily number of cigarettes consumed per smoker.

2.3. Tax Revenue

This model that simulates the impact of tax and price increases required to achieve "less than 5 by 35" by examining the impact on taxation revenues under three different scenarios: (1) excise taxes are increased only to keep up with inflation; (2) "less than 5 by 35" is achieved solely through excise tax increases; and (3) "less than 5 by 35" is achieved through non tax intervention and excise tax increases that raise prices by 5% in real terms annually. We used accepted parameters of elasticity for changes in tobacco prices for adults (−0.4) and twice that for youth [13]. The model accounts for population growth, inflation, and tax evasion. We used data for the province of Ontario to simulate the impact of tax and price increases required to achieve "less than 5 by 35" on tax revenue. At the current tax rates, it is expected that Ontario will collect about $1.5 billion in 2016. All monetary figures below are in constant 2016 dollars. To estimate the changes on tax revenue, we made the following baseline model parameters and assumptions.

Own-price elasticity: There is overwhelming evidence that individuals respond to changes in tobacco prices. In high-income countries such as Canada and the United States, it is generally accepted that a 10% increase in prices would reduce total consumption by about 4%; and that half of the reduction comes from a reduction in the number of smokers and half from a reduction in consumption among continuing smokers [13]. It is also generally accepted that youth respond more to changes in prices—about twice as much as older adults [13]. Consequently, as a baseline assumption for own-price elasticity for cigarettes, we used −0.4 for adults (20 years of age and above) (−0.2 for own-price prevalence elasticity and −0.2 for own-price consumption elasticity), and twice that for youth (12 to 19 years of age).

Pass-through rate: Tax changes do not necessarily lead to price changes as manufacturers are rarely required to pass on the full extent of tax increases to consumers. Manufacturers often under- or over-shift tax changes. In mature cigarette markets such as Canada, manufacturers typically over-shift tax increases [24]. As a baseline assumption, we assumed that tobacco manufacturers over-shift tax increases by 10%.

Prices: In order to estimate the effect of tax changes on smoking, it is necessary to first estimate the effect of tax changes on current prices. We used $0.40 per cigarette stick.

Expected inflation: As a measure of expected inflation, we used 2% annual increases to reflect the Bank of Canada's 2% inflation-control target [25].

Cigarette tax evasion: Although cigarette tax evasion has many causes, high taxes undeniably create an incentive for tobacco users and manufacturers to elaborate ways to evade tobacco taxes. While the illegal nature of cigarette tax evasion makes it intrinsically difficult to measure accurately, cigarette tax evasion in some Canadian regions such as southern Ontario is not negligible [26]. Our model allows for a portion of the effect of tax and price increases on tobacco use and consumption to be directed towards contraband cigarettes.

3. Results

3.1. Smoking Prevalence Modelling

Results from the Ontario SimSmoke simulation model indicate that each of the Tobacco Endgame strategies predicts a greater reduction in smoking prevalence by 2035 compared to the status quo scenario (Table 1 and Figure 1).

Table 1. SimSmoke Model Predicted Smoking Prevalence, for Both Sexes, Ages 15–85, With and Without Tobacco Endgame Policies, Ontario, 2018–2035.

Policy	2018	2019	2020	2025	2030	2035
Status Quo Policies [a]	15.5%	15.4%	15.2%	14.4%	13.6%	12.9%
Independent Policy Effects						
Plain packaging	15.5%	14.8%	14.7%	13.8%	13.0%	12.3%
Free cessation services in primary care and hospitals	15.5%	14.8%	14.7%	13.8%	13.0%	12.2%
Free cessation services everywhere	15.5%	14.7%	14.5%	13.6%	12.8%	12.1%
Decreased tobacco availability	15.5%	13.7%	13.5%	12.7%	12.0%	11.4%
Increased taxation	15.5%	12.7%	12.5%	11.7%	10.8%	10.1%
Combined Policy Effects						
All above	15.5%	10.9%	10.7%	9.9%	9.1%	8.5%

[a] Status quo represents the policy levels prior to the first projection year (2019). Source: Ontario SimSmoke.

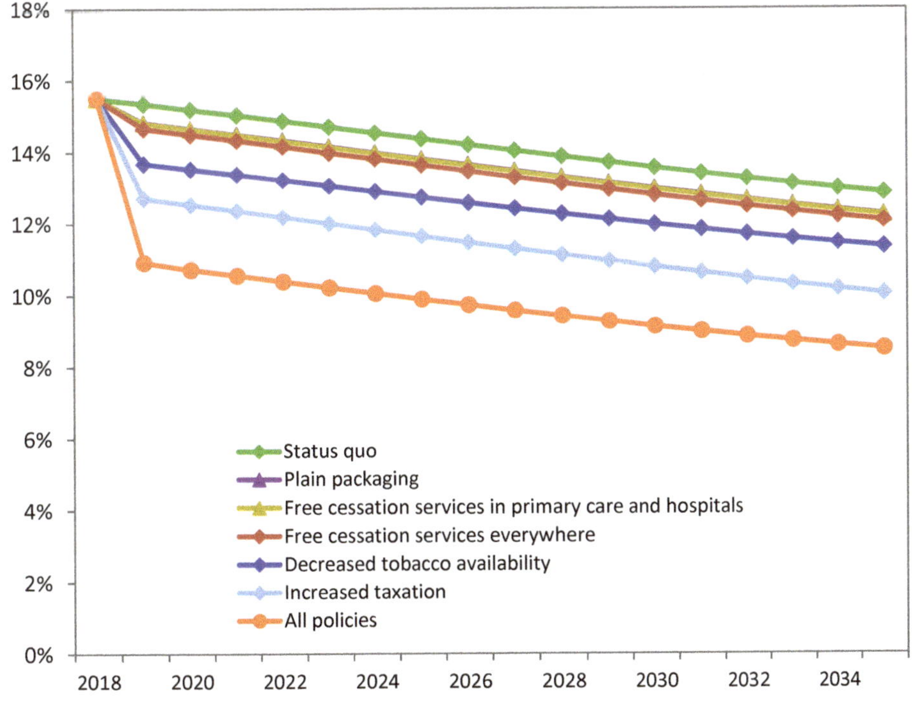

Figure 1. SimSmoke Model Predicted Smoking Prevalence, for Both Sexes, Ages 15–85, With and Without Tobacco Endgame Policies, Ontario, 2018–2035. Status quo represents the policy levels prior to the first projection year (2019). Note: Full data table for this graph provided in the Appendix A (Table A1) Source: Ontario SimSmoke.

Increased taxation had the greatest independent impact on smoking prevalence. By 2035, smoking prevalence is projected to reach 10.1% with increased tobacco taxes, while the status quo prevalence is projected to be 12.9% in 2035 (a 2.8 percentage point reduction).

Decreased tobacco availability is projected to reduce smoking prevalence by 1.5 percentage points in 2035, from 12.9% with the status quo scenario to 11.4% with fewer tobacco outlets.

Offering free cessation services in primary care and hospital settings (i.e., Ottawa Model of Smoking Cessation model) is projected to reduce smoking prevalence to 12.2% in 2035, while free cessation services offered in primary care, hospitals, offices of health professionals, community and 'other' settings is projected to further reduce smoking prevalence to 12.1% in 2035. Both cessation policy models project lower smoking prevalence in 2035 compared to the status quo scenario (12.9% in 2035; a 0.61 and 0.78 percentage point reduction, respectively).

Plain packaging is projected to reduce smoking prevalence by 0.6 percentage points in 2035, from 12.9% with the status quo scenario to 12.3% with plain packaging.

The combined effect of all four Tobacco Endgame strategies modelled in Ontario SimSmoke is projected to reduce smoking prevalence to 8.5% in 2035, a 4.4 percentage point reduction compared to the status quo scenario (12.9% in 2035).

In the model assessing the impact of a higher minimum age for legal purchase, population smoking prevalence was expected to decline 3.7 percentage points by 2035 to 13.2% from an imputed value of 16.9% under the baseline status quo scenario. Increasing the minimum legal purchase age to 21 would be expected to reduce smoking prevalence to 10.5% (8.0% among the 20–34 year olds; 2.7 and 5.2 percentage point decrease, respectively). Eliminating the effect on cessation in the model would predict a 2035 prevalence of 11.2% (10.8% among the 20–34 year olds; 2.0 and 2.4 percentage point decrease, respectively) (Figure 2).

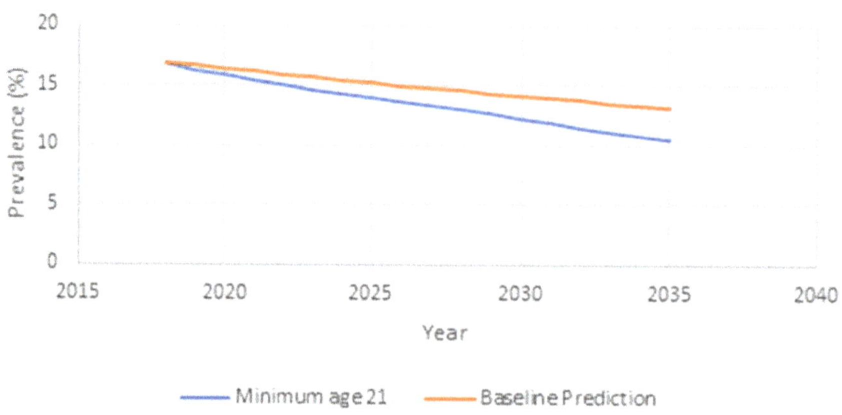

Figure 2. Model Predicted Smoking Prevalence, for Both Sexes, With and Without Increased Minimum Age Tobacco Purchasing Law, Ontario, 2018–2035.

3.2. Taxation Revenue Models

Average number of cigarettes per day was expected to be 4.0 cigarettes smoked per day among the 5% who were expected to continue smoking on average by 2035 down from 13.3 cigarettes a day in 2014.

Scenario 1. "Less than 5 by 35" achieved through non-tax interventions (excise taxes assumed to keep up with inflation):

- Tax revenue, 2035: $163 million
- Tax revenue, 2016–2035: $12,605 million
- Tax revenue, average, 2016–2035: $630 million

Scenario 2. "Less than 5 by 35" achieved solely through excise tax increases (assuming an underlying annual downward trend in smoking prevalence and consumption of 2.5%). Note that such a scenario requires that taxes increase annually by more than 20%:

- Tax revenue, 2035: $5054 million
- Tax revenue, 2016–2035: $68,884 million
- Tax revenue, average, 2016–2035: $3444 million

Scenario 3. "Less than 5 by 35" achieved through non-tax interventions and excise tax increases that raise prices by 5% in real terms, annually:

- Tax revenue, 2035: $673 million
- Tax revenue, 2016–2035: $24,261 million
- Tax revenue, average, 2016–2035: $1213 million

4. Discussion

The modelling results presented in this report highlight the effects of five key Tobacco Endgame strategies to reduce the smoking prevalence in Ontario by the year 2035. Increasing the tobacco taxes had the greatest independent predicted decrease in smoking prevalence by the year 2035 (2.8%), followed by increasing the minimum age for legal purchase to 21 years (2.4%) and decreasing the number of tobacco outlets (1.5%). Offering free cessation services and introducing plain packaging on all tobacco products each reduced the smoking prevalence by less than 1% compared to the status quo. Notably, none of the Tobacco Endgame strategies (either independently or combined) projected a smoking prevalence that was less than 5% by 2035.

Regarding impact of tax interventions on government revenue, our model shows that if Canada achieves "less than 5 by 35" through non-tax interventions, annual tobacco tax receipts would decrease from about $1.5 billion to about $160 million in 2035. However, if tax rates increase such that prices increase by 5% annually (in excess of inflation)—a policy pursued by France from 1991 to the early 2000s—average annual tax revenue would amount to about $1.2 billion and the cumulative taxes collected between 2016 and 2035 would near $25 billion.

The scenario 2 model showing the potential prices needed to achieve "less than 5 by 35" through taxation alone demonstrates the need for a comprehensive policy for the Tobacco Endgame that relies on both tax and non-tax interventions. Allowing for a portion of the effect of tax and price increases on tobacco use and consumption to be directed towards contraband cigarettes, as expected, reduces tax receipts, but does not invalidate any of the key findings. Similarly, our results are not sensitive to the use of a more conservative own-price elasticity estimates of -0.3. Taxation revenue should not be a barrier to the endgame. The analysis shows that with a sensible taxation policy, fiscal cost impact over the period of implementation is minimal compared to the health care and social costs of tobacco which currently are estimated at $16.2 billion per year [27]. Ultimately, however, it is important to recognize that the massive health and mortality burden due to tobacco is not worth sustaining for any amount of profit or revenue.

Caution should be taken when interpreting the projections presented in this report as they depend on the reliability of the data, and the estimated parameters and assumptions used in the models. A reduction in smoking prevalence and consumption in excess of current trends would inevitably lead to future populations that are larger than projected by Statistics Canada's medium growth population projections. There is strong evidence that higher incomes increase the demand for tobacco products [13]. However, income growth in Canada is projected to be relatively low [28]. Consequently, income effects are unlikely to affect the above results. Our approach examines the effect of changes in tobacco excise rates on tobacco excise revenue and not on harmonized sales tax (HST) which is a non-tobacco specific tax applicable on any taxable supplies in Canada, as ex-smokers and continuing smokers that reduce their consumption will very likely divert their spending towards goods and services that are also subject to HST. Our approach does not address the issue of tax avoidance such as brand switching. Because governments in Canada rely entirely on tobacco specific excise taxes and not on specific ad valorem taxes, which differs between brands of tobacco products. More broadly, the endgame potential interventions here are only a possible subset of innovative strategies that could change the landscape of tobacco control. For instance, this study does not consider the role of e-cigarettes, reduced nicotine, or structural changes to the tobacco industry. These other interventions may have a greater impact on smoking prevalence or health burden than the intervention set considered here.

5. Conclusions

Simulation models project that increasing tobacco taxes would result in the greatest decrease in smoking prevalence, and that reducing smoking prevalence to "less than 5 by 35" by both non-tax interventions and excise tax increase would result in minimal impact on government tax revenue. However, despite significant projected decrease in smoking prevalence, achieving "less than 5 by 35" might not be possible through the five key Tobacco Endgame strategies, either independently or combined.

Author Contributions: Conceptualization, M.C., G.E.G., R.S.; methodology, G.E.G., M.C.; formal analysis, G.E.G., J.D. writing—original draft preparation, G.E.G. writing—review and editing, M.C., J.D., G.E.G., R.S.; funding acquisition, R.S. All authors have read and agreed to the published version of the manuscript.

Funding: Health Canada Substance Use and Addiction Program.

Data Availability Statement: Model for forecasting tax simulation available on request.

Acknowledgments: The Ontario Population model was developed for the Tobacco Endgame Summit and we appreciate the contributions of the steering committee and summit attendees.

Conflicts of Interest: The authors declare no conflict of interest. The funders had no role in the design of the study; in the collection, analyses, or interpretation of data; in the writing of the manuscript; or in the decision to publish the results.

Appendix A

Table A1. SimSmoke Model Predicted Smoking Prevalence, for Both Sexes, Ages 15–85, With and Without Tobacco Endgame Policies, Ontario, 2018–2035.

Policy	2018	2019	2020	2021	2022	2023	2024	2025	2026	2027	2028	2029	2030	2031	2032	2033	2034	2035
Status Quo Policies [a]	15.5%	15.4%	15.2%	15.0%	14.9%	14.7%	14.5%	14.4%	14.2%	14.0%	13.9%	13.7%	13.6%	13.4%	13.3%	13.1%	13.0%	12.9%
Independent Policy Effects																		
Plain packaging	15.5%	14.8%	14.7%	14.5%	14.3%	14.2%	14.0%	13.8%	13.6%	13.5%	13.3%	13.1%	13.0%	12.8%	12.7%	12.5%	12.4%	12.3%
Free cessation services in primary care and hospitals	15.5%	14.8%	14.7%	14.5%	14.3%	14.2%	14.0%	13.8%	13.6%	13.5%	13.3%	13.1%	13.0%	12.8%	12.7%	12.5%	12.4%	12.2%
Free cessation services everywhere	15.5%	14.7%	14.5%	14.3%	14.2%	14.0%	13.8%	13.6%	13.5%	13.3%	13.1%	13.0%	12.8%	12.6%	12.5%	12.3%	12.2%	12.1%
Decreased tobacco availability	15.5%	13.7%	13.5%	13.4%	13.2%	13.1%	12.9%	12.7%	12.6%	12.4%	12.3%	12.1%	12.0%	11.8%	11.7%	11.6%	11.5%	11.4%
Increased taxation	15.5%	12.7%	12.5%	12.4%	12.2%	12.0%	11.8%	11.7%	11.5%	11.3%	11.1%	11.0%	10.8%	10.6%	10.5%	10.3%	10.2%	10.1%
Combined Policy Effects																		
All above	15.5%	10.9%	10.7%	10.6%	10.4%	10.2%	10.1%	9.9%	9.7%	9.6%	9.4%	9.3%	9.1%	9.0%	8.9%	8.7%	8.6%	8.5%

[a] Status quo represents the policy levels prior to the first projection year (2019). Note: Data table is for Figure 1.

References

1. World Health Organization. *Report on implementation of the Framework Convention on Tobacco Control*; World Health Organization: Geneva, Switzerland, 2014.
2. Statistics Canada. Canadian Community Health Survey, 2014. Cansim Table 105-0501. Available online: https://www150.statcan.gc.ca/t1/tbl1/en/tv.action?pid=1310045101 (accessed on 13 March 2021).

3. Rehm, J.; Baliunas, D.; Brochu, S.; Fischer, B.; Gnam, W.; Patra, J.; Popova, S.; Taylor, B. The Costs of Substance Abuse in Canada 2002 Highlights. *Can. Cent. Subst. Abus.* **2006**, *1*, 1–14.
4. Queen's University. *A Tobacco Endgame for Canada: Summit Background Paper*; Queen's University: Kingston, ON, Canada, 2016.
5. Physicians for a Smoke-Free Canada. Tax Revenues from Tobacco Sales. Ottawa. 2015. Available online: https://www.smoke-free.ca (accessed on 13 March 2021).
6. Zhang, B.; Cohen, J.; Ferrence, R.; Rehm, J. The Impact of Tobacco Tax Cuts on Smoking Initiation Among Canadian Young Adults. *Am. J. Prev. Med.* **2006**, *30*, 474–479. [CrossRef] [PubMed]
7. Hamilton, V.H.; Levinton, C.; St-Pierre, Y.; Grimard, F. The Effect of Tobacco Tax Cuts on Cigarette Smoking in Canada. *Cmaj* **1997**, *156*, 187–191. [PubMed]
8. Guindon, G.E.; Paraje, G.R.; Chaloupka, F.J. The Impact of Prices and Taxes on the Use of Tobacco Products in Latin America and the Caribbean. *Am. J. Public Health* **2018**, *108*, S492–S502. [CrossRef]
9. Gruber, J.; Sen, A.; Stabile, M. Estimating Price Elasticities When There Is Smuggling: The Sensitivity of Smoking to Price in Canada. *J. Health Econ.* **2003**, *22*, 821–842. [CrossRef]
10. Gallet, C.A.; List, J.A. Cigarette Demand: A Meta-Analysis of Elasticities. *Health Econ.* **2003**, *12*, 821–835. [CrossRef] [PubMed]
11. Galbraith, J.W.; Kaiserman, M. Taxation, Smuggling and Demand for Cigarettes in Canada: Evidence from Time-Series Data. *J. Health Econ.* **1997**, *16*, 287–301. [CrossRef]
12. Chaloupka, F.J.; Warner, K.E.; Chaloupka, F.; Warner, K.E. The Economics of Smoking. *Handb. Health Econ.* **2000**, *1*, 1539–1627.
13. International Agency for Research on Cancer. *IARC Handbooks of Cancer Prevention: Tobacco Control. Volume 14. Effectiveness of Price and Tax Policies for Control of Tobacco*; International Agency for Research on Cancer: Lyon, France, 2011.
14. Government of Canada. *Public Accounts of Canada 2015. Volume II. Details of Expenses and Revenues. Prepared by the Receiver General for Canada*; Public Works and Government Services Canada, Government of Canada: Ottawa, ON, Canada, 2015.
15. Levy, D.T.; Meza, R.; Zhang, Y.; Holford, T.R. Gauging the Effect of U.S. Tobacco Control Policies from 1965 Through 2014 Using SimSmoke. *Am. J. Prev. Med.* **2016**, *50*, 535–542. [CrossRef] [PubMed]
16. Levy, D.T.; Nikolayev, L.; Mumford, E. Recent Trends in Smoking and the Role of Public Policies: Results from the SimSmoke Tobacco Control Policy Simulation Model. *Addiction* **2005**, *100*, 1526–1536. [CrossRef] [PubMed]
17. Levy, D.T.; Cho, S.I.; Kim, Y.M.; Park, S.; Suh, M.K.; Kam, S. SimSmoke Model Evaluation of the Effect of Tobacco Control Policies in Korea: The Unknown Success Story. *Am. J. Public Health* **2010**, *100*, 1267–1273. [CrossRef]
18. Zhang, B.; Schwartz, R. *Technical Report of the Ontario SimSmoke: The Effect of Tobacco Control Strategies and Interventions on Smoking Prevalence and Tobacco Attributable Deaths in Ontario, Canada*; Ontario Tobacco Research Unit: Toronto, ON, Canada, 2013.
19. Chaiton, M.O.; Mecredy, G.; Cohen, J. Tobacco Retail Availability and Risk of Relapse among Smokers Who Make a Quit Attempt: A Population-Based Cohort Study. *Tob. Control* **2018**, *27*, 163–169. [CrossRef]
20. Callaghan, R.C.; Sanches, M.; Gatley, J.; Cunningham, J.K.; Chaiton, M.O.; Schwartz, R.; Bondy, S.; Benny, C. Impacts of Canada's Minimum Age for Tobacco Sales (MATS) Laws on Youth Smoking Behaviour, 2000–2014. *Tobacco Control* **2018**, *27*, e105–e111. [CrossRef]
21. Statistics Canada. Projected Population, by Projection Scenario, Age and Sex, as of July 1. Available online: https://www150.statcan.gc.ca/t1/tbl1/en/tv.action?pid=1710005701 (accessed on 15 March 2021).
22. Ontario Ministry of Finance. Ontario Population Projections, 2019–2046—Table 9. Available online: https://www.fin.gov.on.ca/en/economy/demographics/projections/table9.html (accessed on 15 March 2021).
23. Tobacco Use in Canada: Historical Trends in Smoking Prevalence. Available online: https://uwaterloo.ca/tobacco-use-canada/adult-tobacco-use/smoking-canada/historical-trends-smoking-prevalence (accessed on 15 March 2021).
24. Ross, H.; Tesche, J.; Vellios, N. Undermining government tax policies: Common legal strategies employed by the tobacco industry in response to tobacco tax increases. *Prev. Med.* **2017**, *105*, S19–S22. [CrossRef]
25. Bank of Canada. Understanding Inflation Targeting. 2020. Available online: https://www.bankofcanada.ca/2020/08/understanding-inflation-targeting/ (accessed on 15 March 2021).
26. Haché, T. *Commercial Tobacco in First Nations & Inuit Communities*; Non-Smokers' Rights Association/Smoking and Health Action Foundation: Toronto, ON, Canada, 2009.
27. Conference Board of Canada, "The Costs of Tobacco Use in Canada, 2012". 2017. Available online: https://www.conferenceboard.ca/e-library/abstract.aspx?did=9185 (accessed on 15 March 2021).
28. Government of Canada. Update of Long-Term Economic and Fiscal Projections 2018. Available online: https://www.canada.ca/en/department-finance/services/publications/long-term-projections/2018.html (accessed on 15 March 2021).

Article

Load Forecasting in an Office Building with Different Data Structure and Learning Parameters

Daniel Ramos [1,2], Mahsa Khorram [1,2], Pedro Faria [1,2] and Zita Vale [2,*]

[1] GECAD—Research Group on Intelligent Engineering and Computing for Advanced Innovation and Development, Rua DR, Antonio Bernardino de Almeida 431, 4200-072 Porto, Portugal; dados@isep.ipp.pt (D.R.); makgh@isep.ipp.pt (M.K.); pnf@isep.ipp.pt (P.F.)

[2] Polytechnic of Porto, Rua DR, Antonio Bernardino de Almeida 431, 4200-072 Porto, Portugal

[*] Correspondence: zav@isep.ipp.pt; Tel.: +351-2-2834-0511; Fax: +351-2-2832-1159

Abstract: Energy efficiency topics have been covered by several energy management approaches in the literature, including participation in demand response programs where the consumers provide load reduction upon request or price signals. In such approaches, it is very important to know in advance the electricity consumption for the future to adequately perform the energy management. In the present paper, a load forecasting service designed for office buildings is implemented. In the building, using several available sensors, different learning parameters and structures are tested for artificial neural networks and the K-nearest neighbor algorithm. Deep focus is given to the individual period errors. In the case study, the forecasting of one week of electricity consumption is tested. It has been concluded that it is impossible to identify a single combination of learning parameters as different parts of the day have different consumption patterns.

Keywords: building energy management; forecast; neural network; SCADA; user comfort

Citation: Ramos, D.; Khorram, M.; Faria, P.; Vale, Z. Load Forecasting in an Office Building with Different Data Structure and Learning Parameters. *Forecasting* **2021**, *3*, 242–255. https://doi.org/10.3390/forecast3010015

Received: 30 January 2021
Accepted: 17 March 2021
Published: 20 March 2021

Publisher's Note: MDPI stays neutral with regard to jurisdictional claims in published maps and institutional affiliations.

1. Introduction

Energy consumption forecast is very important in the context of energy consumption management towards improved energy efficiency. The forecast's accuracy may be improved based on retraining with a fixed size of training, discarding older information while retaining new information. The selection of sensors from smart technologies is another aspect that provides more training data that are expected to decrease the forecast errors [1].

The electricity markets face possible generation costs caused by environmental issues [2,3]. Smart grids are implemented in many of these markets, supporting efficient energy use [4]. Solutions involving smart grids consist of an adequate consumer schedule aimed to reduce the electricity consumption in particular periods [5]. These solutions are contextualized when markets launch demand response programs to make the consumption schedule adequate to reduce electricity costs interpreted by peaks [6].

Smart buildings play an important role in the electricity sector to satisfy occupants' electric needs and exploit operational flexibilities. Therefore, the launch of model optimization evidences the need to control the microgrids' power flows [7]. To deal with the situation, it requires solutions from demand response programs, reducing the energy costs using the smart grid opportunities to readapt the consumption to play an important role in load management and energy efficiency [8].

The optimization of electrical energy is possible with data monitored from a measurement system that captures real-time data and automatic forecasting [9,10]. With regard to forecasting, several machine learning algorithms can be used [11–14]. An artificial neural network (ANN) is described by layers containing neurons with weighted connections starting in an input layer, at least one hidden layer, and an output layer [15]. An alternative technique, K-nearest neighbor (KNN), performs data searches and associations in a large resource space with non-linear mapping support [16].

Various types of time-scaled forecast data may be evidenced in the field of energy, with Short Time Load Forecasting (STLF) being a good option. ANN is recommended for many short-term applications, including the prediction of daily peaks by using the training data with past data framed on past years [17]. KNN is suggested for both classification and regression tasks, and in the suggested approach, it is used for regression problems that involve energy predictions. The reduction of data complexity is a relevant aspect evidenced in the algorithm, possible with the nearest neighbors' readaptation to several subsets of data [18]. An even more innovative algorithm is suggested in [19] featuring a KNN-ANN model that uses the K-nearest neighbors process while adding a backpropagation function known to be a particular aspect of an artificial neural network (ANN). The application of the KNN-ANN model is suggested for a stock price prediction problem. The NPower Forecasting Challenge, taken in the year 2015 edition evidenced in [20], challenges the participants to perform daily energy predictions of a customer group. Several algorithms, including artificial neural network and Random Forest, are suggested. In another study, students' classification in algorithms like artificial neural network and Support Vector Machines is analyzed and their limitations are studied [21].

A research area of high interest is the energy efficiency of buildings—more specifically, the power distribution network that connects the equipment to end-users. The energy efficiency is highlighted on several worldwide applications including Supervisory Control And Data Acquisition (SCADA) and IoT systems [22]. These technologies allow the monitoring and management of consumption data on all the types of building from residential to commercial level. Thess data are relevant for the forecasting of data in the field of energy that are associated with electricity markets and policy formulations [23].

The forecasting of energy consumption with daily profile data usually improves the financial profit of consumers considering the monthly electricity bills reducing the peaks of energy detected in particular periods. The accuracy of energy forecasting algorithms depends on infrastructure and planning [24]. There are three ways to model an energy forecasting system mentioned in [25], including physics-based, data-driven, and hybrid models. While pros and cons are in question, the data-driven method has been proven as the best option for merging buildings in the smart grids. An additional factor that may improve the forecast reliability is to use sensor data that performs different measures according to each device according to smart meters [26,27]. The validation of forecasting models is another factor that should be taken into account in several smart buildings [28]. Real-time automatic energy forecasts with access to electric energy are recommended to be performed with data monitored in a building to achieve energy management optimization [29]. In [30], the component estimation technique is used for electricity consumption forecasts; historic consumption data were used. In [31], the impact of data quality in the electricity consumption forecast is discussed. The main focus is given to the dataset cleaning.

This paper provides a methodology to improve electricity consumption forecasting accuracy with sensor data measured by different devices, including presence, temperature, consumption, and humidity. The forecasting algorithms, namely ANN [32] and KNN [33], are implemented as a service and are the recommended options for the decision-making approaches to be used in the present paper. The innovative scientific aspect relies on the specific manipulations of data to overcome anomalies in data, including missing and excess occurrences. Second, the systematic analysis of different learning parameters is implemented to define the most relevant parameters in different periods of the day. This major aspect is usually treated in the literature by analyzing overall average forecasting errors without looking in detail at particular periods [34]. This aspect refers to a limitation in the recent literature, including the one published by the authors of the present paper in [1]. The forecasts are done for intervals (referred to as periods) of 5 min.

After this introduction, the proposed method is explained in Section 2, describing what is done at each stage. Proceeding to Section 3, the results of using the method are presented. The discussion is made in Section 4, and the main conclusions are presented in Section 5.

2. Materials and Methods

This section illustrates and explains the different phases of a method. The parameterization definition, the data reduction, the training and forecasting tasks, and the error calculation are parts of the tasks presented in Figure 1. The presented method is very important to support a building's participation, namely an office building, in demand response programs [35]. Addressing consumer comfort, a SCADA system can make autonomous decisions for participation in demand response programs issued by the distribution network operator [36].

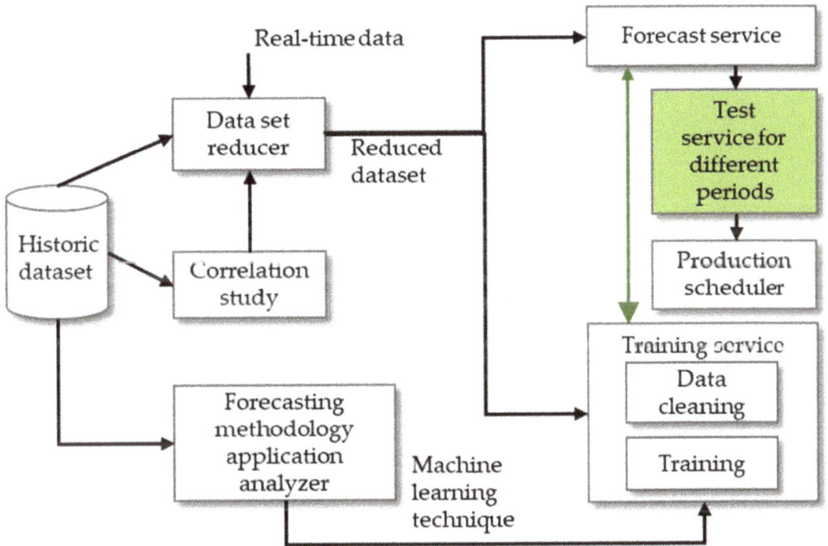

Figure 1. Proposed methodology diagram.

The innovative aspect of the present method is highlighted in green in Figure 1. As can be seen in the green arrow, the forecasting provides feedback to the training service regarding the accuracy of different learning parameters in different periods of the day. The test service is adapted to accommodate the fact that different periods of the day are related to different consumption patterns, so the test service must be run for each period. Different time frames are considered in the "Test service for different periods", namely: weekly Symmetric Mean Absolute Percentage Error (SMAPE) accuracy; daily SMAPE accuracy; period of day SMAPE accuracy; specific period accuracy. SMAPE is defined in Equation (3). The periods in a day for SMAPE in this paper are considered to be three periods: 00:00 to 08:00; 08:00 to 17:00; and 17:00 to 24:00.

The tuning process performs parametrization of data required for later use on forecasting tasks with the support of analysis, studies, optimizations, and data manipulations. Two main aspects describe this process. The first one evaluates the data content analyzing the best possible forecasting technique that should provide better results in that specific situation. The second one performs data transformations to the initial dataset reducing the original version of data to a more accurate version fed by the forecasting technique that should provide more accurate forecasts. There is a balance between the completion and simplicity of data to avoid wrong interpretations. Therefore, data structure and reliability are two main aspects to improve the accuracy of the algorithm.

The real-time data consist of all monitored and persistent data that the building technologies track in the system more concretely with consumption and sensors data. The correlation process has the goal of analyzing which sensors are more associated with consumption. Both the tasks of providing a sample and the correlation study influence the participation towards reducing the dataset.

Despite reducing the dataset to the entire historic series, the same rules apply for real-time data. The forecasting methodology studies which technique is better for the sampling of data. Both the reduced version of the dataset and the forecasting method are sent to the training service.

The cleaning operation makes data more accurate for further use on forecasting tasks. It goes through several phases, starting with reorganizing all data in a unique spreadsheet with data split into several fields, including year, month, day of the month, days of the week, hours, and minutes. The criterion applied for missing information is to make sequential copies of previous records.

Outliers treatments are applied to detect erroneous readings made by technology devices. The outlier's detection occurs with the support of the mean and standard deviation operations, as seen in Equations (1) and (2). The conditions implicit in the outlier's detection with the support of the mean and standard deviation are presented in Equation (3), suggesting scenarios where a point is outside of an interval between two values: the average minus or plus of a product between the error factor and the standard deviation. In the present paper, consumptions above 4800 W or below 300 W are considered outliers. These values have been established according to the authors' knowledge about building consumption.

$$A = \frac{\sum_{t=n-F}^{n} P(t)}{F} \tag{1}$$

- A—average consumption in F;
- n—current moment;
- P—consumption;
- t—index of time;
- F—frame (time interval) used for calculation.

$$S = \sqrt{\frac{1}{F} \times \sum_{t=n-F}^{n} (P(t) - A)^2} \tag{2}$$

- S—standard deviation consumption in F;
- F—frame used for calculation.

The service ends by extracting the cleaned data into a suitable structure that is understandable by the forecasting technique.

The forecast service is triggered the first time after the end of the training service. There are alternative ways, including testing requests or scheduling a new iteration after the error calculation process. The forecasting service reads the test parameters that are synchronous with each iteration with the support of a schedule that forecasts different contexts according to the forecasting technique [11–16] determined in the tuning service representing the total target consumptions. The test service is triggered the first time by default after the forecasting service ending. This service goal is to calculate the forecasting errors in each context which interprets how distance is the actual value from the forecast counterpart. The errors are calculated based on three possible metrics: Weight Absolute Percentage Error (WAPE), Symmetric Mean Absolute Percentage Error (SMAPE), and Root Mean Square Percentage Error (RMSPE). This paper highlights the use of SMAPE, as seen in Equation (3), as it has been identified as the adequate one for this application [37].

$$SMAPE = \frac{1}{F} * \sum_{t=n-F}^{n} \frac{|PF(t) - P(t)|}{(P(t) + PF(t))/2} \tag{3}$$

- PF—forecast consumption;
- F—frame used for calculation;
- t—period.

Following this, a trigger is activated, sending a new retrain request [1] to rerun the training service with more updated information that will discard previous data while also retaining new ones until the trigger point while keeping the same size data. In the present paper, artificial neural network (ANN) and K-nearest neighbor (KNN) forecasting algorithms are used [23]. ANN features a set of artificial neurons connected and structured in layers with a learning process that resembles the biological brain. The layers' structures describe an input and output layer separated by a hidden layer that performs calculations iteratively, learning a logic that associated the input to output data. The neurons transmit data to other neurons with signals according to the edges and layers' structures. The data received from the neurons are propagated afterward to other neurons following a process where the output of each neuron is computed through a non-linear function of the sum of inputs. All the combinations composed of neurons and edges are associated with a weight that adjusts during the learning process [15]. An alternative technique, K-nearest neighbor (KNN), performs data searches and associations in a large resource space with the support of non-linear mapping. This alternative is a method used both for classification and regression applications. In both cases, the input consists of different subsets named neighbors described by the historical data's closest examples.

The output differs from the classification and regression applications following different logics. For classification, the output consists of a class component that associates the nearest neighbor with the most common features. For regression, the output consists of a property of an object value calculated through the average of the set of nearest neighbors [16]. In [1] and [15], the authors have explored using different algorithms in the forecasting of office building consumption, namely ANN, KNN, Random Forest, and SVM. It has been concluded that ANN and KNN are adequate for the specific application under study in this paper. Other deep learning and ensemble learning algorithms can be explored in future work. Nonetheless, the present paper's main idea is to show that different algorithms can be more advantageous in different periods of the day or the week.

3. Results

This section presents the case study, including scenarios and the respective results. The building's historical data have been used as input data, so that the building has been divided into three zones [1]. In Figure 2, the topology of the building can be seen, with the respective three zones and the nine rooms (R1 to R9). In the bottom-right of Figure 2 is shown the detail of Zone 1. The zones of the building have been defined according to the sub-metering installed in the building. It matches the electrical switchboard coverage zones. In this way, the sensors data and consumption data are aggregated according to these zones. For this case study, the historical data of Zone 1 are selected. The selected historical data span the period from 22 May 2017 to 17 November 2019 with 5 min time intervals. It should be noted that the building is equipped with energy meters to record the consumption data and PV generation data as well. Additionally, there are different building sensors such as seven light power indicators, four movement sensors, three door status indicators, one air quality sensor, one temperature sensor, one humidity sensor, and one CO_2 sensor.

The input data are a matrix structure composed of twelve columns evidencing attributes associated to specific five-minute periods. A total of 262,060 rows evidencing the total number of observations from 22 May 2017 to 17 November 2019 were separated by five-minute intervals. The historic dataset represented by 22 May 2017 to 8 November 2019 contains 260,054 rows while the target week represented by 11 to 17 November contains 2006 rows. The initial ten columns identify consumption values, while the remaining two identify additional values obtained from enhanced sensors data, more specifically CO_2 and light intensity. The ten-input consumption featuring five-minute field values that precede the output counterpart corresponds to a period of fifty minutes. The CO_2 and light intensity resemble a single value placed in the five minutes preceding the output consumption. This dataset has been categorized based on the weeks, so focused time period includes

130 weeks. Figures 3–5 show the building's present input data in 130 weeks, related to the power consumption, CO_2 concentration, and intensity of lights, respectively. It means that each line represents the consumption data of one specific week in 2016 periods (5 min time interval).

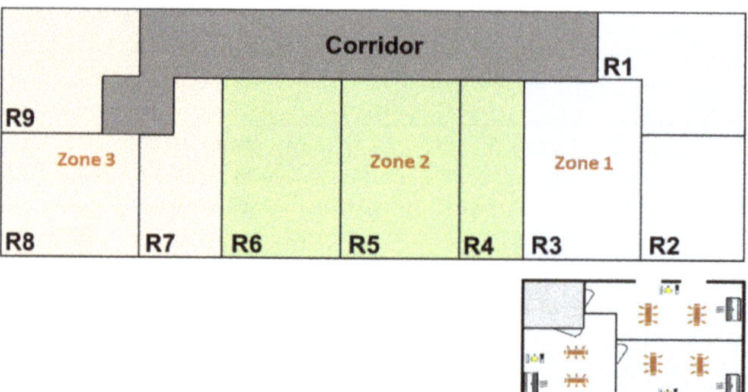

Figure 2. Building zones.

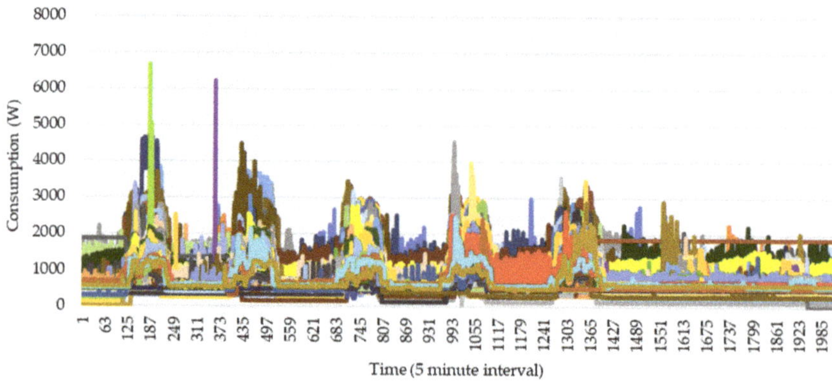

Figure 3. Power consumption of building from 22 May 2017 to 17 November 2019 is categorized based on the weeks.

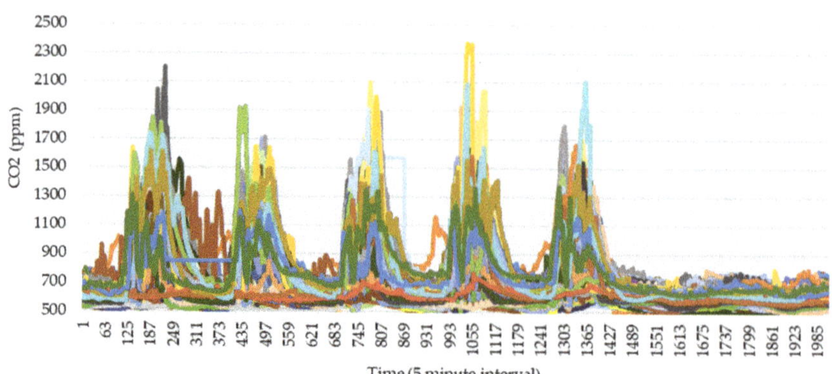

Figure 4. CO_2 concentration data from 22 May 2017 to 17 November 2019 are categorized based on the weeks.

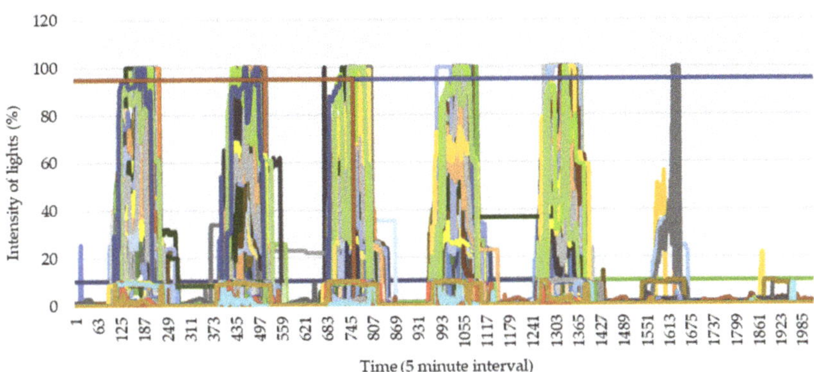

Figure 5. Light intensity data from 22 May 2017 to 17 November 2019 are categorized based on the weeks.

Several other environment data and parameters, such as the weather data, can impact the forecasting model's accuracy; the authors have discussed this in [1]. It has been concluded that, for the office building under study, as the researchers have a very specific routine, weather data do not contribute to improving the accuracy of the forecasting. This case study's main purpose is to forecast the consumption of 7 days based on the proposed training dataset. Additionally, 60 scenarios have been tested on different parameters such as number of entries, learning rate, number of neurons, clipping ratio, epochs, early stopping, and validation split on the forecasting results. Figure 6 shows the real consumption of 7 days of the test dataset. It should be noted that each day includes 288 periods (5 min interval), and each color represents one day.

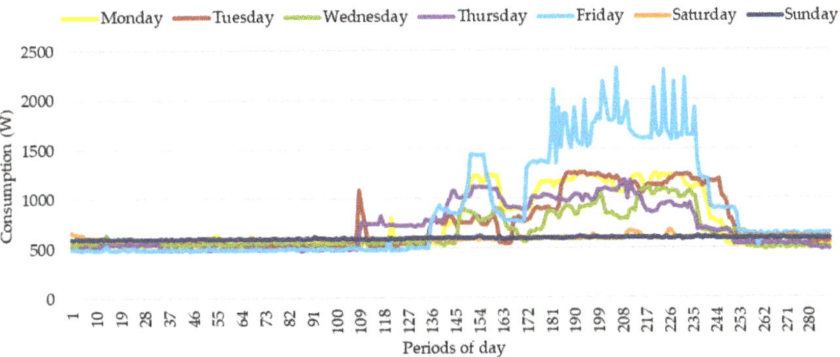

Figure 6. Actual power consumption of 7 days of the week with 5-minute time intervals.

The CO_2 concentration and intensity of lights have been presented in Figures 7 and 8, respectively, to propose the real data in the last week.

Table 1 introduces the characteristics of 60 scenarios with different parameters. Additionally, the calculated error of each forecasting can be seen on the right side of the table based on the ANN and KNN approaches. As shown in Table 1, the rank of calculated errors has been presented by dark color to bright color so that dark green cells show the lower error and white cells present the higher errors. To present the details of these error calculations, three scenarios (A, B, and C) have been selected to be illustrated by figures. The characteristics of these three cases can be seen in Table 1. The characteristics of scenarios A and C are equal. However, the applied techniques for the forecast are different.

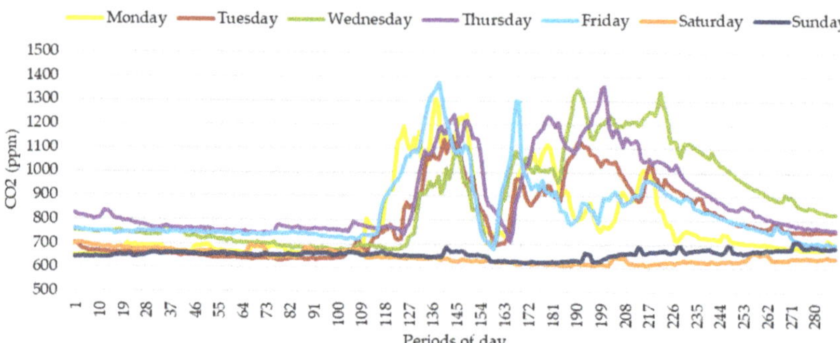

Figure 7. CO2 concentration data of 7 days of the week with 5 min time intervals.

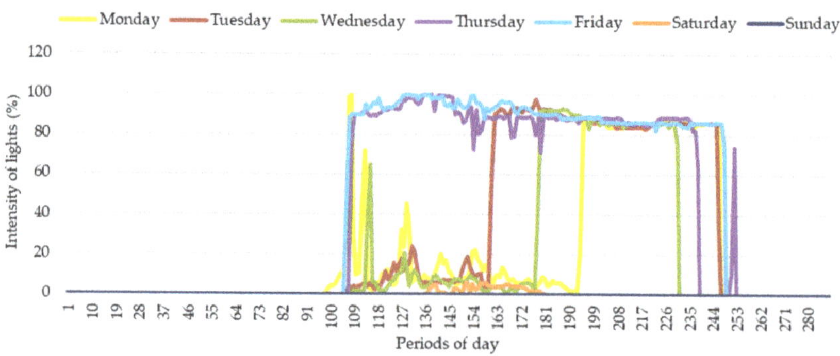

Figure 8. Light intensity data of 7 days of the week with 5 min time intervals.

Table 1. Error calculation based on artificial neural network (ANN) and K-nearest neighbour (KNN) approaches for 60 different scenarios.

Learn. Rate	# Neurons	Clipping Ratio	Epochs	Early Stopping	Validation Split	Days of the Week	SMAPE_ANN (Entries)			SMAPE_KNN (Entries)		
							10	50	100	10	50	100
0.001	32	5	500	20	0.2	−	2.77 *	2.75	4.14	3.60 ***	5.27	7.57
0.001	32	5	500	20	0.2	x	3.37	2.73	5.83	3.61	5.27	7.57
0.001	32	6	200	10	0.3	−	2.75	5.75	3.29	3.60	5.27	7.57
0.001	32	6	200	10	0.3	x	2.53 **	3.63	5.24	3.61	5.27	7.57
0.001	128	5	500	20	0.2	−	3.63	3.52	5.97	3.60	5.27	7.57
0.001	128	5	500	20	0.2	x	2.56	2.72	3.72	3.61	5.27	7.57
0.001	128	6	200	10	0.3	−	4.17	3.07	3.98	3.60	5.27	7.57
0.001	128	6	200	10	0.3	x	3.38	3.10	3.44	3.61	5.27	7.57
0.005	32	5	500	20	0.2	−	6.26	3.97	5.41	3.60	5.27	7.57
0.005	32	5	500	20	0.2	x	2.78	8.64	5.29	3.61	5.27	7.57
0.005	32	6	200	10	0.3	−	5.31	6.42	7.76	3.60	5.27	7.57
0.005	32	6	200	10	0.3	x	3.66	2.74	6.94	3.61	5.27	7.57
0.005	128	5	500	20	0.2	−	4.31	4.66	3.99	3.60	5.27	7.57
0.005	128	5	500	20	0.2	x	4.04	4.21	6.74	3.61	5.27	7.57
0.005	128	6	200	10	0.3	−	4.26	4.24	8.11	3.60	5.27	7.57
0.005	128	6	200	10	0.3	x	6.36	5.06	7.91	3.61	5.27	7.57
0.005	64	5	500	20	0.2	−	5.10	4.52	5.64	3.60	5.27	7.57
0.005	64	5	500	20	0.2	X	3.03	3.44	5.94	3.61	5.27	7.57
0.005	64	6	200	10	0.3	−	5.40	7.00	6.48	3.60	5.27	7.57
0.005	64	6	200	10	0.3	x	3.49	4.79	11.38	3.61	5.27	7.57

* Scenario A; ** Scenario B; *** Scenario C.

Each scenario focuses on seven days, shown by three figures based on the focused time. Figure 9 indicates 96 periods related to the 00:00 to 08:00 (5 min time interval), Figure 10 focuses on 108 periods from 08:00 to 17:00 (5 min time interval), and Figure 11 is related to the 84 periods from 17:00 to 24:00 (5 min time interval). The three referenced figures are related to scenario A. In Appendix A, the figures are presented related to scenario B (Figures A1–A3) and the figures related to scenario C (Figures A4–A6). The values selected for each parameter have been defined by the authors based on the experiments made on the ranges of each parameter that affect the results of forecasting. Additionally, the authors wanted to determine the influence of using the day-of-the-week information as input data to decide if it contributes or not to improving the accuracy.

Figure 9 presents the calculated SMAPE of scenario A in the first part of the day: 96 periods of 5 min are presented, related to the period between 00:00 and 08:00.

Each period of 5 min includes seven points in the graph, corresponding to the consumption for seven days of the week. Figure 10 presents the calculated SMAPE of scenario A in the second part of the day (from 08:00 to 17:00). Figure 11 presents the calculated SMAPE of scenario A in the third part of the day.

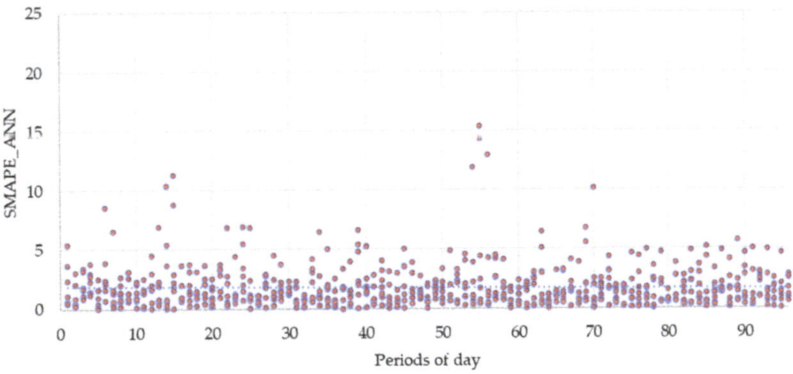

Figure 9. Forecast errors based on ANN approach in scenario A from 00:00 to 08:00.

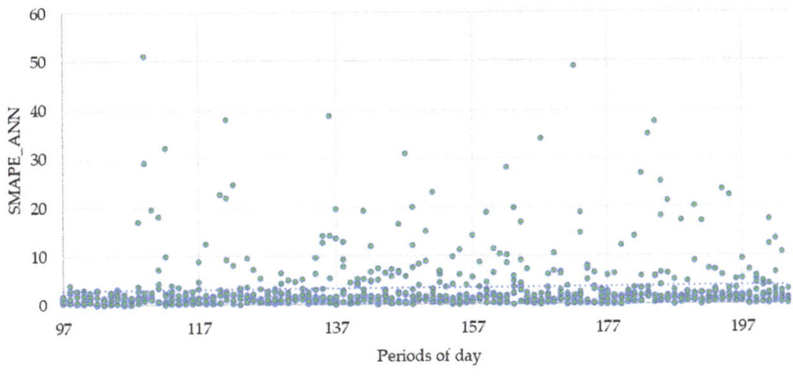

Figure 10. Forecast errors based on ANN approach in scenario A from 08:00 to 17:00.

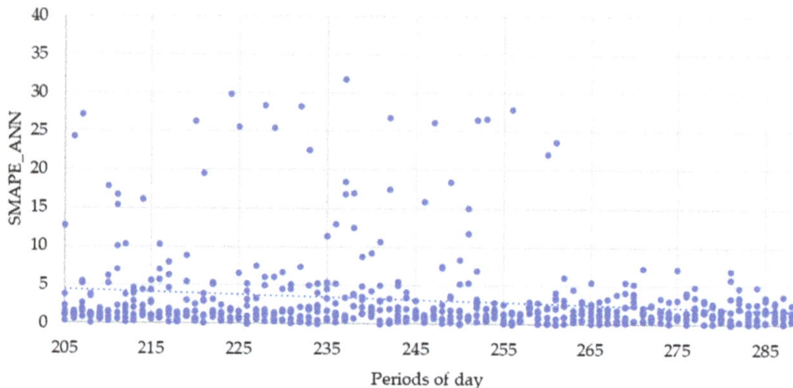

Figure 11. Forecast errors based on ANN approach in scenario A from 17:00 to 24:00.

The discussion of the results obtained will be presented in Section 4, focusing on the results already presented and Appendix A.

Regarding the error analysis in each day, Table 2 presents the SMAPE errors for each method. The data used in Table 2 relate to ten entries: learning rate (0.005), number of neurons in intermediate layers (64), clipping ratio (5.0), number of epochs (500), early stopping (20), validation split (0.2). The day of the week is not considered.

Table 2. SMAPE of ANN and KNN methods for each day.

Method	Full Period	Monday	Tuesday	Wednesday	Thursday	Friday	Saturday	Sunday
ANN	2.69	2.61	3.04	3.45	2.62	5.16	1.13	0.81
KNN	3.95	3.41	4.94	4.67	5.52	6.85	1.38	0.94

It can be seen that for every single day, ANN is always providing a more accurate forecast. However, as can be seen in the period-by-period analysis, KNN can have better accuracy in specific periods of the day or week.

4. Discussion

Looking at Figures 9–11 and Figures A1–A6 it is possible to see that the same method with the same parameters is not more accurate for all the periods. Focusing on the first period of the day, from 00:00 to 08:00, it can be seen that scenario C is the one with the highest dispersion of SMAPE for each period. Looking at Table 1, scenario C is the one with higher SMAPE between the three scenarios. However, for the period between 08:00 and 17:00, scenario C's results are not the worst ones, mainly compared with scenario A (Figures 9 and A2). Finally, regarding the third part of the day, from 17:00 to 24:00, scenario C is the worst one. Scenario B has a regular behavior along this period. However, scenario A is the best one at the end of this period (in the last third of this period). Comparing ANN and KNN, it can be seen that it is impossible to decide on the best one as scenario C is very accurate in a specific period of the day.

It has been found that, generally, the number of entries should be 10, as increasing the number of entries does not provide better results. Regarding the learning rate, it has been found that lower learning rates were more accurate in the results. The same comment applies to the number of neurons. Regarding the clipping ratio and the epochs, the early stopping, the validation split, and the days of the week, it is not possible to make a selection, as both values provide good results in different scenarios.

These results and discussion lead us to conclude that the definition of the ANN and KNN features must be done contextually, as different contexts bring different consumption patterns, and therefore, deserve different configurations in algorithms.

5. Conclusions

This paper has presented a forecasting service used in an office building aiming to support decisions regarding energy management towards efficiency. Two algorithms for forecasting have been used, namely artificial neural network and K-nearest neighbor, testing different algorithms and data features. It has been found that, for different periods of the day, which means different contexts regarding consumption patterns, different algorithm parameters can have higher accuracy levels. This means that it is not possible to say that a single algorithm is more accurate for the office building under study. In other words, one should select KNN for some periods of the day and ANN for other periods of the day, as discussed in Section 4.

Author Contributions: Conceptualization, P.F., and Z.V.; methodology, P.F., Z.V.; software, D.R., M.K.; validation, D.R., P.F., Z.V.; formal analysis, D.R.; investigation, D.R., M.K., P.F., Z.V.; resources, P.F., Z.V.; data curation, D.R., M.K., P.F., Z.V.; writing—original draft preparation, D.R., M.K., P.F., Z.V.; writing—review and editing, D.R., M.K., P.F., Z.V.; visualization, D.R., M.K., P.F.; supervision, P.F., Z.V.; project administration, P.F., Z.V.; funding acquisition, P.F., Z.V. All authors have read and agreed to the published version of the manuscript.

Funding: This work has received funding from FEDER Funds through COMPETE program and from National Funds through (FCT) under the projects UIDB/00760/2020, MAS-Society (PTDC/EEI-EEE/28954/2017) and CEECIND/02887/2017.

Data Availability Statement: The data used in this study are available in [1].

Conflicts of Interest: The authors declare no conflict of interest.

Appendix A

This appendix presents six figures that are added to the results.

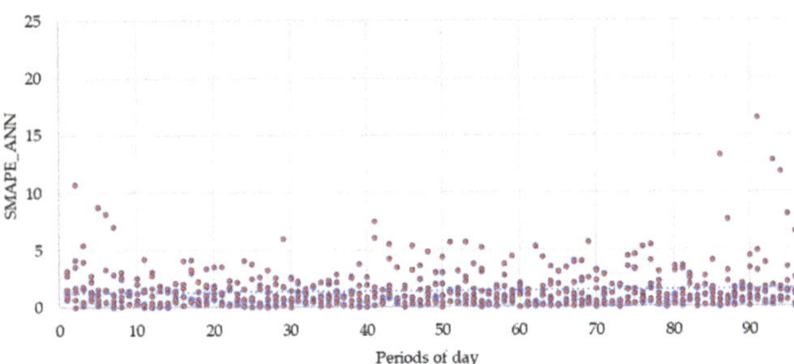

Figure A1. Forecast errors based on ANN approach in scenario B from 00:00 to 08:00.

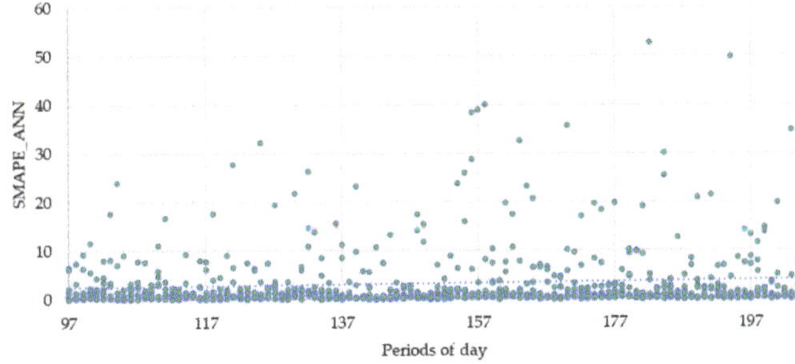

Figure A2. Forecast errors based on ANN approach in scenario B from 08:00 to 17:00.

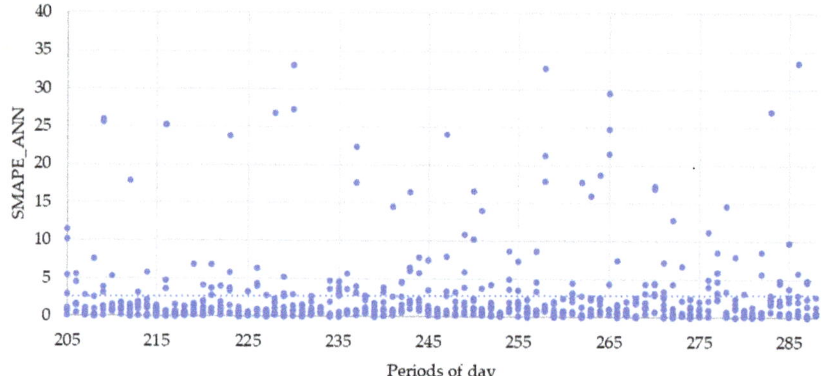

Figure A3. Forecast errors based on ANN approach in scenario B from 17:00 to 24:00.

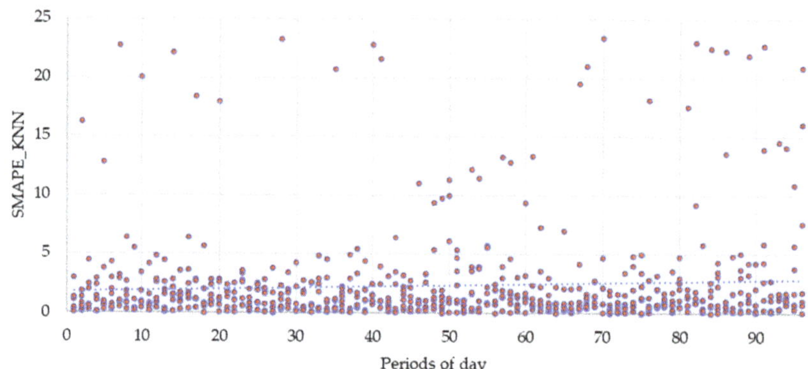

Figure A4. Forecast errors based on the KNN approach in scenario C from 00:00 to 08:00.

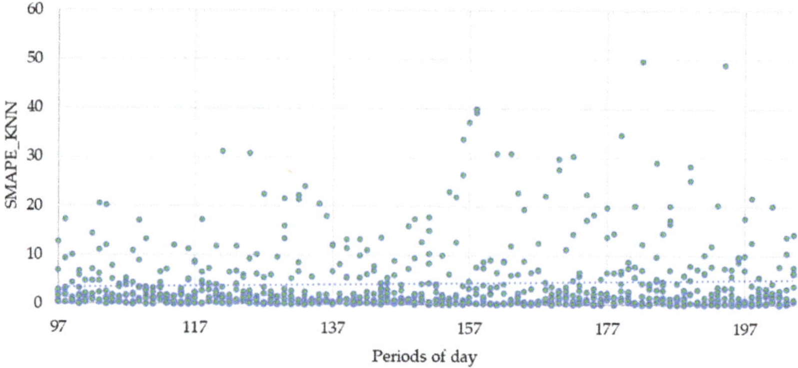

Figure A5. Forecast errors based on the KNN approach in scenario C from 08:00 to 17:00.

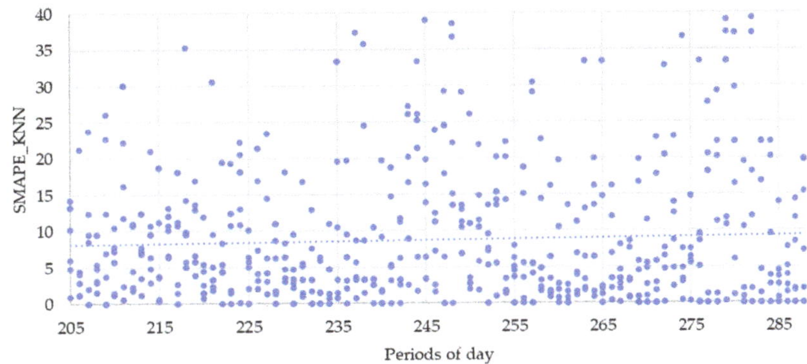

Figure A6. Forecast errors based on the KNN approach in scenario C from 17:00 to 24:00.

References

1. Ramos, D.; Teixeira, B.; Faria, P.; Gomes, L.; Abrishambaf, O.; Vale, Z. Use of Sensors and Analyzers Data for Load Forecasting: A Two Stage Approach. *Sensors* **2020**, *20*, 3524. [CrossRef]
2. Bless, K.; Furong, L. Allocation of Emission Allowances to Effectively Reduce Emissions in Electricity Generation. In Proceedings of the 2009 IEEE Power & Energy Society General Meeting, Calgary, AB, Canada, 26–30 July 2008; pp. 1–8. [CrossRef]
3. Rudnick, H. Environmental impact of power sector deregulation in Chile. In Proceedings of the 2002 IEEE Power Engineering Society Winter Meeting. Conference Proceedings (Cat. No.02CH37309), New York, NY, USA, 27–31 January 2002; Volume 1, p. 392. [CrossRef]
4. Faria, P.; Vale, Z. A Demand Response Approach to Scheduling Constrained Load Shifting. *Energies* **2019**, *12*, 1752. [CrossRef]
5. Pop, C.; Cioara, T.; Antal, M.; Anghel, I.; Salomie, I.; Bertoncini, M. Blockchain Based Decentralized Management of Demand Response Programs in Smart Energy Grids. *Sensors* **2018**, *18*, 162. [CrossRef] [PubMed]
6. Faria, P.; Vale, Z. Demand response in electrical energy supply: An optimal real time pricing approach. *Energy* **2011**, *36*, 5374–5384. [CrossRef]
7. Cao, Y.; Du, J.; Soleymanzadeh, E. Model predictive control of commercial buildings in demand response programs in the presence of thermal storage. *J. Clean. Prod.* **2019**, *218*, 315–327. [CrossRef]
8. Law, Y.; Alpcan, T.; Lee, V.; Lo, A. Demand Response Architectures and Load Management Algorithms for Energy-Efficient Power Grids: A Survey. In Proceedings of the 2012 7th International Conference on Knowledge, Information and Creativity Support Systems, KICSS 2012, Melbourne, Australia, 8–10 November 2012; pp. 134–141. [CrossRef]
9. Abrishambaf, O.; Faria, P.; Vale, Z. Application of an optimization-based curtailment service provider in real-time simulation. *Energy Inform.* **2018**, *1*, 1–17. [CrossRef]
10. Marzband, M.; Ghazimirsaeid, S.S.; Uppal, H.; Fernando, T. A real-time evaluation of energy management systems for smart hybrid home Microgrids. *Electr. Power Syst. Res.* **2017**, *143*, 624–633. [CrossRef]
11. Allen, J.; Snitkin, E.; Nathan, P.; Hauser, A. Forest and Trees: Exploring Bacterial Virulence with Genome-wide Association Studies and Machine Learning. *Trends Microbiol.* **2021**. [CrossRef] [PubMed]
12. Kilincer, I.; Ertam, F.; Sengur, A. Machine Learning Methods for Cyber Security Intrusion Detection: Datasets and Comparative Study. *Comput. Netw.* **2021**, *188*, 107840. [CrossRef]
13. Trivedi, S. A study on credit scoring modeling with different feature selection and machine learning approaches. *Technol. Soc.* **2020**, *63*, 101413. [CrossRef]
14. Merghadi, A.; Yunus, A.; Dou, J.; Whiteley, J.; ThaiPham, B.; Bui, D.; Avtar, R.; Abderrahmane, B. Machine learning methods for landslide susceptibility studies: A comparative overview of algorithm performance. *Earth Sci. Rev.* **2020**, *207*, 103225. [CrossRef]
15. Jozi, A.; Ramos, D.; Gomes, L.; Faria, P.; Pinto, T.; Vale, Z. Demonstration of an Energy Consumption Forecasting System for Energy Management in Buildings. In *Progress in Artificial Intelligence, EPIA 2019, Lecture Notes in Computer Science*; Springer: Cham, Switzerland, 2019; p. 11804.
16. Chen, G.H.; Shah, D. Explaining the success of nearest neighbor methods in prediction. *Found. Trends Mach. Learn.* **2018**, *10*, 337–588. [CrossRef]
17. Matsumoto, T.; Kitamura, S.; Ueki, Y.; Matsui, T. Short-term load forecasting by artificial neural networks using individual and collective data of preceding years. In Proceedings of the Second International Forum on Applications of Neural Networks to Power Systems, Yokohama, Japan, 19–22 April 1993; pp. 245–250. [CrossRef]
18. Barrash, S.; Shen, Y.; Giannakis, G.B. Scalable and Adaptive KNN for Regression Over Graphs. In Proceedings of the 2019 IEEE 8th International Workshop on Computational Advances in Multi-Sensor Adaptive Processing (CAMSAP), Le Gosier, Guadeloupe, 15–18 December 2019; pp. 241–245.

19. Liu, S.; Zhou, F. On stock prediction based on KNN-ANN algorithm. In Proceedings of the 2010 IEEE Fifth International Conference on Bio-Inspired Computing: Theories and Applications (BIC-TA), Changsha, China, 23–26 September 2010; pp. 310–312.
20. Xie, J.; Liu, B.; Lyu, X.; Hong, T.; Basterfield, D. Combining load forecasts from independent experts. In Proceedings of the 2015 North American Power Symposium (NAPS), Charlotte, NC, USA, 4–6 October 2015; pp. 1–5.
21. Imdad, U.; Ahmad, W.; Asif, M.; Ishtiaq, A. Classification of students results using KNN and ANN. In Proceedings of the 2017 13th International Conference on Emerging Technologies (ICET), Islamabad, Pakistan, 27–28 December 2017; pp. 1–6.
22. González-Vidal, A.; Jiménez, F.; Gómez-Skarmeta, A.F. A methodology for energy multivariate time series forecasting in smart buildings based on feature selection. *Energy Build* **2019**, *196*, 71–82. [CrossRef]
23. Ahmad, T.; Zhang, H.; Yan, B. A review on renewable energy and electricity requirement forecasting models for smart grid and buildings. Sustain. *Cities Soc.* **2020**, *55*, 102052. [CrossRef]
24. Ahmad, T.; Huanxin, C.; Zhang, D.; Zhang, H. Smart energy forecasting strategy with four machine learning models for climate-sensitive and non-climate sensitive conditions. *Energy* **2020**, *198*, 117283. [CrossRef]
25. Bourdeau, M.; Zhai, X. qiang, Nefzaoui, E.; Guo, X.; Chatellier, P. Modeling and forecasting building energy consumption: A review of data-driven techniques. Sustain. *Cities Soc.* **2019**, *48*, 101533. [CrossRef]
26. Yaïci, W.; Krishnamurthy, K.; Entchev, E.; Longo, M. Internet of Things for Power and Energy Systems Applications in Buildings: An Overview. In Proceedings of the 2020 IEEE International Conference on Environment and Electrical Engineering and 2020 IEEE Industrial and Commercial Power Systems Europe (EEEIC/I&CPS Europe), Madrid, Spain, 9–12 June 2020; pp. 1–6. [CrossRef]
27. Marinakis, V.; Doukas, H. An Advanced IoT-based System for Intelligent Energy Management in Buildings. *Sensors* **2018**, *18*, 610. [CrossRef] [PubMed]
28. Jayasuriya, D.; Rankin, M.; Jones, T.; Hoog, J.; Thomas, D.; Mareels, I. Modeling and validation of an unbalanced LV network using Smart Meter and SCADA inputs. In Proceedings of the IEEE 2013 Tencon—Spring, Sydney, NSW, Australia, 17–19 April 2013; pp. 386–390. [CrossRef]
29. Gomes, L.; Sousa, F.; Vale, Z. An Intelligent Smart Plug with Shared Knowledge Capabilities. *Sensors* **2018**, *18*, 3961. [CrossRef] [PubMed]
30. Shah, I.; Iftikhar, H.; Ali, S. Modeling and Forecasting Medium-Term Electricity Consumption Using Component Estimation Technique. *Forecasting* **2020**, *2*, 163–179. [CrossRef]
31. Nespoli, A.; Ogliari, E.; Pretto, S.; Gavazzeni, M.; Vigani, S.; Paccanelli, F. Electrical Load Forecast by Means of LSTM: The Impact of Data Quality. *Forecasting* **2021**, *3*, 91–101. [CrossRef]
32. Keras. Available online: https://www.tensorflow.org/guide/keras (accessed on 4 May 2020).
33. K-Neighbors. Available online: https://scikit-learn.org/stable/modules/generated/sklearn.neighbors.KNeighborsRegressor.html (accessed on 4 May 2020).
34. Leva, S. Editorial for Special Issue: "Feature Papers of Forecasting". *Forecasting* **2021**, *3*, 135–137. [CrossRef]
35. Faria, P.; Vale, Z.; Baptista, J. Constrained consumption shifting management in the distributed energy resources scheduling considering demand response. *Energy Convers. Manag.* **2015**, *93*, 309–320. [CrossRef]
36. Vale, Z.; Morais, H.; Faria, P.; Ramos, C. Distribution system operation supported by contextual energy resource management based on intelligent SCADA. *Renew. Energy* **2013**, *52*, 143–153. [CrossRef]
37. Armstrong, J.S. *Long-Range Forecasting from Crystal Ball to Computer*; Wiley: Hoboken, NJ, USA, 1985.

forecasting

Article

A Model Predictive Control for the Dynamical Forecast of Operating Reserves in Frequency Regulation Services

Pavlos Nikolaidis * and Harris Partaourides

Department of Electrical Engineering, Cyprus University of Technology, P.O. Box 50329, 3603 Limassol, Cyprus; c.partaourides@cut.ac.cy
* Correspondence: pavlos.nikolaidis@cut.ac.cy; Tel.: +357-25-002-041; Fax: +357-25-002-635

Abstract: The intermittent and uncontrollable power output from the ever-increasing renewable energy sources, require large amounts of operating reserves to retain the system frequency within its nominal range. Based on day-ahead load forecasts, many research works have proposed conventional and stochastic approaches to define their optimum margins for reliability enhancement at reasonable production cost. In this work, we aim at delivering real-time load forecasting to lower the operating-reserve requirements based on intra-hour weather update predictors. Based on critical predictors and their historical data, we train an artificial model that is able to forecast the load ahead with great accuracy. This is a feed-forward neural network with two hidden layers, which performs real-time forecasts with the aid of a predictive model control developed to update the recommendations intra-hourly and, assessing their impact and its significance on the output target, it corrects the imposed deviations. Performing daily simulations for an annual time-horizon, we observe that significant improvements exist in terms of decreased operating reserve requirements to regulate the violated frequency. In fact, these improvements can exceed 80% during specific months of winter when compared with robust formulations in isolated power systems.

Keywords: renewable energy sources; load forecasting; frequency regulation; artificial neural network; model predictive control

Citation: Nikolaidis, P.; Partaourides, H. A Model Predictive Control for the Dynamical Forecast of Operating Reserves in Frequency Regulation Services. *Forecasting* **2021**, *3*, 228–241. https://doi.org/10.3390/forecast3010014

Received: 25 February 2021
Accepted: 16 March 2021
Published: 17 March 2021

Publisher's Note: MDPI stays neutral with regard to jurisdictional claims in published maps and institutional affiliations.

1. Introduction

The power generation sector has seen rapid growth, mainly due to the increasing industrialization, domestic appliances and transportation demand [1]. The global challenge for modern power systems is to satisfy the growing electricity demand, whilst supplying uninterruptible and high-quality services. For several years now, this requirement has been fulfilled mostly by using fossil fuels because of their concentrated energy, which makes their output dispatchable and easy to adjust according to the load needs [2]. Based on well known load curves, the system operators could appropriately plan-ahead adequate operating reserves to allow for deviation corrections between the expected and actual load demand. However, the continuous burning of fossil-fuels poses a serious threat to the global environment and consequent climate change, calling for emission-free and renewable energy sources in the forthcoming years.

On the other hand, the introduction of renewable power generation produces a number of critical changes on the unit commitment and economic dispatch problem formulation. The intermittent and volatile behavior of renewable resources impose further variations on net demand and thus, the clarity of the operating reserves must be carefully scheduled. In addition, their uncontrollable and unpredictable power output increases the reserve requirements and probable deficits are reflected as frequency deviations between the nominal values. Consequently, the simultaneous increase in electricity demand and reduction in contributions of conventional sources create a lot of power integration and fluctuation issues, which undoubtedly disturb the overall system security, stability and reliability. Since the renewable energy sources do not contribute in flexibility, at a relatively low penetration level, they are commonly treated as negative loads providing comparable fluctuations with the existing net load fluctuations. As their penetration level grows, the conventional generating units occur inadequate for load following [3]. Over the last decade, researchers have extensively applied conventional and stochastic optimization techniques to define the optimal operating reserve margins and enhance the overall system reliability at reasonable costs. Based on predefined load curves, the various approaches broadly used can be divided into robust, deterministic and stochastic. The deterministic formulations recommend constant shares to represent the forecast errors in load demand. Without investigating the comparative performance of different risk considerations, the deterministic approaches rely solely on a set of uncertain parameters, offering poor reliability/cost trade-offs. To strengthen the robustness, a conservative formulation may propose a 5% upward and downward deviation space, while more robust approaches involve up to 10% margins for islanded systems [4]. More recently, a variety of solutions have relied on stochastic mechanisms, distinguishing the formulations into random scenario reduction, distributionally robust and uncertainty-set classifications [5,6]. Aiming at the minimization of the expected cost over a probability distribution that is represented by scenarios, these frameworks are versatile [7,8]. However, they require significant computational efforts and it is difficult to retrieve temporal and spatial correlations within scenario-trees [9].

The vast majority of the literature in relating fields concentrates on household or small area level load forecasting (i.e., distribution transformer) due to the significantly limited availability of regular patterns. In their effort to address the imposed uncertainty, the existing methods can be divided into three main categories. The methods of the first category make use of clustering or classification techniques to correlate similar customers, day types or weather conditions, targeting on the reduction of uncertainty variance [10]. A second category focuses on the elimination of the imposed uncertainties at the meter-level by utilizing aggregated smart-meter data [11], whereas the rest of the methods fall in the last category and refer to uncertainty separation within the regular patterns, relying on spectral analysis such as Fourier transformation, wavelet and empirical mode decomposition [12]. Beyond the aggregated level, load forecasting methods are based on sophisticated mechanisms and machine learning techniques. A tutorial review of probabilistic electric load forecasting is provided in [13]. The authors in [14] presented a comparison between hybrid and artificial intelligence models including support vector machines, expert systems, fuzzy logic, regression trees and artificial neural networks, while the notable time series models of long short-term memory (LSTM) systems, recurrent (RNN) and convolutional (CNN) neural networks combined with different regression techniques are discussed in [15]. Although highly flexible and effective, RNN-based approaches outperform traditional forecasting models in terms of root mean square error (RMSE) and mean absolute percentage error (MAPE) [16,17].

The existing methods aim at day-ahead forecasts or make use of RNN systems to only minimize the forecast error against the actual load. To the best of our knowledge, there has not yet been a comprehensive solution that targets real-time forecasts to improve the performance using updated input values. Most approaches utilize temperature as the only weather-dependent variable and no research work is targeted on the real-time estimation of reserve margins. In this work, we propose a radically different framework to determine the operating reserves based on a real-time load forecast. Identifying their vital role in day-ahead power optimization tasks, we aim at the dynamical update of the predefined daily demand based on a model predictive control. Specifically, we make use of independent input predictors to achieve the dependent target, namely the daily load. Based on annual data with respect to some selected predictors, we train a neural network via non-linear regression. During the particular day, the updated values of the predictors are assigned to the model, which assesses their impact and its significance on the output target and re-use them to estimate the new demand ahead. Together with the power balance, they constitute a system-wide constraint that affects the overall system security and total achieved production cost. The obtained results show that significant improvements exist in terms of decreased operating reserve requirements. Considering the performance of the trained neural network, the determined operating reserves account for the mean squared error (MSE) and the actual deviation of the selected predictors. Based on real-time updates, the load forecasting can achieve lower costs, while the system security is preserved.

The rest of the paper is organized as follows. The following section includes the problem formulation and the importance of accurate reserve definition. Section 3 deals with the methodology followed to develop the proposed, real-time load forecast model. All precise descriptions in relation with the different models used are included. In addition, the considered test system is presented along with the main parameters used for predictions. In Section 4, the realizations of our solution are presented and their findings are discussed in detail, while the obtained improvements are listed by their relevance. Finally, the conclusions are drawn in Section 5.

2. Problem Formulation

In order to achieve a comprehensive view regarding the impact of operating reserves on total generation cost, we first define the generic objective function of unit commitment task with the aid of Equation (1).

$$f = min \sum_{t=1}^{T} \sum_{i=1}^{N} \left[F\left(P_i^t\right) + \left(1 - U_i^{t-1}\right) SU_i \right] U_i^t \tag{1}$$

Denoting the total time intervals with T and the total number of available generating units with N, the power contribution of a generator i during the time slot t is expressed via P_i^t. U_i^t defines whether a generator is "on" or "off" during that interval, whereas the cost to start-up is represented by SU_i. The power balance constraint is provided in Equation (2). In general, the summed power of the committed units must satisfy the load demand P_d [18]. Each deviation from the absolute power balance (zero equivalent) violates the nominal frequency (50 or 60 Hz) of the system according to Newton's Second Law of Equation (3).

$$\sum_{i=1}^{N} U_i^t \cdot P_i^t = P_d^t \tag{2}$$

$$T_m - T_e = J\frac{d\omega}{dt} \tag{3}$$

In case of an imbalance between the mechanical torque T_m and electrical torque T_e, the rotating mass will experience an angular acceleration or deceleration $d\omega/dt$, which is reflected as a change in frequency. It is noted that the frequency change is smaller for a system with high inertia (J) compared to a system with low inertia [19]. To guarantee the system stability, different reserve types are needed according to their time of response. For clarification purposes, we express the equation of motion (4) in power terms so that $P = T \cdot \omega$ is preserved.

$$P_m - P_e = M\frac{d\omega}{dt} \tag{4}$$

where $M = J \cdot \omega$ is the angular momentum of the rotating system. Turning to the specification of the minimum technical and operational characteristics that each user connected to the Transmission System must comply, the frequency range during normal conditions is stated between 49.8 and 50.2 Hz and it can be extended to 47–52 Hz during disturbances. A disturbance event is defined as an incident that causes deviations equal or greater than 0.5 Hz from the nominal f_o. The operating reserves are separated into spinning and non-spinning. Spinning reserves are the first acting and derived from the synchronized units to the system [20]. They include the restraint and recovery reserves, which are available within 3 and 20 s and operable for 20 s and 20 min, respectively. Following are the supplemental and replacement reserves which need to be available for 6 h. A last category involves the contingency reserves that are operable within 6–24 h. These categories fall in the non-spinning reserve classification. Day-ahead schedules must satisfy a further system-wide, coupling constraint, namely the spinning reserve margins SR^t. The formulation of such inequality constraints (both upward SR_{up} and downward SR_{down}) is expressed via the following respective equations:

$$\sum_{i=1}^{N} U_i^t \cdot P_{i,max}^t \geq P_d^t + SR_{up}^t \tag{5}$$

$$\sum_{i=1}^{N} U_i^t \cdot P_{i,min}^t \leq P_d^t - SR_{down}^t \tag{6}$$

where $P_{i,min}$ and $P_{i,max}$ denote the minimum and maximum capacity limits of each generator i. Assuming a robust formulation with SR margins in the order of 10% of the instant load, it is worth noting that this expensive requirement forces more generators to start-up, leading to sub-optimal unit commitment schedules and uneconomic power dispatch.

To lower the expensive spinning-reserve requirements, we propose the intra-daily forecast of load demand. In contrast to day-ahead estimations, which may deviate from real-time values, intra-daily forecast with 15 min updates of selected predictors may improve the accuracy and consequent required reserves. Electricity load follows daily patterns, which are repeated according to the human activity and weather conditions. In this regard, we exploit an accurate hours-ahead system for load forecast using neural networks. Our purpose is to enhance the system security and reliability, whilst minimizing the SR requirements by making use of a model predictive control, which performs updates every 15 min to supply the neural networkIn more detail, a number of predictors x are imported in the feed-forward network along with the target y to form our data set $x_i, y_i | i = 1, \ldots, n$. The model is trained using the largest share of the historical data for training, while the rest is equally distributed for validation and testing. The developed model exploits a two-hidden-layer neural network employed as follows:

$$h_1 = \sigma\left(\sum_{k=1}^{K} w_k x_k + \beta_1\right) \tag{7}$$

$$h_2 = \sigma\left(\sum_{l=1}^{L} w_l h_{1_l} + \beta_2\right) \tag{8}$$

$$y = \sum_{m=1}^{M} w_m h_{2_m} + \beta_y \tag{9}$$

where $\sigma(\cdot)$ is the sigmoid activation function and h the output of the hidden layers. K, L, M are the number of predictors, neurons at the first and second hidden layer, respectively [21]. Figure 1 depicts a graphical representation of the proposed network.

During the realization of power dispatch, the selected predictors $\dot{x}(t)$ re-enter the forecast model at t and the remaining $T - t$ sequence is updated based on the model predictive control explained as follows:

$$I = \sum_{j=t}^{T} w_{x_P}[r_p(j) - x_p(j)]^2 + \sum_{j=t}^{T} w_y[\Delta y(j)]^2 \tag{10}$$

The predicted parameters r_j constitute the reference of the model and each deviation from the actual values is recursively corrected to minimize $I \cdot \Delta y$ indicates the impact of the actual deviation on the new, forecasted values when x_j are reused for load forecast. The significance of Δy is regulated by penalizing with w_y, while w_x reflects the importance of each selected predictor p. Finally, the equality constraint of $\Sigma w = 1$ must be preserved [22,23].

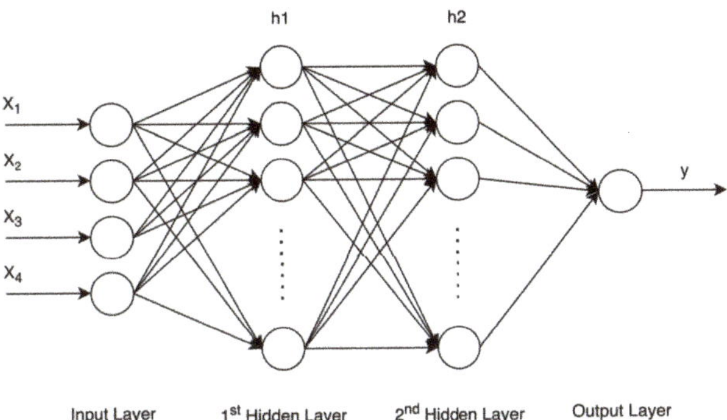

Figure 1. Proposed Neural Network.

3. Test System and Methodology

The considered system concerns the isolated power community of the island of Cyprus. This is a representative, small-to-medium scale network consisting of 20 generators to supply a 1100 MW peak demand (usually occurred in July) with an annual load factor of 56% [24]. Due to its isolation, small area and remoteness, electricity supply for more than 875 thousand people inhabited in the island, mainly relies on imported fossil fuels, the price of which is 3–4 times higher than that in the mainland [4]. As a result, the extremely high SR requirements of up to 10% of the hourly load pose a critical increase on total production cost. To decide which predictors to include in our forecaster, we first tried to extract a physical relationship between them and our target, namely the load demand. Based on actual data obtained from the Cyprus Energy Regulatory Authority (CERA), we demonstrate the hourly load for a representative week for each season in Figure 2.

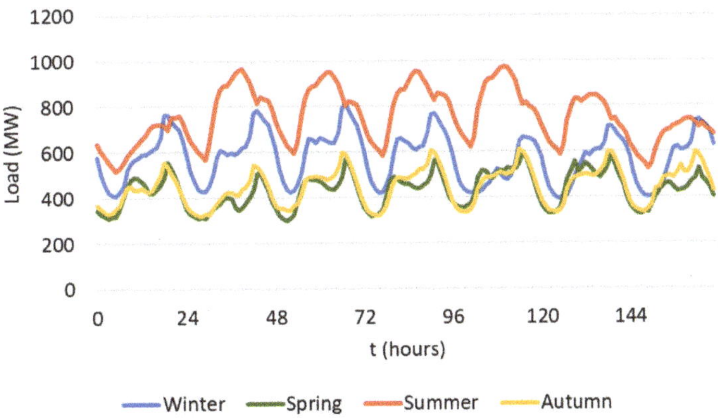

Figure 2. Weekly load demand per season.

Apart from the seasonality and human activity, similar patterns have been observed within the same periods of different years. This way, we choose to express the seasonality by the hour and date, whereas the human activity is represented through the day-type. The repetition of this activity is shown with the aid of three further predictors, such as the daily load of the previous day, week and year. These six predictors form our constant parameters. In Figures 3–5, we provide the fluctuation of temperature and relative humidity which are our further two, variable predictors. Figures 3 and 4 show an hourly histogram relating to the year 2019, while their seasonal values are offered in Figure 5. As can be seen, they both present non-linear relations with time and in order to make easy and accurate predictions, a better resolution is needed. This can be achieved by performing week-to-week comparisons of their hourly variation during different seasons.

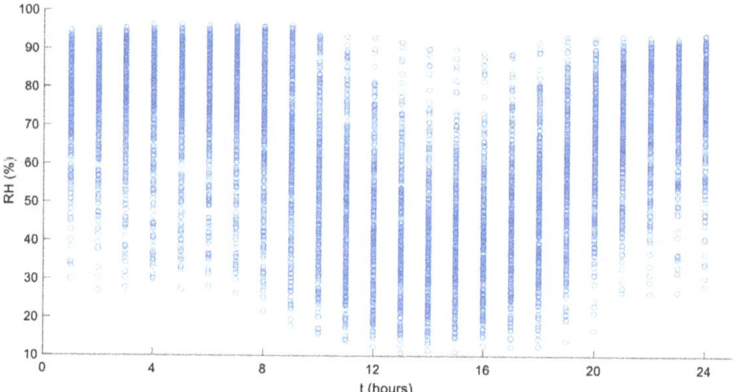

Figure 3. Annual variation of relative humidity.

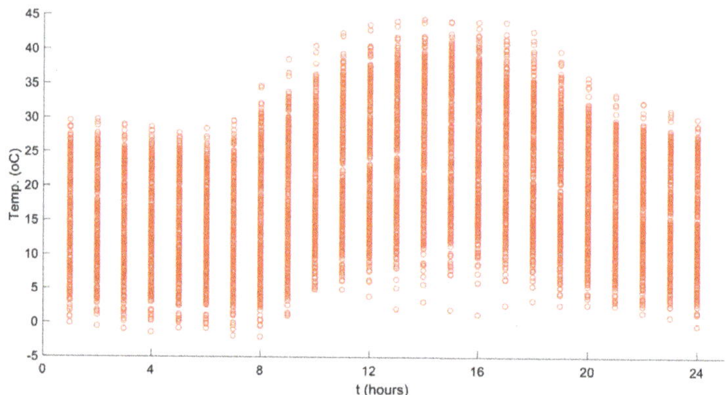

Figure 4. Annual variation in temperature.

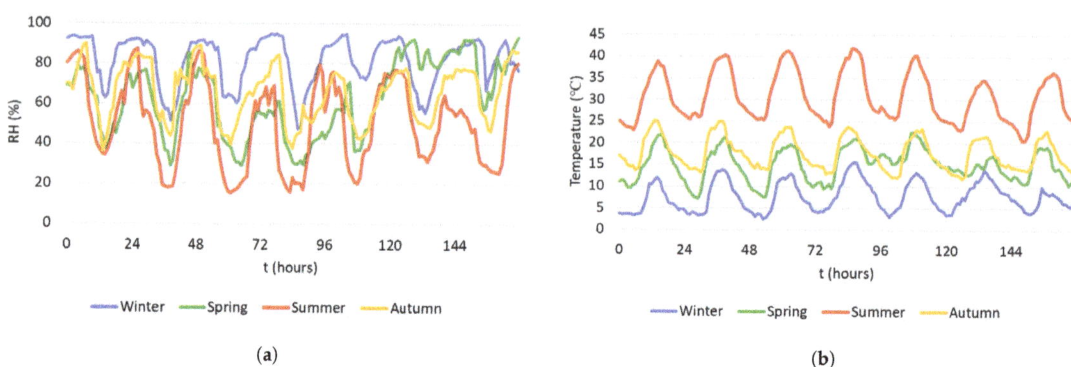

(**a**)   (**b**)

Figure 5. Seasonal variation of (**a**) relative humidity; (**b**) temperature.

Undoubtedly, ambient temperature affects the human comfort and their overall activity. However, relative humidity is the parameter that ultimately determines the rate with which heat is drawn away from the body and thus how does the absolute temperature "feels like" by humans [25,26]. Figure 6 offers the most important values of temperature and relative humidity for the most energy-intensive weeks in 2019's winter and summer.

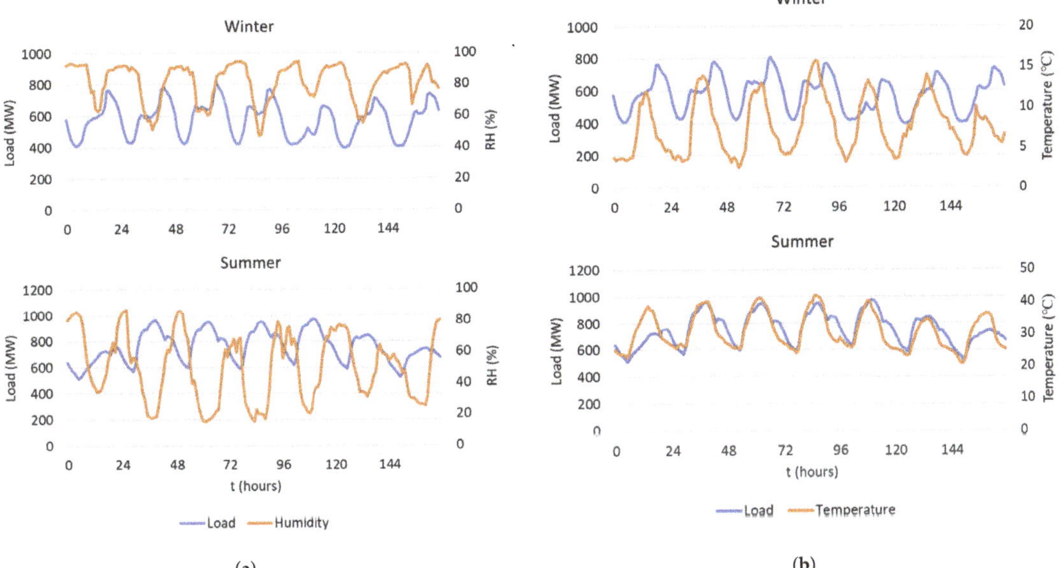

Figure 6. Winter and summer comparisons of hourly load demand and (**a**) relative humidity; (**b**) temperature.

The relative humidity possesses higher values, which tend to decrease during the daylight. On the other hand, the temperature shows an adverse trend, which during the summer shows a linear relationship with load but during winter, it is inversely proportional to the load demand. Therefore, it is obvious that both variables project a fluctuation to load forecast and consequently, they must be updated during the realization of power dispatch. Utilizing actual data from 2010–2019, we train a neural network based on non-linear regression between the following predictors: (1) day (or date), (2) hour, (3) day-type (weekday = 0, weekend = 1, holiday = 2), (4) previous day load, (5) previous week 24h-load, (6) previous year 24h-load, (7) relative humidity and (8) temperature, and the target of actual load demand. The respective settings of our network include 20 neurons per hidden layer. The forecasting model exploits 70% of the historical data for training, 15% for validation and 15% for testing.

Regarding the model used for predictive control, the selected predictors refer to the updated temperature and relative humidity forecasts for the intra-hour periods of 15-minutes, equally weighted by 25%. The remaining 50% is given to the change in the manipulated, depended variable Δy. In contrast to traditional models that regulate their inputs to approximate the referenced values and minimize their impact, in our realization, we set the updated values as the predicted (reference) and we regulate the controlled temperature and humidity to estimate their impact through the forecaster. Then, the model is updated with the new values and dynamically accepts the updates to perform the next cycle until the end of the assessed day. We illustrate our proposed configuration in Figure 7.

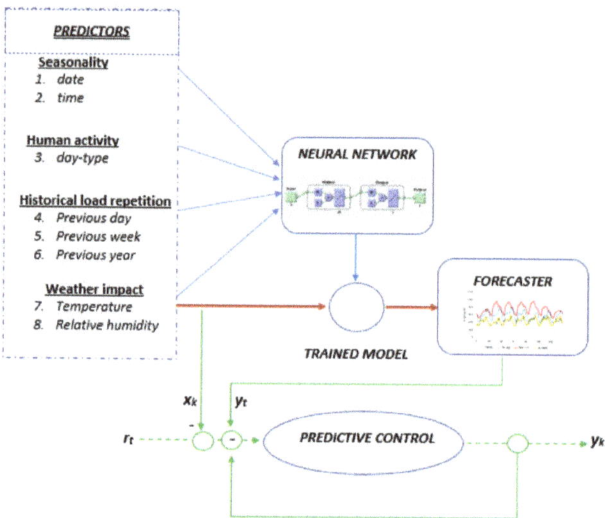

Figure 7. The proposed real-time load forecast model.

4. Results and Discussion

Aiming at the minimization of expensive SR margins for frequency regulation, we apply our proposed solution introducing the actual data obtained from CERA. We make use of a feed-forward neural network with two hidden layers of 20 neurons and a Levenberg–Marquardt algorithm for the curve fitting. This algorithm relies on the minimization of the squared sum of some imposed parameters β [27]. For a given set of n empirical pairs (x_i, y_i), this problem can be formulated as follows:

$$\hat{\beta} = \underset{\beta}{\mathrm{argmin}} \sum_{i=1}^{n} [y_i - f(x_i, \beta)]^2 \tag{11}$$

After the introduction of the predictor matrix x (of $n \times p$ dimensions) and the dependent target y into the model, the achieved performance of the forecaster is calculated in terms of MSE and presented in Figure 8.

$$MSE = \frac{1}{T} \sum_{t=1}^{T} (y_t - \hat{y}_t)^2 \tag{12}$$

As can be observed, the forecasting model shows high performance with R-values above 97.5% in each case and estimated MSE in the order of 2.388%. The regression plots displayed, show that the network outputs with respect to targets for training, validation, and test sets, fall along the 45-degree line, where the network outputs are equal to the targets. This verifies our views on the existence of lower SR requirements. For further verification of the network performance, we illustrate the error histogram in Figure 9.

The outliers' indication shows that most errors fall between −75 and +75. The respective training, validation and test error appear in Figure 10. Since the test set error presents similar characteristics with the validation set error, as well as the final mean squared error being small, the obtained result is quite reasonable.

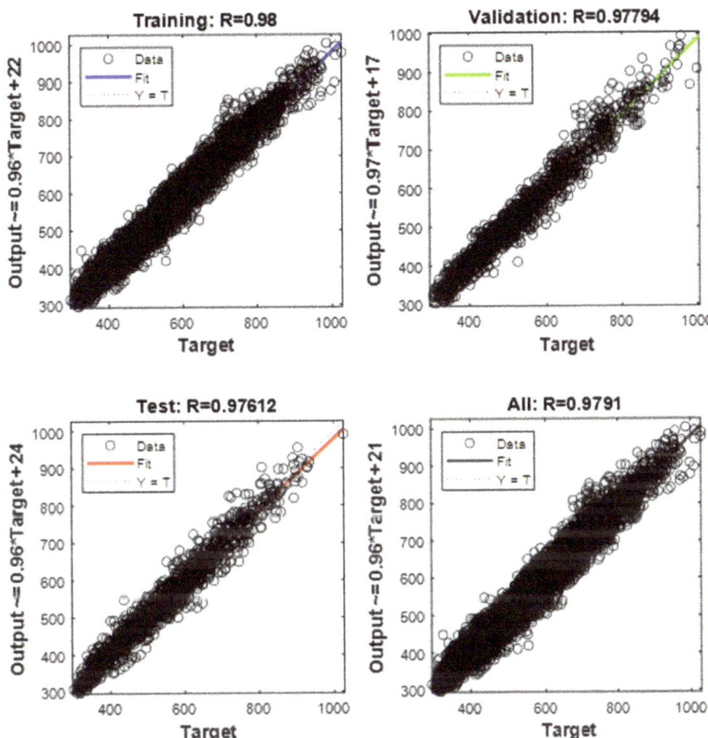

Figure 8. Performance of the trained model for load forecasting.

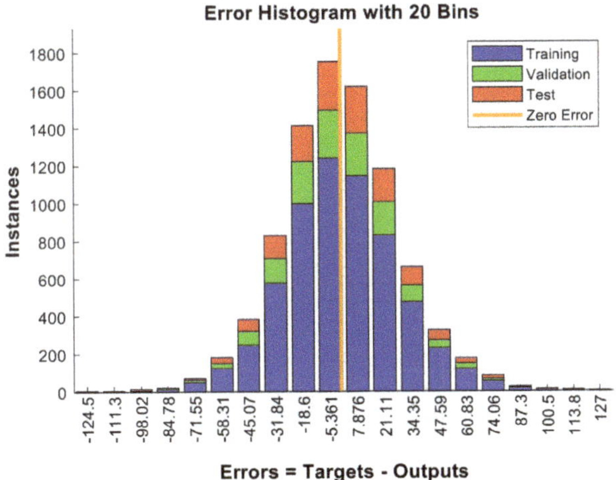

Figure 9. The error histogram of the load forecast model.

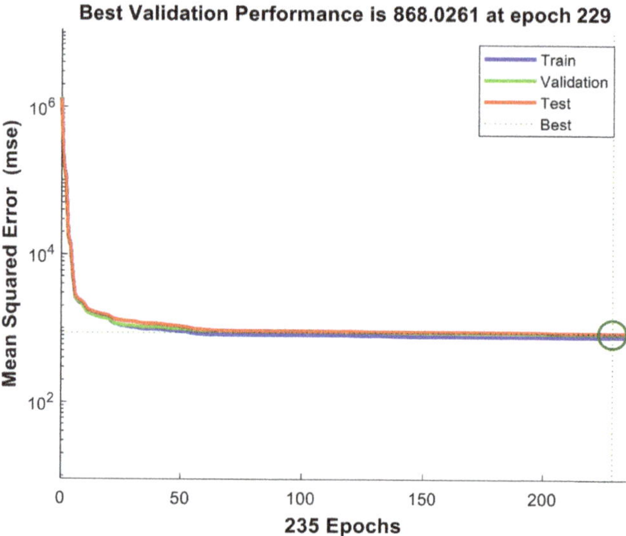

Figure 10. A graphical representation of the training errors, validation errors, and test errors.

To gain a broader overview of the efficacy of our approach, we compare our proposed solution with a benchmark optimizer, namely Gradient Descent. Based on Equations (13) and (14), the achieved RMSE and MAPE are 10.6227 and 0.0105, respectively, when Levenberg–Marquardt is used, against Gradient Descent, which accounts for 168.4502 RMSE and 0.2875 MAPE. Figure 11 demonstrates the load forecast recommendation for the considered optimizers. Selecting Levenberg–Marquardt as the optimizer for curve fitting, we illustrate the performance of the proposed neural network over the alternative regression trees in Figure 12. Although the proposed solution almost perfectly fits the actual load demand, the alternative regression tree-based approach deviates considerably, providing the respective 68.8261 and 0.0907 RMSE and MAPE.

$$\text{RMSE} = \sqrt{\frac{1}{T} \sum_{t=1}^{T} (y_t - \hat{y}_t)^2} \tag{13}$$

$$\text{MAPE} = \frac{1}{T} \sum_{t=1}^{T} \left| \frac{y_t - \hat{y}_t}{y_t} \right| \tag{14}$$

Applying daily simulations for the entire year of 2020, we estimate the deviation errors between the day-ahead, forecasted load and actual, real-time values during the assessed dates. The input of the model predictive control is updated using intra-hour (15-minutes sampling rates) data regarding the forecasted ambient temperature and relative humidity. The worst deviations are found to be during summer and their actual representation is shown in Figure 13. It is noted that there imposed 24 updates which represent the most prevalent of the 4 intra-hour ones. We depict the most relevant deviations which accounts for over 2% error.

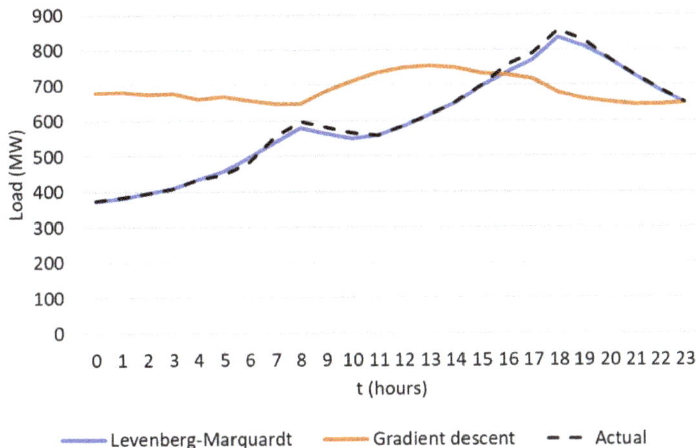

Figure 11. Implications of different optimizers on the feed-forward neural network performance.

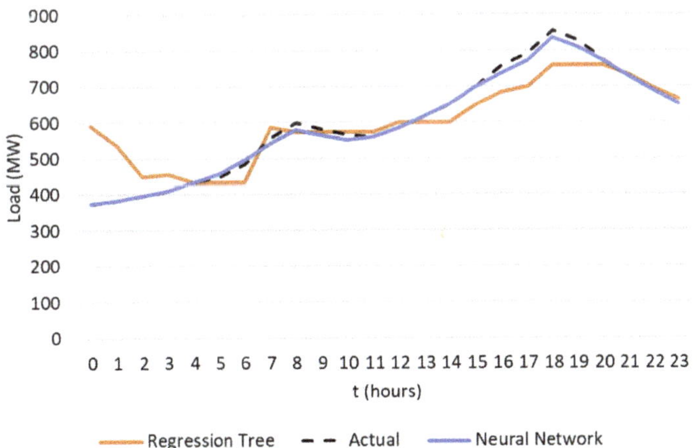

Figure 12. Performance of neural network against regression tree with best-fit optimizer.

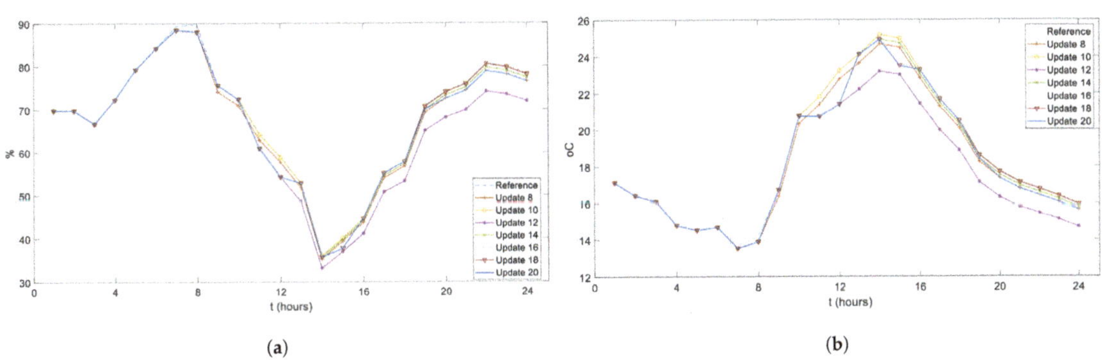

(**a**)

(**b**)

Figure 13. Real-time deviations from the day-ahead forecast of (**a**) relative humidity; (**b**) temperature.

These deviations have a daily impact on the forecasted load, which is reflected as frequency violations. To correct the deviations, more generators are required to serve the varying demand or spinning reserves are called upon. Any generation deficits may lead into load interruptions, while excess generation can cause active power curtailment. In any case, the unexpected deviations increase the total production cost and force the system operators to plan-ahead bulk operating reserves to appropriately regulate the system frequency. In our paradigm, the SR minimization relies on the high-performance neural network and the real-time corrections based on the updated forecasts of temperature and humidity. In contrast to traditional alternatives, which associate the SR requirements solely with the forecaster performance, performing real-time, intra-hour load forecast, these requirements are reasonably mitigated.

We provide the realization of our proposed solution to an energy-intensive winter day in Figure 14. In this case, one can observe how the negative temperature deviations between 10:00 and 16:00 affect the hourly-load forecast. Considering that $E = P \cdot t$, this deviation corresponds to a daily power of 146.867 MWh or 35.864 MW instant power equivalent in the worst case. To recover this imbalance, a spinning reserve of up to 4.67% would be adequate if planned ahead.

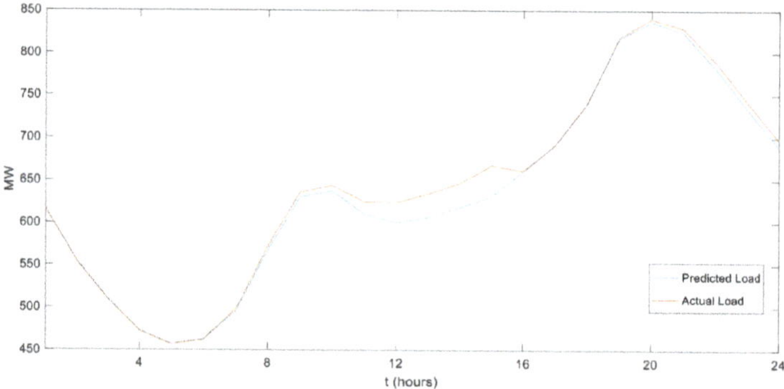

Figure 14. A realization of the real-time load forecast model for an energy-intensive winter day.

Finally, we depict similar configurations for the more mitigated load curves in spring and autumn, together with the most energy-intensive day in summer, in Figure 15. For completeness sake, we list the comparative results with respect to the achieved SR capacity per month in Table 1, considering the real-time weather impact and overall performance of our load forecasting model.

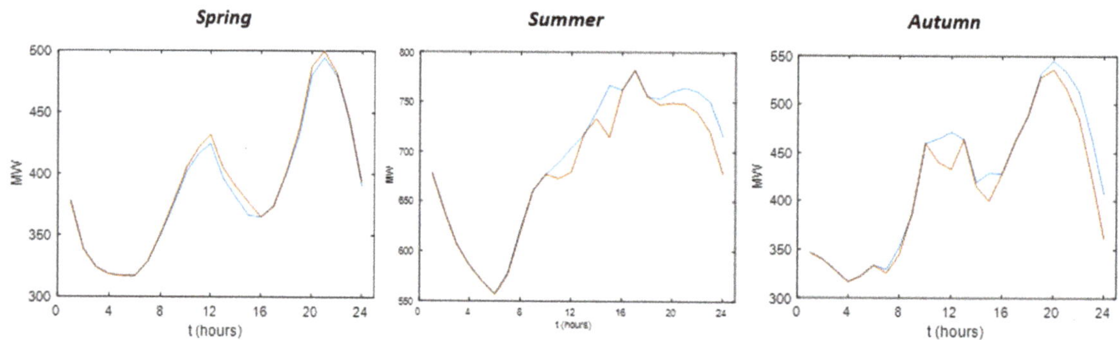

Figure 15. Real-time deviations from the day-ahead forecast concerning specific, energy-intensive days in spring, summer and autumn.

Table 1. Spinning reserve comparisons pertaining our proposed solution and robust alternatives.

Month	Load Demand (GWh)	Robust Formulation (GWh)	Real-Time Solution (GWh)
January	448.06	22.4	2.98
February	404.7	20.24	2.69
March	278.99	13.95	8.5
April	288.29	14.42	8.79
May	285.97	14.3	8.58
June	498.21	24.91	13.86
July	514.82	25.74	14.32
August	502.37	25.12	14.21
September	304.39	15.22	15.1
October	314.54	15.73	15.61
November	312.0	15.6	15.48
December	444.02	22.2	2.76

5. Conclusions

The continuous increase in the renewable energy contribution deteriorates the flexibility and stability of modern power systems calling for bulk spinning reserve margins. In this work, we proposed a dynamical forecaster to ameliorate the expensive requirements of spinning reserves based on real-time updates. Utilizing neural networks, we trained artificial models to forecast the load ahead with great accuracy, based on critical predictors and without using any model development structure to individuate and select the appropriate input parameters. Instead, we exploited eight predictors and distinguished them into constant and variable inputs by making use of a model predictive control. Apart from the most actively used data for historical load, seasonality and human activity, we also considered relative humidity as one of our main variable inputs. We performed real-time applications with the aid of a model predictive control, developed to update the recommendations intra-hourly and further correct the imposed deviations. Exploiting actual data regarding an isolated power system, the experimental results show that improvements exist in terms of decreased spinning reserve requirements to regulate the violated frequency. These findings strongly collaborate our claims and strengthen the arsenal of independent system operators with an effective tool for real-time load forecasting and total generation cost minimization.

As for future directions of research, we highlight the consolidation of more predictors correlated with renewable generation such as wind and solar. This way, a global forecaster could recommend the residual load target by making use of multi-input/multi-output neural networks. In addition, the fuel-dependent electricity prize may also take place as a real-time update, affecting both the human activity and hourly load demand.

Author Contributions: Conceptualization, P.N.; methodology, P.N.; software, P.N.; validation, H.P.; writing—original draft preparation, P.N. All authors have read and agreed to the published version of the manuscript.

Funding: This research received no external funding.

Institutional Review Board Statement: Not applicable.

Informed Consent Statement: Not applicable.

Acknowledgments: The authors would like to thank the Cyprus Energy Regulatory Authority for the provision of the needed annual data.

Conflicts of Interest: The authors declare no conflict of interest.

References

1. Anand, H.; Narang, N.; Dhillon, J. Profit based unit commitment using hybrid optimization technique. *Energy* **2018**, *148*, 701–715. [CrossRef]
2. Nikolaidis, P.; Poullikkas, A. Sustainable Services to Enhance Flexibility in the Upcoming Smart Grids. In *Sustaining Resources for Tomorrow*; Springer: Berlin/Heidelberg, Germany, 2020; pp. 245–274.
3. Nikolaidis, P.; Poullikkas, A. Cost metrics of electrical energy storage technologies in potential power system operations. *Sustain. Energy Technol. Assess.* **2018**, *25*, 43–59. [CrossRef]
4. Nikolaidis, P.; Chatzis, S.; Poullikkas, A. Renewable energy integration through optimal unit commitment and electricity storage in weak power networks. *Int. J. Sustain. Energy* **2019**, *38*, 398–414. [CrossRef]
5. Zhu, R.; Wei, H.; Bai, X. Wasserstein metric based distributionally robust approximate framework for unit commitment. *IEEE Trans. Power Syst.* **2019**, *34*, 2991–3001. [CrossRef]
6. Kazemzadeh, N.; Ryan, S.M.; Hamzeei, M. Robust optimization vs. stochastic programming incorporating risk measures for unit commitment with uncertain variable renewable generation. *Energy Syst.* **2019**, *10*, 517–541. [CrossRef]
7. Shahbazitabar, M.; Abdi, H. A novel priority-based stochastic unit commitment considering renewable energy sources and parking lot cooperation. *Energy* **2018**, *161*, 308–324. [CrossRef]
8. Yu, Y.; Luh, P.B.; Litvinov, E.; Zheng, T.; Zhao, J.; Zhao, F.; Schiro, D.A. Transmission contingency-constrained unit commitment with high penetration of renewables via interval optimization. *IEEE Trans. Power Syst.* **2016**, *32*, 1410–1421. [CrossRef]
9. Lorca, A.; Sun, X.A. Multistage robust unit commitment with dynamic uncertainty sets and energy storage. *IEEE Trans. Power Syst.* **2016**, *32*, 1678–1688. [CrossRef]
10. Kong, W.; Dong, Z.Y.; Jia, Y.; Hill, D.J.; Xu, Y.; Zhang, Y. Short-term residential load forecasting based on LSTM recurrent neural network. *IEEE Trans. Smart Grid* **2017**, *10*, 841–851. [CrossRef]
11. Kong, W.; Dong, Z.Y.; Hill, D.J.; Luo, F.; Xu, Y. Short-term residential load forecasting based on resident behaviour learning. *IEEE Trans. Power Syst.* **2017**, *33*, 1087–1088. [CrossRef]
12. Shi, H.; Xu, M.; Li, R. Deep learning for household load forecasting—A novel pooling deep RNN. *IEEE Trans. Smart Grid* **2017**, *9*, 5271–5280. [CrossRef]
13. Hong, T.; Fan, S. Probabilistic electric load forecasting: A tutorial review. *Int. J. Forecast.* **2016**, *32*, 914–938. [CrossRef]
14. Zhang, J.; Wei, Y.M.; Li, D.; Tan, Z.; Zhou, J. Short term electricity load forecasting using a hybrid model. *Energy* **2018**, *158*, 774–781. [CrossRef]
15. Wang, Y.; Gan, D.; Sun, M.; Zhang, N.; Lu, Z.; Kang, C. Probabilistic individual load forecasting using pinball loss guided LSTM. *Appl. Energy* **2019**, *235*, 10–20. [CrossRef]
16. Amarasinghe, K.; Marino, D.L.; Manic, M. Deep neural networks for energy load forecasting. In Proceedings of the 2017 IEEE 26th International Symposium on Industrial Electronics (ISIE), Edinburgh, UK, 19–21 June 2017; pp. 1483–1488.
17. Chen, K.; Chen, K.; Wang, Q.; He, Z.; Hu, J.; He, J. Short-term load forecasting with deep residual networks. *IEEE Trans. Smart Grid* **2018**, *10*, 3943–3952. [CrossRef]
18. Nikolaidis, P.; Poullikkas, A. Enhanced Lagrange relaxation for the optimal unit commitment of identical generating units. *IET Gener. Transm. Distrib.* **2020**, *14*, 3920–3928. [CrossRef]
19. Hansen, A.; Sørensen, P.; Zeni, L.; Altin, M. *Frequency Control Modelling—Basics*; DTU Wind Energy: Copenhagen, Denmark, 2016.
20. Kirby, B.J. *Frequency Regulation Basics and Trends*; U.S. Department of Energy Office of Scientific and Technical Information: Oak Ridge, TN, USA, 2005. [CrossRef]
21. Sadollah, A.; Sayyaadi, H.; Yadav, A. A dynamic metaheuristic optimization model inspired by biological nervous systems: Neural network algorithm. *Appl. Soft Comput.* **2018**, *71*, 747–782. [CrossRef]
22. Arnold, M.; Andersson, G. Model predictive control of energy storage including uncertain forecasts. In Proceedings of the Power Systems Computation Conference (PSCC), Stockholm, Sweden, 22–26 August 2011; Voume 23, pp. 24–29.
23. Bennett, C.; Stewart, R.A.; Lu, J. Autoregressive with exogenous variables and neural network short-term load forecast models for residential low voltage distribution networks. *Energies* **2014**, *7*, 2938–2960. [CrossRef]
24. Nikolaidis, P.; Chatzis, S.; Poullikkas, A. Optimal planning of electricity storage to minimize operating reserve requirements in an isolated island grid. *Energy Syst.* **2019**, *10*, 1–18. [CrossRef]
25. Ustaoglu, B.; Cigizoglu, H.; Karaca, M. Forecast of daily mean, maximum and minimum temperature time series by three artificial neural network methods. *Meteorol. Appl. A J. Forecast. Pract. Appl. Train. Tech. Model.* **2008**, *15*, 431–445. [CrossRef]
26. Castañeda-Miranda, A.; de Icaza-Herrera, M.; Castaño, V.M. Meteorological temperature and humidity prediction from fourier-statistical analysis of hourly data. *Adv. Meteorol.* **2019**, *2019*, 4164097. [CrossRef]
27. Transtrum, M.K.; Sethna, J.P. Improvements to the Levenberg-Marquardt algorithm for nonlinear least-squares minimization. *arXiv* **2012**, arXiv:1201.5885.

Article

The Wisdom of the Data: Getting the Most Out of Univariate Time Series Forecasting

Fotios Petropoulos [1,*] and **Evangelos Spiliotis** [2]

[1] School of Management, University of Bath, Bath BA2 7AY, UK
[2] Forecasting and Strategy Unit, School of Electrical and Computer Engineering, National Technical University of Athens, 15780 Athens, Greece; spiliotis@fsu.gr
[*] Correspondence: f.petropoulos@bath.ac.uk

Abstract: Forecasting is a challenging task that typically requires making assumptions about the observed data but also the future conditions. Inevitably, any forecasting process will result in some degree of inaccuracy. The forecasting performance will further deteriorate as the uncertainty increases. In this article, we focus on univariate time series forecasting and we review five approaches that one can use to enhance the performance of standard extrapolation methods. Much has been written about the "wisdom of the crowds" and how collective opinions will outperform individual ones. We present the concept of the "wisdom of the data" and how data manipulation can result in information extraction which, in turn, translates to improved forecast accuracy by aggregating (combining) forecasts computed on different perspectives of the same data. We describe and discuss approaches that are based on the manipulation of local curvatures (theta method), temporal aggregation, bootstrapping, sub-seasonal and incomplete time series. We compare these approaches with regards to how they extract information from the data, their computational cost, and their performance.

Keywords: information; combination; uncertainty; theta; temporal aggregation; bagging; sub-seasonal series

Citation: Petropoulos, F.; Spiliotis, E. The Wisdom of the Data: Getting the Most Out of Univariate Time Series Forecasting. *Forecasting* **2021**, *3*, 478–497. https://doi.org/10.3390/forecast3030029

Academic Editor: Sonia Leva

Received: 17 May 2021
Accepted: 21 June 2021
Published: 23 June 2021

Publisher's Note: MDPI stays neutral with regard to jurisdictional claims in published maps and institutional affiliations.

1. Introduction

Univariate time series forecasting is the creation of extrapolations for a single variable based on past, time-ordered observations of the same variable. Despite the geometric increase in data availability, univariate forecasts are even today the basis for the decision making in many organisations. Improvements in the performance of such forecasts are crucial for reducing costs associated with operational, tactical, and strategic planning [1].

Nowadays, automatic time series forecasting can be easily achieved using dedicated forecasting software or open source packages. Examples include ForecastPro®, SAS Forecasting Server®, and the *forecast* package for R statistical software. Such software and packages offer tools for batch and automatic forecasting with minimal to zero manual input. They integrate families of models, like exponential smoothing [2] and autoregressive integrated moving average, ARIMA [3], that can capture a wide range of data patterns and produce extrapolations with ease. However, such families of models rely on assumptions that are barely met in practice, and struggle to select the most appropriate model for a given time series due to the uncertainties involved: identifying the optimal model form, estimating the optimal set of parameters, and dealing with the inherent uncertainty in the data [4].

The purpose of this article is to provide an overview of approaches that can be used to enhance the performance of univariate forecasting methods. There are four common characteristics that govern the approaches covered in this article. First, all approaches attempt to distil as much information from the original time series data as possible by exploring them through alternative lenses. This is achieved through amplification of specific time series features and transformation of the original time series. Second, the approaches build on the success of forecast combinations to offer improved forecasting performance while tackling uncertainties regarding model form and parameter specification. Third, all approaches are model-free in the sense that do not rely on a particular family or pool of models. Four, each of the approaches manage to handle at least one of the uncertainties associated with fitting forecasting models: model form, parameter, and data.

In summary, we consider, present, and discuss the following five approaches:

- Theta method, where the seasonally adjusted data are decomposed into theta lines with different curvatures that are suitable to handle local and global movements in the data [5–10];
- Multiple temporal aggregation (MTA), where the original time series is transformed into multiple new series of lower frequencies (higher temporal aggregation levels) [11–14];
- Bagging (bootstrapping and aggregation), in which the remainder of a time series is bootstrapped towards the creation of new series with the same underlying structure (trend and seasonality) but different random components [4,15,16];
- Forecasting with sub-seasonal series (FOSS), in which a seasonal time series is split into multiple new series where only particular seasons (or sets of seasons) are observed and modelled [17];
- Forecasting from multiple starting points, where the least recent observations of a time series are not used when estimating forecasting models [18,19].

We should clarify that although the literature involves several other univariate approaches in addition to the aforementioned ones for extracting more information from the original data and mitigating the drawbacks that forecasts from single forecasting methods may involve, these are not considered in the present study as they are not characterised by the four key attributes discussed earlier. For instance, when the forecast errors of a method display strong auto-correlations (e.g., because the method fails to fully capture seasonality or trend), a common approach is to adjust the forecasts originally produced according to their expected error, specified using a second univariate forecasting method on the residuals of the first one. TBATS [20] exponential smoothing state space model with Box-Cox transformation, ARMA errors, Trend and Seasonal components, and Theta with ARMA errors [21], are just some examples of this approach which, although enhances forecasting performance, does not rely on combinations. Similarly, decomposition techniques that allow for complex, multiple seasonal patterns to be captured [22], can be regarded as "wisdom of the data" approaches, but do not involve combinations, depending also on particular models and, in many cases, explanatory variables.

The next five sections expand on each of the above approaches: We offer a summary of the related research studies, we describe how these approaches handle and manipulate the original time series data, and we discuss the advantages gained from their application. Section 7 offers a cross-comparison of the approaches, with an emphasis on the uncertainties that each handles, as well as their computational cost. Finally, Section 8 offers our conclusions and insights for future research.

2. Theta Method

The theta method was the top-performing submission in the M3 forecasting competition [23]. Its name originates from the first letter of the Greek word for "temperature", θ. Similarly to how a decrease or increase in the temperature would result in contraction or expansion, the theta method amplifies or smooths the local curvatures of a time series, i.e., the distances between the points of the series with those of a simple linear regression line, computed over its observations against time. The result of this local-curvatures manipulation process is the creation of new series that are called "theta lines". The degree of amplification or reduction in the local variations is controlled by a parameter, θ, where a value of 1 corresponds to the original data with the original local curvatures. If $\theta > 1$, then the local variations are amplified; if $\theta < 1$, then the resulting theta line is smoother than the original data.

In its simplest form, the theta method decomposes the original data into two theta lines with parameters $\theta = 0$ and $\theta = 2$ [5]. The theta line with $\theta = 0$ corresponds to linear regression on a time-trend indicator. This is a straight line that captures the long term trend of the data and has no local variations. The theta line with $\theta = 2$ displays double the curvatures of the data. It is argued that this second theta line is able to better capture the short term variations in the data. Each of these two theta lines are extrapolated separately. Assimakopoulos and Nikolopoulos [5] used the forecasts of the linear regression on trend to extrapolate theta line with $\theta = 0$ and the simple exponential smoothing (SES) method to produce forecasts for the other theta line ($\theta = 2$). Once the forecasts from the two theta lines have been produced, then these are combined with equal weights to form the forecast for theta line with $\theta = 1$ that corresponds to the data with the original curvatures.

The above process works directly on data that do not exhibit seasonality. However, if the original data are seasonal, then they need to be adjusted for seasonality before the application of the theta decomposition. Assimakopoulos and Nikolopoulos [5] proposed the use of the classical decomposition method with the assumption that the seasonal pattern is multiplicative in nature; a not unreasonable assumption for real life applications. As an alternative, Spiliotis et al. [9] proposed using shrinkage estimators of time series seasonal indices to avoid cases where their values are exaggerated. In both cases, a simple statistical test based on the autocorrelation coefficient with a lag that matches the periodicity of the data is used to decide on the existence of a (sufficiently strong) seasonal pattern, typically considering a confidence level of 90%. This test is described in detail in [8]. If the theta decomposition is applied on the seasonally adjusted data, then the resulting forecasts are not seasonal, and a seasonal re-adjustment is needed. This is simply done by multiplying the combined forecasts with the respective seasonal indices computed earlier by the decomposition method. A visual example of producing theta lines from seasonal time series data is presented in Figure 1.

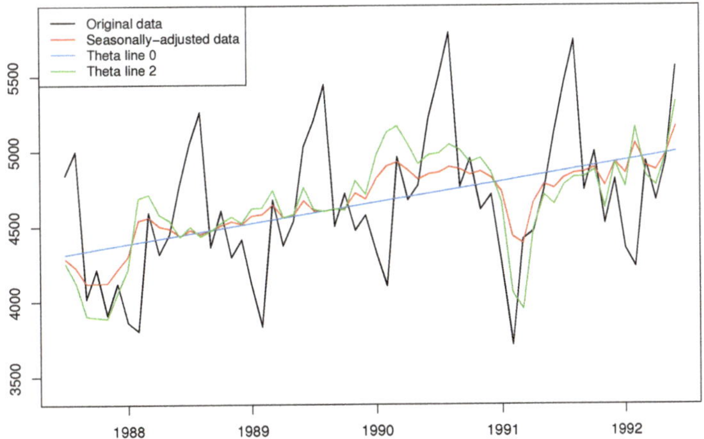

Figure 1. An illustrative example of producing theta lines for the theta method. The original data (black line) are de-seasonalised (red line). Then, a linear regression on trend produces the theta line with $\theta = 0$ (blue line). The theta line with $\theta = 2$ (green line) has double the curvatures of the seasonally adjusted data.

When theta is restricted to the simple form of two theta lines (0 and 2) that are extrapolated by the linear regression line and SES, then its application on some seasonally adjusted data is mathematically equivalent to SES with drift [6]. However, it would be more appropriate if theta is seen as a decomposition framework rather than a forecasting method. One can decide on the number of theta lines, their theta parameters, the forecasting methods to be applied on each of them, and the combination weights, among other modelling choices. In fact, as explained by Spiliotis et al. [9], *"the advantage of theta derives exactly from (its) "divide and conquer" property: There is no single forecasting model capable of effectively capturing all possible time series patterns. Yet, if the series is decomposed into multiple lines of a reduced amount of information, improvements in forecasting accuracy are possible even for the case of conventional models"*.

Several studies have worked on expanding the theta method to the aforementioned directions. Petropoulos and Nikolopoulos [24] examine the use of equal and unequal weights for the combinations of the two theta lines forecasts, and conclude that optimally choosing the combination weights per series may result in performance benefits. Petropoulos [25] proposes the addition of a third theta line with $\theta = 1$ that is extrapolated by the damped-trend exponential smoothing method. He also suggests the addition of a second short term trend-line that is fitted on the most recent observations, which is closely connected with the concept of multiple starting points (see Section 6). Fioruci et al. [26] and Fiorucci et al. [8] offer generalised rolling origin evaluation methods and state space models for optimising the theta parameter of the second theta line, showcasing the benefits in the out-of-sample accuracy of the method. Thomakos and Nikolopoulos [27] expand the application of the theta method in a multivariate setting and show the conditions under which this is expected to work better than its standard, univariate implementation.

Two recent extensions on the theta method are particularly interesting. Following the work of Spiliotis et al. [9], Spiliotis et al. [10] offer a taxonomy of theta models that can capture several forecasting profiles regarding the type of trend (additive or multiplicative) and seasonality (none, additive, or multiplicative). This is a significance advancement since the original theta method was designed on the assumptions of a linear trend and multiplicative seasonality. The authors propose non-linear trends, but also alternative seasonal profiles in a framework that resembles that of the exponential smoothing family of models [6]. Moreover, they define a process for selecting an optimal theta method and offer a simple way to empirically estimate the prediction intervals. Their "AutoTheta" method shows improved performance over the standard theta method for both point forecast accuracy but also the estimation of uncertainty. Legaki and Koutsouri [28] deal with non-linear trends in an alternative fashion. They apply a Box-Cox transformation on the seasonally adjusted data prior to the theta decomposition and extrapolation. The value of the Box-Cox transformation parameter, λ, is selected so that the profile log-likelihood of a linear model fitted to the seasonally adjusted series is maximised, with the choice of λ being restricted in $[0, 1]$. A Box-Cox transformation allows the application of theta on data with non-linear trends but also results in stabilisation of the variance. The "Box-Cox Theta" was one of the solutions submitted in the M4 forecasting competition [29], resulting in very good point forecast accuracy with very low computational cost [30].

The theta method has performed well in a variety of settings that involve financial [31], tourism [32], and inventory forecasting [33]. It is not a surprise that nowadays it is considered to be one of the default time series forecasting benchmarks along with the automatic implementations of exponential smoothing and ARIMA [34], as showcased by the M4 forecasting competition [29]. The theta method is attractive for its simplicity, robust performance, and computational efficiency. The book of Nikolopoulos and Thomakos [35] exclusively focuses on the theory and applications of the theta method, highlighting the conditions under which it will outperform other forecasting methods. Several open source implementations of the theta method exist. We would like to spotlight the *forecTheta* package for R statistical software, as well as the functions `thetaf()` and `theta()` of the packages *forecast* and *tsutils*, respectively. Finally, Petropoulos and Nikolopoulos [36] offer a step-by-step tutorial of the standard theta method coupled with an implementation in just 10 lines of R code.

3. Multiple Temporal Aggregation

The theta method extracts more information from the data by amplifying or deflating the local curvatures. In other words, the theta method manipulates the data on the vertical axis of a standard time series plot. The next approach we explore manipulates the data on the horizontal axis, i.e., the time. Temporal aggregation refers to a time series transformation where a higher frequency series is translated into a series of lower frequency see Section 2.9.2 in [37]. For example, a time series on the daily frequency can be converted into a weekly-frequency series when considering non-overlapping time buckets of 7 days each. Different levels of temporal aggregation result in new, shorter series where the high frequency components (seasonality and noise) are filtered out while level and trends are made easier to discern and model. Moreover, when temporal aggregation is applied on very granular, intermittent data, then we observe a decrease on the degree of intermittence, i.e., the number of zero observations included in the series, thus facilitating the overall forecasting process. An example of the temporal aggregation process applied on fast moving data is presented in Figure 2.

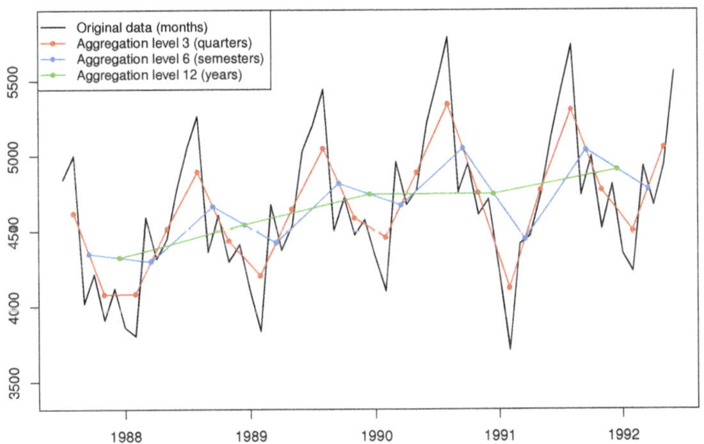

Figure 2. A visual example where multiple new temporally aggregated time series are created based on the original data. The monthly data (black line) are temporally aggregated to quarterly (red line), semesterly (blue line), and yearly (green line) data.

Although it is possible that one focuses on modelling a single aggregation level, even if this is not the original level on which the data are recorded [38–40], more benefits will usually arise from modelling multiple temporal aggregation (MTA) levels and combining the resulting forecasts. Kourentzes et al. [11] offer one of the first systematic studies to explore the beneficial effects of MTA. Focusing on exponential smoothing models [2], they propose that model selection should be applied on each temporally aggregated series separately. The exponential smoothing model components (level, trend, and seasonality) are estimated per aggregation level and their additive-transformed estimates are averaged across levels. The summation of the three average components is the final forecast. This approach is known as the "multiple aggregation prediction algorithm" (MAPA). The need for averaging at a component level rather than at a forecast level was driven by the fact that seasonality may not be possible to estimate in some levels (consider, for instance, monthly data and an aggregation level of five periods). Combining at a component level avoids the excessive shrinkage of the seasonal pattern [41,42]. MAPA showcased improved performance over the exponential smoothing benchmark that was applied on the original data only [11]. The improvements of MAPA over the benchmarks were more obvious on the longer forecasting horizons.

MAPA, as introduced by Kourentzes et al. [11] and implemented with exponential smoothing, is a great solution for amplifying and smoothing data patterns for fast moving series. However, when the series become intermittent, with the presence of many zeros among the non-zero demand observations, then the toolbox of forecasting models applied across the various aggregation levels can be updated to include specialised methods for intermittent demands. Such methods include the Croston's method [43] and the Syntetos-

Boylan approximation (SBA) [44]. Petropoulos and Kourentzes [13] suggest the use of multiple temporal aggregation levels for the case of slow-moving demand series, where a selection between the Croston's method and the SBA is made based on the degree of intermittence and the variability of the non-zero values [45]. Finally, if the average inter-demand interval becomes equal to unity (i.e., the intermittent data are sufficiently temporally aggregated to become non-intermittent), then Petropoulos and Kourentzes [13] suggest replacing specialised methods for intermittent demand with SES. The empirical results from the application of the MAPA version for intermittent demand data showed superior forecasting performance on a variety of metrics that included proxies for the inventory performance.

Another extension to MAPA was introduced by Kourentzes and Petropoulos [46] to allow the algorithm handle exogenous variables which are estimated as additional components. The concept is similar to how exponential smoothing models (ETS) are extended to include exogenous variables (ETSx). However, the multivariate version of MAPA (MAPAx) performs temporal transformation on the exogenous variables too. This, by turn, tackles the uncertainty associated with estimating not only the effects of such predictors, but also their timing, i.e., leading and lagging effects. Applied on demand volumes affected by promotions, MAPAx offered a performance that was better to either ETSx or ARIMAx (ARIMA models with exogenous variables), both in terms of accuracy and bias, across multiple planning horizons.

One of the most important milestones in the development of MTA has been its conceptualisation in a hierarchical fashion. This enabled to directly apply the advances of the rich hierarchical literature [47–50] to the MTA application, that includes the estimation of coherent forecasts from the base forecasts of each hierarchical node. In essence, each hierarchy consists of observations at the most granular frequency at the bottom level, which are then added up to higher hierarchical levels, with the top level usually being a full periodic cycle. For example, monthly observations are added to bi-monthly, quarterly, four-monthly, semesterly, and yearly. Temporal hierarchies were first proposed by Athanasopoulos et al. [14], who showed that such structures allow MTA to be applied to a wide range of forecasts that are not limited to exponential smoothing ones and could even include judgment. The authors performed a large simulation study to better understand why temporal hierarchies work better than simply modelling the original data. Finally, they discussed the managerial implications of MTA through a case study of accident and emergency demand data.

Since the work of Athanasopoulos et al. [14], there has been a spark of research studies around forecasting with temporal hierarchies (THIEF). We now provide some highlights. Spiliotis et al. [41] proposed three simple ways to improve performance of temporal hierarchies: (i) model combinations to the base forecasts prior reconciliation, (ii) additive and multiplicative bias adjustments to the base forecasts, and (iii) a selective application of temporal hierarchies so that unnecessary seasonal shrinkage is avoided for the time series that exhibit strong seasonality, closely related also to the work of Kourentzes et al. [51] on optimal selection of temporal aggregation levels. Jeon et al. [52] expanded temporal hierarchical forecasting from point forecast reconciliation to probabilistic coherent forecasts, showcasing its benefits on high frequency wind power production and electricity load data. Additionally, focusing on short term electricity load data, Nystrup et al. [53] showed that temporal hierarchical forecasting can be significantly improved when auto- and cross-correlations are taken into account in the reconciliation stage of the base forecasts. Finally, Kourentzes and Athanasopoulos [54] applied temporal hierarchies on intermittent demand data, arguing that some data patterns (trend and seasonality) are difficult to discern on low levels of aggregation where the degree of intermittence is high. They selectively used Teunter-Syntetos-Babai (TSB) [55] method for intermittent demand or ETS based on an intermittence threshold, which acts as a hyperparameter. Generally, the accuracy improvements were higher for lower intermittence thresholds, i.e., TSB switches to ETS when the intermittence is low, as investigated on 5000 time series depicting the demand of aerospace spare parts.

A fertile field for research is the integration of temporal aggregation forecasting with the more traditional cross-sectional one, towards what is dubbed as "cross-temporal forecasting". To the best of our knowledge, Spiliotis et al. [42] were the first to investigate this issue, focusing on hourly electricity consumption data from a bank, disaggregated into branches and further disaggregated into energy uses. They proposed a sequential process where a simplified version of MAPA is first applied on the seasonally adjusted data, followed by reseasonalisation of the temporally combined forecasts and consequent application of cross-sectional hierarchical forecasting for the production of coherent forecasts across all cross-sectional levels. Kourentzes and Athanasopoulos [56] approached cross-temporal aggregation from a hierarchical approach, instead of using MAPA. Although they defined full cross-sectional hierarchies, they still used a sequential approach where they first apply temporal hierarchies for each cross-sectional node followed by cross-sectional reconciliation at each aggregation level with the resulting forecast being combined using equal weights towards a "consensus reconciliation matrix". The authors showed that this approach resulted in improvements when applied on Australian tourism data. Yagli et al. [57] explored further the sequential implementation of cross-temporal hierarchies by comparing the appropriate order of application, i.e., spatial then temporal, or temporal then spatial. Using photovoltaic power generation data, they showed greater benefits when temporal aggregation is applied prior to cross-sectional (spatial), while they also provided evidence that temporal aggregation may not be needed at all levels of the cross-sectional hierarchy.

Overall, we can see a large number of studies over the last few years that focus on issues surrounding MTA. MTA is attractive as it offers significant performance improvements that are coupled with aligned decision making [14]. Forecasts are produced at different frequencies and are then reconciled, rendering them suitable for use in several functions within companies and organisations, including operational, tactical, and strategic planning. Although, normally, different teams and departments within organisations would produce their own sets of forecasts, MTA brings us one step closer to the concept of "one number forecast", where the same sets of forecasts can be used for logistics, manufacturing, scheduling, budgeting, etc. Various implementations of MTA are available in open source forecasting packages that include the `mapa()` function of the *MAPA* package (MAPA and MAPAx), the `imapa()` function of the *tsintermittent* package (MAPA for intermittent demand data), and the `thief()` function of the *thief* package (temporal hierarchies) for R statistical software.

4. Bagging

The next approach that we investigate is called "bagging", which is short for "bootstrapping and aggregation". In brief, bagging is based on the resampling of the random component of a series towards the creation of new series with the same underlying patterns (trend and seasonality) but different remainder. Multiple forecasts are produced using the original and the bootstrapped series which are then aggregated (combined) to form the final forecast. In more detail, the steps for the bagging approach are as follows:

1. A Box-Cox transformation is applied on the original series. The λ parameter for the Box-Cox transformation is automatically selected based on the Guerrero's method [58], but other methods such as the maximisation of the profile log likelihood of a linear model fitted to the original data could be used. The purpose of this step is twofold. First, the variance of the series is stabilised. Second, multiplicative patterns are converted into additive ones;

2. The Box-Cox transformed series is decomposed into its components. If the series has periodicity greater than unity (e.g., quarterly or monthly data), then the seasonal and trend decomposition using Loess STL [59] decomposition is applied to separate the transformed series into the trend, seasonal, and remainder components. If the

series has no periodicity (e.g., yearly data), then a Loess decomposition is applied to separate the series into two components: trend and remainder;

3. The remainder component of the above decomposition is bootstrapped towards the creation of new vectors of remainders that follow the empirical distribution of the original remainder vector;

4. The bootstrapped remainder vectors are added to the other extracted components from the decomposition of step 2 (trend and, where applicable, seasonality) to form new bootstrapped series. These series have the same underlying structural patterns with the original series;

5. An inverse Box-Cox transformation is applied on each of the bootstrapped series, using the same λ parameter of step 1. This transformation brings the bootstrapped series back to the same scale as the original data. Figure 3 shows the estimated bootstrapped series based on the original (seasonal) data;

6. The original and the bootstrapped series are extrapolated using an automatic forecasting process, which may result on the use of the same or different model forms and parameters. In any case, many sets of forecasts are produced at the end of this step;

7. The forecasts from the original and bootstrapped series of the previous step are aggregated (averaged).

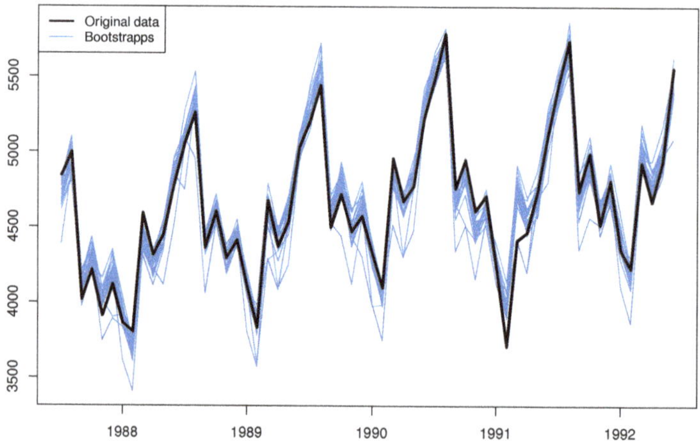

Figure 3. The original data (black line) together with 30 bootstrapped series (blue lines).

Effectively, bagging should be seen as a data augmentation (or oversampling) approach applied in univariate settings, in the sense that the amount of modified series added over the existing one for training the forecasting methods and producing the final forecasts are solely based on the series being predicted. This is a key difference compared to the multivariate data augmentation approaches used in the literature for successfully implementing "cross-learning" (or "global") forecasting methods [60], where the synthetic data share the underlying patterns of multiple series found in a broad set of series.

Bagging was first proposed by Bergmeir et al. [15], who applied it to improve the performance of exponential smoothing. They used the moving block bootstrapping MBB [61] algorithm to produce bootstrapped vectors of the remainder, and produced 99 bootstrapped series. The best ETS model was fitted on each of the original series and the 99 bootstrapped series in order to produce point forecasts. The final forecasts were obtained using the median operator, while the authors discuss that they also tried mean and trimmed means. Bagging on ETS offered improved performance over ETS simply applied on the original data. The authors also tried replacing Box-Cox and Loess decomposition with decomposition based on the components of the best ETS model fitted on the original data. The authors also explored replacing MBB with the sieve bootstrap method [62]. However, both these modifications resulted in, overall, inferior results.

In a follow-up study, Petropoulos et al. [4] sailed to explore the reasons behind the good performance of the bagging approach. They argued that bagging succeeds in tackling, at the same time, three sources of forecast uncertainty: (*i*) the uncertainty in selecting the correct model form, (*ii*) the uncertainty in estimating the model's parameters, and (*iii*) the inherent uncertainty of the data. They devised three simple experiments to disintegrate the benefits of bagging:

- After producing the bootstrapped series the usual way, the authors identified the optimal models on these bootstrapped series. Instead of using the forecasts from these models directly, their model forms were applied back to the original data, for which a different optimal model form may have been identified. Effectively, the bootstrapped series provided the frequencies with which each model form was identified as 'optimal', and these frequencies were then translated into combination weights for averaging the point forecasts of the different model forms when applied on the original data. All model parameters and forecasts were estimated using the original data only. This variation of bagging is known as "bootstrap model combination" (BMC) and handles only the uncertainty in the model form;
- The optimal model form was identified using the original data only. Subsequently, this optimal model form was applied on the original data and the bootstraps to obtain multiple independent estimates of the model parameters. The combination of each set of model parameters with the unique optimal model form was then applied again on the original data to produce multiple sets of point forecasts. As with bootstrap model combination, the bootstrapped series were not used to produce forecasts directly. This variation solely handles the uncertainty in estimating the model parameters;
- The optimal model form and set of parameters were estimated using the original series only. Subsequently, this unique model form and set of parameters were applied on all bootstrapped series to produce multiple sets of point forecasts. This variation solely tackles the uncertainty associated with the data, as the bootstrapped series are not used for selecting between models or estimating their parameters.

Using the data from M and M3 forecasting competitions, the results of Petropoulos et al. [4] showed that, on average, tackling model uncertainty alone through bootstrap model combination offers benefits that are higher than bagging itself. Simply addressing the uncertainty in estimating the parameters of the applied model is overall worse than either bagging or bootstrap model combination but still slightly better than forecasting without bootstraps. Tackling only the data uncertainty does not offer notable gains. The authors went one step further towards generalising bagging by considering replacing the estimator (ETS) with ARIMA. The results were consistent, with bootstrap model combination being the best approach overall. Finally, they replaced MBB with two other bootstrapping approaches, circular block bootstrap CBB [63] and linear process bootstrap LPB [64], showing that the relative average ranks of the various approaches would not significantly change.

Although the last two studies focused on the performance of bagging when applied on families of models (ETS and ARIMA), bagging can also lead in improvements in the forecasting performance when applied on single methods. Dantas et al. [65] showed that bagging with the Holt-Winters method, an exponential smoothing method that is able to capture the trend and the seasonality in the data, results in better performance than either ETS, ARIMA, or bagged ETS when forecasting the demand for air transportation.

To control the effect of the covariance on the combination step of the bagging approach, Dantas and Cyrino Oliveira [16] proposed the use of clusters of similar forecasts. Instead of aggregating across all forecasts, a diverse set of forecasts are selected from each cluster and then these selected forecasts are combined across clusters. This simple trick leads to reduced variance of the forecasts and, as a result, in reduced forecast error. They tested their cluster-modified bagging approach using ETS and data from the M3 forecasting competition and showed improvements in the point forecast accuracy.

Meira et al. [66] extended the previous works towards allowing the various bagging approaches to produce robust prediction intervals. They proposed "treating and pruning" strategies to selectively exclude models from the pool of candidate models such that models with explosive or outlying prediction interval values are not considered. This not only improved the performance of bagging and its variations, but also offered improvements upon the standard ETS. Overall, the authors demonstrated that bootstrap model combination offered very competitive performance compared to other bagging variations, both in terms of point forecast accuracy but also uncertainty estimation.

Research around bagging is much more scarce compared to theta or MTA. However, it is a robust alternative to deal with the various sources of uncertainty; arguably, though, an expensive one. Most published studies use between 50 and 100 bootstraps per series, with the computational cost need to fit all models and produce forecast being increased with the same rate. Open source implementations of the bagging approach include the functions `baggedETS()` and `baggedClusterETS()` of the R packages *forecast* and *tshacks*, respectively.

5. Sub-Seasonal Series

Instead of transforming a series to another of lower frequency through temporal aggregation using all observations, the next approach we review applies sub-sampling such that the resulting series includes only some of the periods within a periodic cycle. Consider, for instance, the case of daily data and focus on the weekly periodic pattern (weekly seasonality) of length 7. One (the traditional) option would be to consider all observations and model a series with a seasonal cycle equal to 7 periods. However, we could also consider only the values for a specific day of the week (such as Monday) and create a new time series which will not be seasonal and model it independently; and we could repeat this for every single day. Expanding this idea, we could also consider pairs, triplets, quadruplets, etc., of adjacent days (such as Monday-Tuesday or Monday-Tuesday-Wednesday, etc.) and form even more series of varying degrees of periodicity. In other words, we do not do any transformation per se, but systematically remove (through subsampling) specific periods of the series to create new ones of lower periodicity. Figure 4 shows an example of this sub-sampling process assuming some data originally recorded in the monthly frequency.

Forecasting with sub-seasonal series (FOSS) allows for simplified modelling of the patterns in the original series as different seasons are excluded every time [17]. This offers a more robust estimation of the trends but also the seasonal patterns in the data, with FOSS serving as a "magnifying glass" to the forecasting models used for their extraction. FOSS uses combination, and its welcome side effects, to aggregate the forecasts produced using the sub-seasonal series. Assuming a time series with periodicity s ($s = 7$ of daily data; $s = 12$ for monthly data), then FOSS entails the creation and modelling of $s^2 - s + 1$ series. However, most of these series have periodicity that is much lower than s and are relatively short, so the increase in the computational cost is not linearly associated with the increase in the number of models to fit. Each set of series produced by FOSS that has the same periodicity is referred to as "level of information". In its simplest form, FOSS models all such levels of information and combines the forecasts with equal weights.

Li et al. [17] offer a large empirical evaluation of FOSS using data from the M3 and M4 forecasting competitions. They showed that FOSS acts as a self-improving approach for the state-of-the-art batch forecasting benchmarks ETS and ARIMA. The improvements achieved are amplified when the periodicity of the original series is higher but also when the forecast horizon increases, i.e., when forecasting becomes more challenging. In addition, the authors applied FOSS on double seasonality, high frequency load data, and showed that FOSS is also a useful tool in the presence of complex seasonal patterns.

FOSS is publicly available through the *foss* package for R. However, research in this area is still premature. We can see several avenues for future exploration that include the selective use of levels of information, the use of unequal combination weights, and the creation of series using non-adjacent periods.

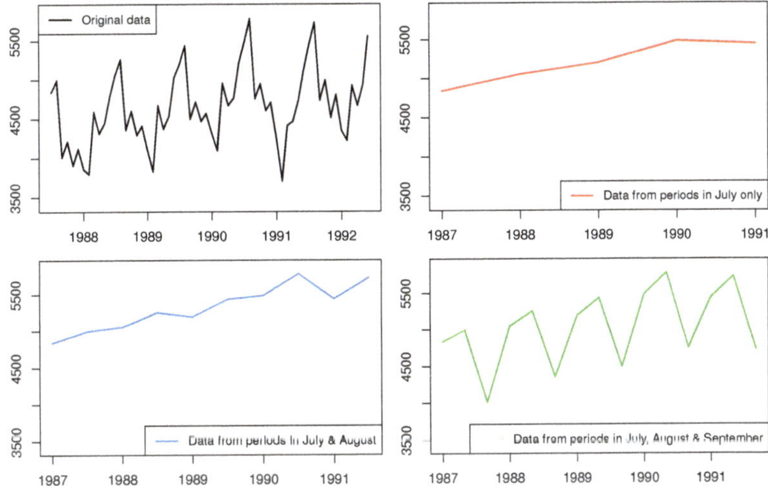

Figure 4 An illustrative example of producing sub-seasonal series by sub-sampling the original monthly data (first panel). In the second panel, we have produced a non-seasonal series that consists only of the periods in July of each year. The third and fourth panels show two more sub-sampled series with periodicity 2 and 3, respectively. Note that by considering particular subsamples, the level as well as other patterns change significantly.

6. Multiple Starting Points

In the era of big data, retaining long histories of time series values is quite inexpensive. However, would using as many data as possible for producing forecasts warrant the best performance? Although increasing the number of the available observations is expected to lead to better accuracy, such a result is subject to a certain degree of determinism in the data. If the data exhibit structural changes (level shifts, changes in the trend and seasonal patterns, etc.) or contain outlying values, then it may be better to use the most recent window of the data that would not be subject to such data irregularities [67]. Another extreme way to handle changes in the structure of the data would be to only retain the most recent window that contains enough observations that are necessary to produce forecasts. For example, the "Demand Planning" functionality of the SAP APO retains only three years of monthly data, discarding the least recent history.

Determining the optimal window of data on which forecasting models are fitted is not a straightforward exercise. Instead, one can consider multiple windows. Assume that a time series consist of n observations. A first set of forecasts can be produced using all n observations. A second set of forecasts can be produced using the most recent $n - 1$ observations. This process can be repeated $m + 1$ times, such that $n - m$ would still be enough data points for producing forecasts, i.e., at least two seasonal cycles for periodic data. Finally, the multiple sets of forecasts can be combined to obtain the final forecasts. This approach does not transform nor manipulate the original series, but simply trims the beginning of the data to produce multiple overlapping in-sample windows of different lengths based on which forecasts are produced. This approach is known as "forecasting using multiple starting points" (MSP). Figure 5 demonstrates the process of trimming the original series to create new series from multiple starting points.

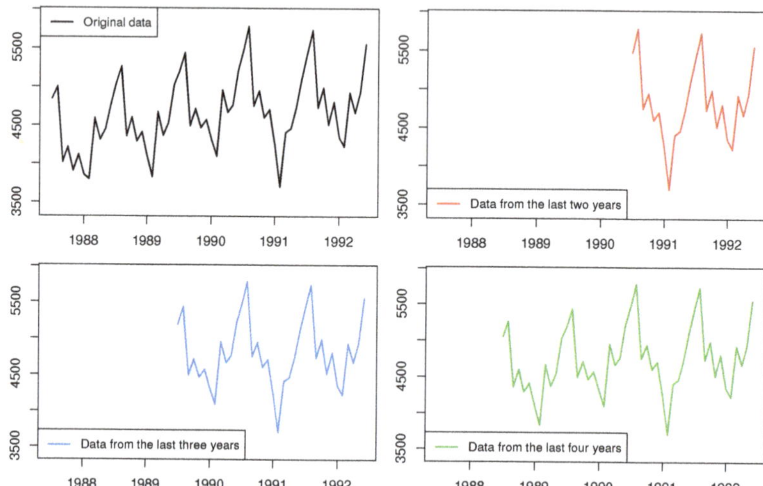

Figure 5. An illustrative example of producing series from multiple starting points. The original data (first panel) are trimmed so that the periods from only the last two years (second panel), the last three years (third panel), or the last four years (fourth panel) are considered.

Research in this stream is limited. To our knowledge, Disney and Petropoulos [18] were the first to empirically examine the approach based on multiple starting points. They applied it on data from the M3 forecasting competition using simple averaging operators (mean, median, and mode), which resulted in improved forecasting performance especially for the yearly frequency. They showed that the improvements generally increase as the number of starting points also increases. They also presented a case study based on the demand of 23 different types of spare parts, showing that forecasting from multiple starting points improves the accuracy in about three-fourths of the cases, with average improvements of about 10%. Bai et al. [19] also empirically investigated this approach, comparing equal versus optimal weights when combining across the forecasts but also considering non-consecutive starting points for their in-sample windows.

We believe that there is scope for more research in this area. Future studies could focus on applying formal techniques for detecting structural changes, which then can be used to select the starting points in a more systematic manner. Another possibility for future investigation could be the application of the concept of multiple starting points within cross-sectional hierarchical structures, where it is usually assumed that every node in the hierarchy has the same number of historical observations. Finally, understanding the circumstances under which forecasting from multiple starting points works best is vital towards implementing it in practice. To our knowledge, there does not exist an open source implementation for forecasting from multiple starting points.

7. Cross-Comparison

The five approaches that were described in the previous five subsections attempt to extract more information from the original time series by performing various forms of data modifications, adjustments, manipulations, and transformations. These can be summarise in three larger categories: *random component, frequency,* and *length.* Table 1 summarises how the extraction of information works for each of these five approaches. The theta method retains the frequency and length of the data, but amplifies the local curvatures which are represented as the residuals of a linear regression on trend. MTA transforms the original series through temporal aggregation to new shorter series of lower frequency; inevitably, the upsampling also results in lower noise [40]. Bagging is based on the bootstrapped series that are produced through re-sampling of the remainder from a decomposition process.

FOSS focuses on the subsampling of the original series resulting, similar to MAPA, in new series that are shorter and have lower periodicity. Finally, forecasting from multiple starting points is based on trimming the original series by removing the least recent values, retaining the frequency and random component intact.

Table 1. How does extraction of the information work?

Approach	Random Component	Frequency	Length
Theta	✓		
MTA	✓	✓	✓
Bagging	✓		
FOSS		✓	✓
MSP			✓

In Table 2, we map the five approaches with regards to how they handle the three sources of uncertainty: data uncertainty, model form uncertainty, and model parameters uncertainty. Our mapping involves two levels: ✓ denotes full account of that type of uncertainty, while ✓ denotes partial account. The theta method handles the uncertainty in the data in the sense that the local curvatures are amplified or reduced to better identify short and long term movements in the data. MTA also handles data uncertainty as temporal aggregation results in smoothing the noise in the data [40]. However, MTA also addresses the uncertainty in the model form, as different models may be identified as optimal at different temporal levels: a dominating seasonal pattern may lead to the selection of a seasonal-only model at the lowest aggregation level. However, as seasonality is smoothed out by temporal aggregation, a trend pattern may become apparent in a higher aggregation level [11]. Even if the same models are identified as optimal in various temporal levels, then MTA is still likely to help by partially addressing parameters' uncertainty.

Table 2. How do the five approaches handle the sources of uncertainty?

Approach	Sources of Uncertainty		
	Data	Model	
		Form	Parameters
Theta	✓		
MTA	✓	✓	✓
Bagging	✓	✓	✓
FOSS		✓	✓
MSP	✓	✓	✓

Bagging is the only approach that is able to tackle all three types of uncertainty, something that was extensively discussed by Petropoulos et al. [4]. However, some bagging variations focus on particular sources of uncertainty, as discussed in Section 4. FOSS is the only approach that does not explicitly handle the data uncertainty, but directly focuses on the model form uncertainty (and the model parameters). Finally, MSP tackles data uncertainty in the sense that, by trimming series, outliers or structural changes are removed. However, the new (shorter) series might also result in alternative model forms and sets of parameters.

Next, we consider the computational cost required by each of the approaches to produce forecasts. For simplification, instead of recording computational time per se (as this would depend on length of the series, among others) we compare the various approaches in terms of models required to be fitted. As a benchmark, it is noteworthy that the `ets()` function of the *forecast* package for R statistical software fits 19 models (8 for non-seasonal data) before a final model is selected and its forecasts are produced. The theta method is arguably one of the most inexpensive robust time series forecasting methods. In its standard implementation, it requires the fitting of just 2 models, one for each theta line (a simple linear regression model and SES). Even theta variations that consider more than two theta lines, the number of models required is small. The robust implementation by Legaki and Koutsouri [28] that uses a Box-Cox transformation offered, arguably, one of the best trade-offs in performance versus cost in the M4 competition [30].

Compared to theta, all other approaches are more costly. MTA requires forecasts for each aggregation level: 12 for monthly data; 4 for quarterly data. However, this could be slightly reduced when one uses temporal hierarchies (6 for monthly; 3 for quarterly). It is common that in each level an automatic algorithm, like ETS or ARIMA, is used. This means that the number of models required to be fitted increases a lot. Using temporal hierarchies with ETS results in fitting 103 exponential smoothing models for a monthly time series (5 seasonal levels $\times$ 19 models $+$ 1 non-seasonal level $\times$ 8 models). Empirical evidence https://kourentzes.com/forecasting/2014/10/31/guest-post-on-the-robustness-of-bagging-exponential-smoothing/ (accessed on 1 June 2021) has shown that Bagging's performance converges when at least 50 bootstrap series are aggregated—while most of the studies consider 100 bootstrap series. This means that Bagging with ETS requires fitting as little as 950 models (50 bootstraps $\times$ 19 models) for a single seasonal series and 400 models for a non-seasonal series, rendering it one of the most expensive approaches in this review study. Forecasting with sub-seasonal series is also very costly. From the $s^2 - s + 1$ series created, s of them have a periodicity of 1 with the potentially displaying seasonal patterns. Again assuming ETS, FOSS entails fitting and parametrising 165 models when modelling a series on the quarterly frequency ($(s^2 - 2s + 1) \times 19$ models for the sub-series with $s > 1$, plus $s \times 8$ models for the rest) rising to 2395 models for a monthly time series. The cost for the forecasting from multiple starting points heavily depends on the length of the series. Assuming a monthly time series ($s > 12$) with length $n = 50$, we would require at least $2s = 24$ periods to produce forecasts, which allows us to consider at most 27 starting points, translating to fitting 513 models when using ETS.

Lastly, we consider the performance of the various approaches as published in various studies so far. We focus on the data used in two forecasting competitions, M3 [23] and M4 [29], and particularly the yearly, quarterly, and monthly frequencies. It is important to note that our summary results, presented in Table 3, are based on the empirical evidence presented on other studies, which are identified next to each numerical result. We also limit our results to the values of the symmetric mean absolute percentage error (sMAPE) as reporting the mean absolute scaled error (MASE) was not possible (different researchers apply the scaling differently). For some studies that only provided relative improvements over a benchmark, such as [14], did not differentiate between the results of each competition, such as [41,42], or were limited to one of the two competitions considered, such as [28], we have reproduced the results using the code provided by the corresponding authors. Overall, we observe that some of these approaches are more suited in forecasting non-seasonal patterns (see, for instance, the very good performance of the Box-Cox Theta on the yearly frequency), while others are better when the series are periodic (see, for instance, FOSS and MTA).

Table 3. The published average performance of the five approaches on the monthly data from the M3 and M4 competitions.

Approach	Variation	M3			M4		
		Yearly	Quarterly	Monthly	Yearly	Quarterly	Monthly
Theta	Standard [5]	16.90 [23]	8.96 [23]	13.85 [23]	14.59 [29]	10.31 [29]	13.00 [29]
	Optimised θ [8]	15.94 [8]	9.28 [8]	13.74 [8]	13.68	10.09	13.32
	Box-Cox [28]	16.20	9.13	13.55	13.37 [29]	10.15 [29]	13.00 [29]
	AutoTheta [42]	16.02	9.18	13.86	13.80	10.13	13.13
	Theta-EXP [41]	16.48	8.99	13.44	14.11	10.37	13.12
MTA	MAPA-ETS [11]	18.37 [11]	9.63 [11]	13.69 [11]	14.88	10.27	12.97
	THIEF-ETS [14]	17.00	9.38	13.58	15.36	10.40	12.89
	THIEF-ARIMA [14]	17.10	9.79	14.49	15.15	10.61	13.39
Bagging	MBB-ETS [15]	17.89 [15]	10.13 [15]	13.64 [15]	14.47 [66]	10.23 [66]	13.30 [66]
	BMC-ETS [4]	17.15 [4]	9.56 [4]	13.79 [4]	14.94 [66]	10.08 [66]	13.07 [66]
	Pruned and Treated [66]	17.36 [66]	9.74 [66]	13.61 [66]	14.49 [66]	10.22 [66]	13.27 [66]
FOSS	FOSS-ETS [17]		9.24 [17]	13.56 [17]		10.15 [17]	12.84 [17]
	FOSS-ARIMA [17]		9.68 [17]	14.01 [17]		10.41 [17]	12.87 [17]
MSP	MSP-ETS [18]	16.90 [18]	9.79 [18]	14.00 [18]			

THIEF is applied using the "structural" reconciliation approach, while MAPA using the "hybrid" approach with a mean combination operator for aggregating the ETS components at different temporal aggregation levels. Optimised θ refers to the "Dynamic Optimised Theta Model". Results are replicated, where required, using the "thief", "MAPA", "forecast", and "forecTheta" packages for R, of versions 0.3, 2.0.4, 8.14, and 2.2, respectively.

Given the high-representativeness of the data in the M3 and M4 datasets [68], we believe that the results can be safely generalised in other settings and contexts, where the presented approaches are expected to work well. However, we will highlight here some particular applications on different contexts. Nikolopoulos et al. [31] apply the theta method on finance data, demonstrating its good performance over other benchmarks. Athanasopoulos et al. [14] offer a case study for the application of MTA (in the form of temporal hierarchies) for forecasting the demand of the Accident and Emergency departments in the UK. Additionally, working with MTA, Yagli et al. [57] improved the performance of solar forecasts. De Oliveira and Cyrino Oliveira [69] demonstrate the effectiveness of the bagging approach on energy consumption data. Finally, the case study of Li et al. [17] also involves high-frequency energy consumption data and shows the good performance of FOSS when complex patterns exist. The application of MSP on different contexts is limited, as this approach has not been—to our knowledge—widely applied yet.

8. Conclusions and a Look to the Future

Univariate time series forecasting can be challenging, especially since real life data do not comply with the assumptions and do not follow data generating processes usually assumed by models that can be found implemented in the forecasting support systems. At the same time, improving forecast accuracy can be crucial, as even a small decrease in the forecast error may translate to significant gains in terms of the utility of the forecasts see, for example, references [33,70], who discuss the case of forecasting for inventories. In this paper, we reviewed five approaches that can enhance the performance of univariate time series forecasting methods. These approaches are based on two basic principles: (*i*) manipulation of the original data to extract as much information from them as possible, and (*ii*) forecast combination which has been proved to be extremely beneficial in the forecasting field see, for example, references [71,72].

The five approaches that we presented can be applied on top of established time series forecasting models, such as ETS or ARIMA. In fact, we can argue that all these five approaches work as self-improving mechanisms to boost the performance of the underlying forecasting methods. Although the term "self-improving mechanism" was originally used by Nikolopoulos et al. [38] to describe the performance gained by applying temporal aggregation, we argue that this is a good descriptor for all the approaches discussed in our study. It is important to highlight that the improvements achieved by the application of these approaches do not entail the collection of additional data, such as explanatory variables, that usually come with an additional cost, as well as uncertainty in a sense that, in most cases, the future values of these variables must also be predicted for supporting forecasting methods in a regression fashion. The input for all approaches described is simply the past values of the dependent variable of interest.

When a large number of data are available, then empirical evidence from the latest forecasting competitions [29,73] shows that meta-learning and cross-learning approaches can be used to improve time series forecasting performance. Such "global" approaches are often based of time series features [74] or patterns [75] that may be prevalent and common across many time series. As a result, meta-learning and cross-learning approaches are relevant for companies that require to produce forecasts for myriads of data [76]. Large retailers, such as Walmart, Target, and Carrefour, are representative examples. However, many more companies and organisations are interested in forecasting only a few tens or hundreds of time series to support their operations, marketing, and other functions. As such, "local" solutions, like the ones covered in this study, that use information from singular time series only, are still very useful in practice. More importantly, if one needs to forecast only a small number of series, then it would make sense to invest in the additional computational resources required to handle the most demanding of the approaches (Bagging and FOSS). Regardless, we believe that analysts that wish to apply the approaches presented in this paper should decide based on their added-value across different sampling frequencies (see also the discussions in Section 7) balanced against their relative computational cost.

The various approaches that we presented in this paper have been so far studied in isolation. Although the applying of these approaches in a sequential fashion is entirely feasible, as it is the case with an MTA implementation—the `thief()` function—which offers theta as one of the methods to produce base forecasts, it would be even more interesting to see future studies that focus on the integration of the approaches described here. The only exception that we are aware of is the study by Wang et al. [77] that attempts to structurally integrate the concepts surrounding the theta method (and the manipulation of the local curvatures) with aspects of non-overlapping temporal aggregation. We believe that there is much scope for further research in integrating "wisdom of the data" approaches. For instance, one could consider defining a temporal hierarchy approach in which the base forecasts for the nodes of a certain aggregation level are not produced by considering the entire series consisting of all information at the same aggregation level, but each node is extrapolated separately using sub-seasonal series (FOSS). Another example would be the integration of bagging and multiple starting points approaches, since each of them focuses on a different way in extracting information from the data.

Another interesting path for future investigation would be to explore how these approaches can better support forecasting in practice. For example, consider the extension of these univariate-oriented approaches to fit within a hierarchical framework which contains several series that are cross-sectionally aggregated. Temporal hierarchies naturally extend to cross-temporal hierarchies, see [56], however this is not the case with all other approaches described here. For instance, when using bagging on a particular node of the hierarchy, the bootstrapping of the remainder could be informed by the remainder of the other nodes. Even more interestingly, a bootstrap model combination approach could be based on the models selected as optimal across hierarchical aggregation levels.

In conclusion, univariate time series forecasting benefits from looking the available data through different lenses, attempting to understand them better and model them more efficiently. This is achieved by tackling uncertainties associated with data itself and easing the identification of an 'optimal' model and its parameters. As such, we are looking forward to see more approaches that consider the "wisdom of the data" towards enhancing the forecasting performance.

Author Contributions: Conceptualisation: F.P.; methodology: F.P. and E.S.; software: E.S.; validation: E.S.; visualisation: F.P. and E.S.; formal analysis: F.P. and E.S.; investigation: F.P. and E.S.; resources: F.P. and E.S.; data curation: E.S.; writing—original draft preparation: F.P.; writing—review and editing: E.S.; supervision: F.P.; project administration: F.P.; funding acquisition: F.P. All authors have read and agreed to the published version of the manuscript.

Funding: This research received no external funding.

Institutional Review Board Statement: Not applicable.

Informed Consent Statement: Not applicable.

Data Availability Statement: The data of the M3 and M4 forecasting competitions are publicly available and can be found at https://forecasters.org/resources/time-series-data/, accessed on 1 June 2021.

Acknowledgments: The first author thanks Evelyn and Freddy for their support.

Conflicts of Interest: The authors declare no conflicts of interest.

Abbreviations

The following abbreviations are used in this manuscript:

ARIMA	Auto-regressive Integrated Moving Average
ARIMAx	Auto-regressive Integrated Moving Average with Exogenous Variables
BMC	Bootstrap Model Combination
CBB	Circular Block Bootstrap
ETS	Exponential Smoothing
ETSx	Exponential Smoothing with Exogenous Variables
FOSS	Forecasting with Sub-seasonal Series
LPB	Linear Process Bootstrap
MAPA	Multiple Temporal Aggregation Algorithm
MAPAx	Multiple Temporal Aggregation Algorithm with Exogenous Variables
MASE	Mean Absolute Scaled Error
MBB	Moving Block Bootstrap
MSP	Multiple Starting Points
MTA	Multiple Temporal Aggregation
SBA	Syntetos-Boylan Approximation
SES	Simple Exponential Smoothing
sMAPE	Mean Absolute Percentage Error
STL	Seasonal and Trend decomposition using Loess
THIEF	Temporal Hierarchical Forecasting
TSB	Teunter-Syntetos-Babai (method)

References

1. Fildes, R.; Ma, S.; Kolassa, S. Retail forecasting: Research and practice. *Int. J. Forecast.* **2019**. [CrossRef]
2. Hyndman, R.J.; Koehler, A.B.; Snyder, R.D.; Grose, S. A state space framework for automatic forecasting using exponential smoothing methods. *Int. J. Forecast.* **2002**, *18*, 439–454. [CrossRef]
3. Hyndman, R.; Khandakar, Y. Automatic time series forecasting: The forecast package for R. *J. Stat. Softw.* **2008**, *26*, 1–22.
4. Petropoulos, F.; Hyndman, R.J.; Bergmeir, C. Exploring the sources of uncertainty: Why does bagging for time series forecasting work? *Eur. J. Oper. Res.* **2018**, *268*, 545–554. [CrossRef]
5. Assimakopoulos, V.; Nikolopoulos, K. The Theta model: A decomposition approach to forecasting. *Int. J. Forecast.* **2000**, *16*, 521–530. [CrossRef]
6. Hyndman, R.J.; Billah, B. Unmasking the Theta method. *Int. J. Forecast.* **2003**, *19*, 287–290. [CrossRef]
7. Thomakos, D.; Nikolopoulos, K. Fathoming the theta method for a unit root process. *IMA J. Manag. Math.* **2012**, *25*, 105–124. [CrossRef]
8. Fiorucci, J.A.; Pellegrini, T.R.; Louzada, F.; Petropoulos, F.; Koehler, A.B. Models for optimising the theta method and their relationship to state space models. *Int. J. Forecast.* **2016**, *32*, 1151–1161. [CrossRef]
9. Spiliotis, E.; Assimakopoulos, V.; Nikolopoulos, K. Forecasting with a hybrid method utilizing data smoothing, a variation of the Theta method and shrinkage of seasonal factors. *Int. J. Prod. Econ.* **2019**, *209*, 92–102. [CrossRef]

10. Spiliotis, E.; Assimakopoulos, V.; Makridakis, S. Generalizing the Theta method for automatic forecasting. *Eur. J. Oper. Res.* **2020**, *284*, 550–558. [CrossRef]
11. Kourentzes, N.; Petropoulos, F.; Trapero, J.R. Improving forecasting by estimating time series structural components across multiple frequencies. *Int. J. Forecast.* **2014**, *30*, 291–302. [CrossRef]
12. Petropoulos, F.; Kourentzes, N. Improving forecasting via multiple temporal aggregation. *Foresight Int. J. Appl. Forecast.* **2014**, *34*, 12–17.
13. Petropoulos, F.; Kourentzes, N. Forecast combinations for intermittent demand. *J. Oper. Res. Soc.* **2015**, *66*, 914–924. [CrossRef]
14. Athanasopoulos, G.; Hyndman, R.J.; Kourentzes, N.; Petropoulos, F. Forecasting with temporal hierarchies. *Eur. J. Oper. Res.* **2017**, *262*, 60–74. [CrossRef]
15. Bergmeir, C.; Hyndman, R.J.; Benítez, J.M. Bagging exponential smoothing methods using STL decomposition and Box–Cox transformation. *Int. J. Forecast.* **2016**, *32*, 303–312. [CrossRef]
16. Dantas, T.M.; Cyrino Oliveira, F.L. Improving time series forecasting: An approach combining bootstrap aggregation, clusters and exponential smoothing. *Int. J. Forecast.* **2018**, *34*, 748–761. [CrossRef]
17. Li, X.; Petropoulos, F.; Kang, Y. Improving forecasting with sub-seasonal time series patterns. *arXiv* **2021**, arXiv:2101.00827.
18. Disney, S.M.; Petropoulos, F. Forecast combinations using multiple starting points. In Proceedings of the Logistics & Operations Management Section Annual Conference (LOMSAC 2015), Glasgow, UK, 12–15 July 2015.
19. Bai, Y.; Li, X.; Kang, Y. Improving forecasting with multiple starting points. 2021, Unpublished work.
20. Livera, A.M.D.; Hyndman, R.J.; Snyder, R.D. Forecasting Time Series With Complex Seasonal Patterns Using Exponential Smoothing. *J. Am. Stat. Assoc.* **2011**, *106*, 1513–1527. [CrossRef]
21. Pedregal, D.J.; Trapero, J.R.; Villegas, M.A.; Madrigal, J.J. Submission 260 to the M4 competition. In *Github*; Universidad de Castilla: Ciudad Real, Spain, 2018.
22. Dokumentov, A.; Hyndman, R.J. STR: A Seasonal-Trend Decomposition Procedure Based on Regression. *arXiv* **2020**, arXiv:2009.05894.
23. Makridakis, S.; Hibon, M. The M3-Competition: Results, conclusions and implications. *Int. J. Forecast.* **2000**, *16*, 451–476. [CrossRef]
24. Petropoulos, F.; Nikolopoulos, K. Optimizing Theta Model for Monthly Data. 2013. Available online: https://www.scitepress.org/PublicationsDetail.aspx?ID=PYc+bgnxmJE=&t=1 (accessed on 1 June 2021).
25. Petropoulos, F. θ-reflections from the next generation of forecasters. In *Forecasting with the Theta Method*; Nikolopoulos, K., Thomakos, D.D., Eds.; John Wiley & Sons, Ltd.: Chichester, UK, 2019; pp. 161–175. [CrossRef]
26. Fioruci, J.A.; Pellegrini, T.R.; Louzada, F.; Petropoulos, F. The Optimised Theta Method. *arXiv* **2015**, arXiv:1503.03529.
27. Thomakos, D.D.; Nikolopoulos, K. Forecasting Multivariate Time Series with the Theta Method: Multivariate Theta Method. *J. Forecast.* **2015**, *34*, 220–229. [CrossRef]
28. Legaki, N.Z.; Koutsouri, K. Submission 260 to the M4 competition. In *Github*; National Technical University of Athens: Athens, Greece, 2018.
29. Makridakis, S.; Spiliotis, E.; Assimakopoulos, V. The M4 Competition: 100,000 time series and 61 forecasting methods. *Int. J. Forecast.* **2020**, *36*, 54–74. [CrossRef]
30. Gilliland, M. The value added by machine learning approaches in forecasting. *Int. J. Forecast.* **2020**, *36*, 161–166. [CrossRef]
31. Nikolopoulos, K.; Thomakos, D.; Petropoulos, F.; Assimakopoulos, V. Theta Model Forecasts for Financial Time Series: A Case Study in the S&P500. Technical Report 0033. 2009. Available online: https://ideas.repec.org/p/uop/wpaper/0033.html (accessed on 1 June 2021).
32. Athanasopoulos, G.; Hyndman, R.J.; Song, H.; Wu, D.C. The tourism forecasting competition. *Int. J. Forecast.* **2011**, *27*, 822–844. [CrossRef]
33. Petropoulos, F.; Wang, X.; Disney, S.M. The inventory performance of forecasting methods: Evidence from the M3 competition data. *Int. J. Forecast.* **2019**, *35*, 251–265. [CrossRef]
34. Nikolopoulos, K.; Thomakos, D.D.; Katsagounos, I.; Alghassab, W. On the M4.0 forecasting competition: Can you tell a 4.0 earthquake from a 3.0? *Int. J. Forecast.* **2020**, *36*, 203–205. [CrossRef]
35. Nikolopoulos, K.I.; Thomakos, D.D. *Forecasting with The Theta Method: Theory and Applications*; John Wiley & Sons: Hoboken, NJ, USA, 2019.
36. Petropoulos, F.; Nikolopoulos, K. The Theta Method. *Foresight Int. J. Appl. Forecast.* **2017**, *46*, 11–17.
37. Petropoulos, F.; Apiletti, D.; Assimakopoulos, V.; Babai, M.Z.; Barrow, D.K.; Ben Taieb, S.; Bergmeir, C.; Bessa, R.J.; Bijak, J.; Boylan, J.E.; et al. Forecasting: Theory and Practice. *arXiv* **2021**, arXiv:2012.03854.
38. Nikolopoulos, K.; Syntetos, A.A.; Boylan, J.E.; Petropoulos, F.; Assimakopoulos, V. An aggregate-disaggregate intermittent demand approach (ADIDA) to forecasting: An empirical proposition and analysis. *J. Oper. Res. Soc.* **2011**, *62*, 544–554. [CrossRef]
39. Babai, M.Z.; Ali, M.M.; Nikolopoulos, K. Impact of temporal aggregation on stock control performance of intermittent demand estimators: Empirical analysis. *Omega* **2012**, *40*, 713–721. [CrossRef]
40. Spithourakis, G.; Petropoulos, F.; Nikolopoulos, K.; Assimakopoulos, V. A Systemic View of ADIDA framework. *IMA J. Manag. Math.* **2014**, *25*, 125–137. [CrossRef]
41. Spiliotis, E.; Petropoulos, F.; Assimakopoulos, V. Improving the forecasting performance of temporal hierarchies. *PLoS ONE* **2019**, *14*, e0223422. [CrossRef] [PubMed]

42. Spiliotis, E.; Petropoulos, F.; Kourentzes, N.; Assimakopoulos, V. Cross-temporal aggregation: Improving the forecast accuracy of hierarchical electricity consumption. *Appl. Energy* **2020**, *261*, 114339. [CrossRef]
43. Croston, J.D. Forecasting and Stock Control for Intermittent Demands. *Oper. Res. Q.* **1972**, *23*, 289–303. [CrossRef]
44. Syntetos, A.A.; Boylan, J.E. The accuracy of intermittent demand estimates. *Int. J. Forecast.* **2005**, *21*, 303–314. [CrossRef]
45. Kostenko, A.V.; Hyndman, R.J. A note on the categorization of demand patterns. *J. Oper. Res. Soc.* **2006**, *57*, 1256–1257. [CrossRef]
46. Kourentzes, N.; Petropoulos, F. Forecasting with multivariate temporal aggregation: The case of promotional modelling. *Int. J. Prod. Econ.* **2016**, *181*, 145–153. [CrossRef]
47. Athanasopoulos, G.; Ahmed, R.A.; Hyndman, R.J. Hierarchical forecasts for Australian domestic tourism. *Int. J. Forecast.* **2009**, *25*, 146–166. [CrossRef]
48. Hyndman, R.J.; Ahmed, R.A.; Athanasopoulos, G.; Shang, H.L. Optimal combination forecasts for hierarchical time series. *Comput. Stat. Data Anal.* **2011**, *55*, 2579–2589. [CrossRef]
49. Athanasopoulos, G.; Gamakumara, P.; Panagiotelis, A.; Hyndman, R.J.; Affan, M. Hierarchical Forecasting. In *Macroeconomic Forecasting in the Era of Big Data: Theory and Practice*; Fuleky, P., Ed.; Springer International Publishing: Cham, Switzerland, 2020; pp. 689–719. [CrossRef]
50. Hollyman, R.; Petropoulos, F.; Tipping, M.E. Understanding forecast reconciliation. *Eur. J. Oper. Res.* **2021**, *294*, 149–160. [CrossRef]
51. Kourentzes, N.; Rostami-Tabar, B.; Barrow, D.K. Demand forecasting by temporal aggregation: Using optimal or multiple aggregation levels? *J. Bus. Res.* **2017**, *78*, 1–9. [CrossRef]
52. Jeon, J.; Panagiotelis, A.; Petropoulos, F. Probabilistic forecast reconciliation with applications to wind power and electric load. *Eur. J. Oper. Res.* **2019**, *279*, 364–379. [CrossRef]
53. Nystrup, P.; Lindström, E.; Pinson, P.; Madsen, H. Temporal hierarchies with autocorrelation for load forecasting. *Eur. J. Oper. Res.* **2020**, *280*, 876–888. [CrossRef]
54. Kourentzes, N.; Athanasopoulos, G. Elucidate structure in intermittent demand series. *Eur. J. Oper. Res.* **2021**, *288*, 141–152. [CrossRef]
55. Teunter, R.H.; Syntetos, A.A.; Zied Babai, M. Intermittent demand: Linking forecasting to inventory obsolescence. *Eur. J. Oper. Res.* **2011**, *214*, 606–615. [CrossRef]
56. Kourentzes, N.; Athanasopoulos, G. Cross-temporal coherent forecasts for Australian tourism. *Ann. Tour. Res.* **2019**, *75*, 393–409. [CrossRef]
57. Yagli, G.M.; Yang, D.; Srinivasan, D. Reconciling solar forecasts: Sequential reconciliation. *Sol. Energy* **2019**, *179*, 391–397. [CrossRef]
58. Guerrero, V.M. Time-series analysis supported by power transformations. *J. Forecast.* **1993**, *12*, 37–48. [CrossRef]
59. Cleveland, R.B.; Cleveland, W.S.; McRae, J.E.; Terpenning, I. STL: A seasonal-trend decomposition procedure based on loess. *J. Off. Stat.* **1990**, *6*, 3–73.
60. Bandara, K.; Hewamalage, H.; Liu, Y.H.; Kang, Y.; Bergmeir, C. Improving the Accuracy of Global Forecasting Models using Time Series Data Augmentation. *arXiv* **2020**, arXiv:2008.02663.
61. Kunsch, H.R. The Jackknife and the Bootstrap for General Stationary Observations. *Ann. Stat.* **1989**, *17*, 1217–1241. [CrossRef]
62. Bühlmann, P. Sieve Bootstrap for Time Series. *Bernoulli* **1997**, *3*, 123–148. [CrossRef]
63. Politis, D.N.; Romano, J.P. *A Circular Block-Resampling Procedure for Stationary Data*; Tech. Rep. No. 370; Department of Statistics, Stanford University: Stanford, CA, USA, 1991.
64. McMurry, T.; Politis, D.N. Banded and tapered estimates of autocovariance matrices and the linear process bootstrap. *J. Time Ser. Anal.* **2010**, *31*, 471–482. [CrossRef]
65. Dantas, T.M.; Cyrino Oliveira, F.L.; Varela Repolho, H.M. Air transportation demand forecast through Bagging Holt Winters methods. *J. Air Transp. Manag.* **2017**, *59*, 116–123. [CrossRef]
66. Meira, E.; Cyrino Oliveira, F.L.; Jeon, J. Treating and Pruning: New approaches to forecasting model selection and combination using prediction intervals. *Int. J. Forecast.* **2021**, *37*, 547–568. [CrossRef]
67. Pesaran, M.H.; Timmermann, A. Selection of estimation window in the presence of breaks. *J. Econom.* **2007**, *137*, 134–161. [CrossRef]
68. Spiliotis, E.; Kouloumos, A.; Assimakopoulos, V.; Makridakis, S. Are forecasting competitions data representative of the reality? *Int. J. Forecast.* **2020**, *36*, 37–53. [CrossRef]
69. de Oliveira, E.M.; Cyrino Oliveira, F.L. Forecasting mid-long term electric energy consumption through bagging ARIMA and exponential smoothing methods. *Energy* **2018**, *144*, 776–788. [CrossRef]
70. Syntetos, A.A.; Nikolopoulos, K.; Boylan, J.E. Judging the judges through accuracy-implication metrics: The case of inventory forecasting. *Int. J. Forecast.* **2010**, *26*, 134–143. [CrossRef]
71. Bates, J.M.; Granger, C.W.J. The Combination of Forecasts. *Oper. Res. Soc.* **1969**, *20*, 451–468. [CrossRef]
72. Timmermann, A. Forecast Combinations. *Handb. Econ. Forecast.* **2006**, *1*, 135–196.
73. Makridakis, S.; Spiliotis, E.; Assimakopoulos, V. The M5 Accuracy Competition: Results, Findings and Conclusions. 2020. Available online: https://drive.google.com/drive/u/1/folders/1S6IaHDohF4qalWsx9AABsIG1filB5V0q (accessed on 1 June 2021).
74. Montero-Manso, P.; Athanasopoulos, G.; Hyndman, R.J.; Talagala, T.S. FFORMA: Feature-based forecast model averaging. *Int. J. Forecast.* **2020**, *36*, 86–92. [CrossRef]

75. Kang, Y.; Spiliotis, E.; Petropoulos, F.; Athiniotis, N.; Li, F.; Assimakopoulos, V. Déjà vu: A data-centric forecasting approach through time series cross-similarity. *J. Bus. Res.* **2020**. [CrossRef]
76. Seaman, B. Considerations of a retail forecasting practitioner. *Int. J. Forecast.* **2018**, *34*, 822–829. [CrossRef]
77. Wang, B.; Petropoulos, F.; Jooyoung, J.; Erdogan, G. Integrating theta method and multiple temporal aggregation: Optimising aggregation levels. In Proceedings of the 39th International Symposium on Forecasting ISF 2019, Thessaloniki, Greece, 16–19 June 2019.

MDPI
St. Alban-Anlage 66
4052 Basel
Switzerland
Tel. +41 61 683 77 34
Fax +41 61 302 89 18
www.mdpi.com

Forecasting Editorial Office
E-mail: forecasting@mdpi.com
www.mdpi.com/journal/forecasting